INTRODUCTION TO VECTOR ANALYSIS

INTRODUCTION TO VECTOR ANALYSIS

J. C. TALLACK

Formerly Head of the Mathematics Department,
West Norfolk and King's Lynn High School

CAMBRIDGE
AT THE UNIVERSITY PRESS
1970

CAMBRIDGE UNIVERSITY PRESS
Cambridge, New York, Melbourne, Madrid, Cape Town, Singapore,
São Paulo, Delhi, Dubai, Tokyo

Cambridge University Press
The Edinburgh Building, Cambridge CB2 8RU, UK

Published in the United States of America by Cambridge University Press, New York

www.cambridge.org
Information on this title: www.cambridge.org/9780521124515

This is a revised and enlarged version of
Introduction to Elementary Vector Analysis
first published in 1966
This digitally printed version 2009

A catalogue record for this publication is available from the British Library

Library of Congress Catalogue Card Number: 74-142128

ISBN 978-0-521-07999-0 Hardback
ISBN 978-0-521-12451-5 Paperback

CONTENTS

Preface		*page* vii
1	Introduction to vectors through displacements	1
2	Addition and subtraction of vectors	18
3	Multiplication and division of a vector by a number	32
4	Position vectors and centroids	43
5	Projection and components of a vector	54
6	Applications in mechanics	69
7	Differentiation and integration	88
8	The scalar product	108
Miscellaneous Exercises		134
9	The straight line and the plane	137
10	Other loci	162
11	The vector product	183
12	Product of three vectors	224
13	Further applications of the vector product	250
14	Curves in space	275
Answers to Exercises		287
Bibliography		294
Index		295

PREFACE

The aim of this book is to provide an easy introduction to the algebra of vectors and to the application of vectors in geometry and mechanics.

The idea of a vector is introduced in chapter 1 by discussing displacements. In chapter 2 the algebra of vectors is developed from the definition of the addition of two vectors. Other chapters cover differentiation, integration and the scalar product of vectors. This book replaces *Introduction to Elementary Vector Analysis* (by J. C. Tallack; published in 1966 by Cambridge University Press). The scope of the text has been considerably enlarged by the inclusion of six new chapters. These cover: (i) new techniques (the vector product and the triple products) and (ii) applications (a) in pure mathematics (geometry of the line and plane, plane and simple space loci and the differential geometry of space curves, and (b) in applied mathematics (statics, kinematics, dynamics and electrodynamics).

A feature of the book is the number of worked examples, particularly geometrical, and these it is hoped will be of special value to those who are working on their own.

The book should appeal to the following groups:

(a) scientists and mathematicians in the sixth form;

(b) first year university students of mathematics, science (particularly physics) and engineering, who wish for an easy introduction to the fundamental principles during the first term;

(c) students in training and technical colleges, and

(d) students taking 'Modern' mathematics with vectors in the syllabus.

Although the text has not been written for any particular examination syllabus, it covers the topics for pure and applied mathematics at A-level set by the various Examining Boards.

It is a pleasure to express thanks to the University of Cambridge Local Examinations Syndicate, the University of London, the Joint Matriculation Board and the Mathematical Association for permission to use examination questions. The source of these questions is indicated as follows:

C.	Cambridge University Local Examinations Syndicate. G.C.E. Advanced Level.
L.	London University. Advanced Level.
L.U.	London University. B.Sc.
J.M.B.	Joint Matriculation Board. G.C.E. Advanced Level.
M.A.	Mathematical Association. Diploma in Mathematics.

My thanks are also due to the staff of the Cambridge University Press for their helpful advice at all stages and for the excellence of their printing.

Winchester J.C.T.
November, 1970

1

INTRODUCTION TO VECTORS THROUGH DISPLACEMENTS

Introduction

We shall, in what follows, introduce the idea of a vector and indicate why a special branch of analysis or algebra has to be developed to deal with operations involving vectors. In this chapter the algebra of vectors will be developed by recourse to geometrical drawing and to number-pair form.

Concept of a displacement

An aeroplane starts from London at noon and in half an hour flies 200 km in a straight line. A person wishes to pin-point its exact position at 12.30 p.m. on a map. Now it is obvious that this cannot be done since the information given is insufficient. We have a number 200 representing the distance gone in kilometres but we have no idea of the direction taken by the aeroplane. The information we need is a combined distance–direction or a combined number–direction since the distance in kilometres is expressed as a number. This leads to the idea of displacement as distinct from distance and we may think of a displacement as the distance gone considered together with the direction taken.

We shall now discuss displacements in detail since they will help us to obtain an understanding of vectors in general.

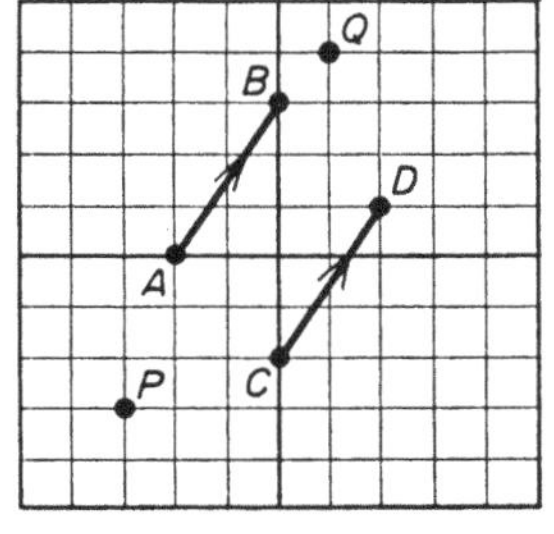

Fig. 1.1

The representation of displacements

In Fig. 1.1 if we start at P and move 1 unit to the right and 3 units up we arrive at A. We can represent this movement in a shorthand way by the ordered number-pair (1, 3). If we start at Q and move 1 unit to the right and 3 units down we arrive at D and we can denote this by $(1, -3)$. Again if we start at Q and move 1 unit to the left and 1 unit down we arrive at B and we denote this by $(-1, -1)$.

The movement or displacement in going from A to B can be described as moving 2 units to the right and 3 units up or (2, 3). The displacement from B to A is the result of moving 2 units to the left and 3 units down or $(-2, -3)$.

The displacement from C to D in number-pair form is (2, 3). This is also the displacement from A to B. Can we make any conclusions about these two displacements? We can say that they are the same since they are equal in length and have the same direction and sense, or we can say they are the same since they are both represented by the same number-pair. However, there is an objection to this equality since it may be argued that they are not the same because they each have different starting-points and ending-points. Thus it seems that when we are considering a displacement we should make it clear whether we wish its location to be included or not. Now if we wish to specify the location of a displacement we use the term 'located' or 'localized' displacement. The displacement itself with no regard to its position is called a 'free' displacement.

We now introduce the notation **PQ** in bold type to denote the directed line segment which represents the free displacement whose magnitude is the length of the line segment PQ, whose direction is that of the line PQ and whose sense is the direction of travel from P to Q, i.e. the sense is given by the order of the letters in **PQ**.

With this notation we can say that for the above displacements **AB, CD**

$$\mathbf{AB} = \mathbf{CD}.$$

Other ways used in printing and writing to indicate the directed line segment **AB** are

$$\overrightarrow{AB},\ \underrightarrow{AB},\ \overline{AB},\ \underline{AB},\ \underset{\sim}{AB}.$$

Magnitude and direction of displacements

In Fig. 1.2 the magnitude of the located displacement **PQ** is the length of PQ which can be associated with the positive number 5. The direction and sense of **PQ** is along PQ and from P to Q.

Suppose **AB** is the free displacement whose number-pair form is (3, 4). **PQ** and **RS** also have the same number-pair form and are therefore the associated located displacements of **AB**. This means that the magnitude, direction and sense of **PQ, RS** and all associated located displacements of **AB** are the same as the magnitude, direction and sense of **AB**. Thus the terms magnitude, direction and sense of **AB** will mean the magnitude, direction and sense of all associated located displacements. In passing we note that we can always associate a displacement with a positive number which defines its magnitude or length.

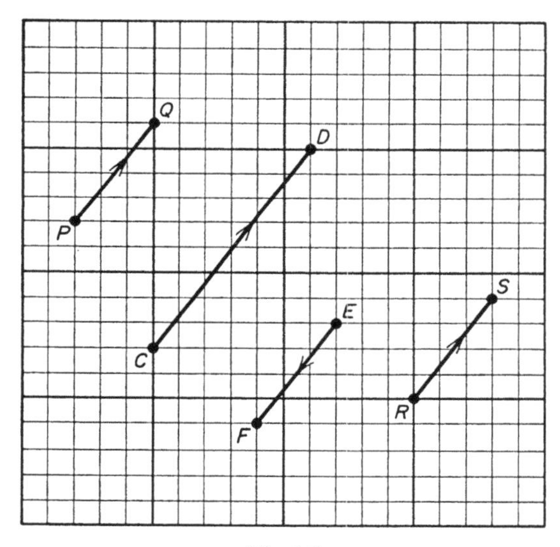

Fig. 1.2

Unless we state otherwise 'displacement' in future will imply 'free displacement'.

The located displacements **PQ, CD, EF** have all the same direction, that is, they are parallel. Writing them in number-pair forms **PQ** = (3, 4), **CD** = (6, 8), **EF** = (−3, −4) we see that the condition for them to be parallel is given by the relation

$$\frac{3}{4} = \frac{6}{8} = \frac{-3}{-4}$$

between the number-pairs. However, the complete reversal of signs of (3, 4), (−3, −4) indicates that **PQ, EF** are opposite in sense.

Successive displacements

Suppose in Fig. 1.3 we start at A and we make a displacement **AB** = (1, 3) followed by a displacement **BC** = (4, 1). We now note that a displacement (5, 4) will take us directly from A to C. Thus we conclude that a displacement from A to B followed by one from B to C is equivalent to one from A to C. This can be symbolized as

AB followed by **BC** = **AC**.

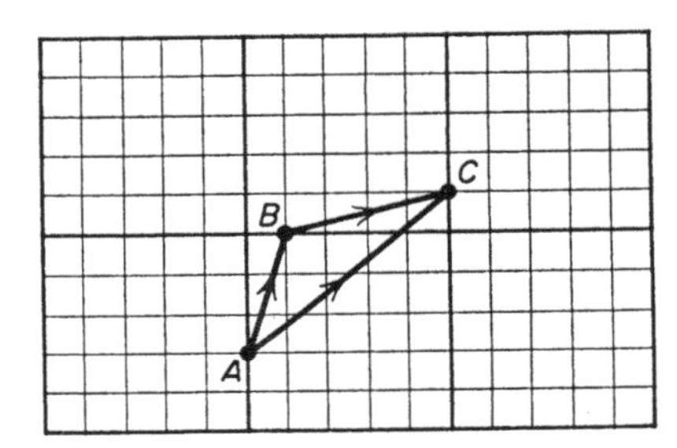

Fig. 1.3

To shorten this further let us use the symbol ⊕ for 'followed by'. Thus

$$\mathbf{AB} \oplus \mathbf{BC} = \mathbf{AC}.$$

However, is it really necessary to introduce a new symbol ⊕? We already have the four symbols $+$, $-$, $\times$, $\div$ which indicate the names of familiar operations in arithmetic and which have been sufficient for doing all that we require. To answer our question we have another look at

$$\mathbf{AB} \oplus \mathbf{BC} = \mathbf{AC}. \qquad (1)$$

We write the displacements in number-pair form

$$\mathbf{AB} = (1, 3),$$

$$\mathbf{BC} = (4, 1),$$

$$\mathbf{AC} = (5, 4).$$

$$\therefore \quad (1, 3) \oplus (4, 1) = (5, 4), \text{ by substituting in (1).}$$

We note that if we add the 1 and the 4 we get the 5, and similarly adding the 3 and 1 gives us the 4.

Let us try again with the following displacements **PQ, QR** as in Fig. 1.4.

The symbolic description of the successive displacements **PQ, QR** is

$$\mathbf{PQ} \oplus \mathbf{QR} = \mathbf{PR}. \qquad (2)$$

Again $\quad \mathbf{PQ} = (3, 5),$

$$\mathbf{QR} = (2, -1),$$

$$\mathbf{PR} = (5, 4).$$

$$\therefore \quad (3, 5) \oplus (2, -1) = (5, 4),$$

by substituting in (2).

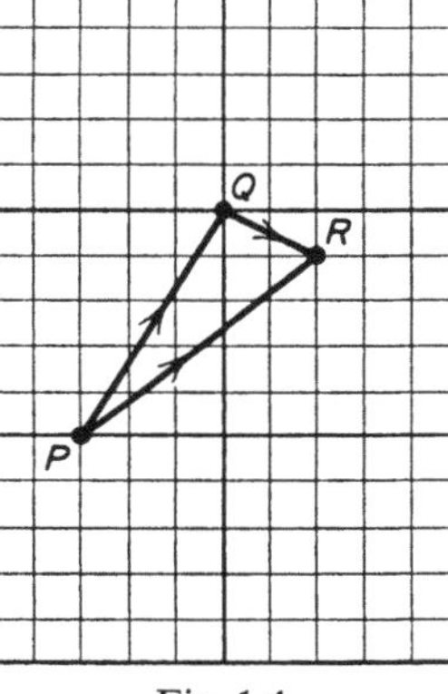

Fig. 1.4

We again see that by adding the 3 and 2 we get the 5 and by adding the 5 and -1 we get the 4.

This suggests the idea that the result of the successive displacements **AB, BC** can be regarded as an addition process, but the operation of addition is to be carried out in a special way. Thus we can discard the symbol $\oplus$ and write $+$ in its place, but we must understand that the $+$ indicates addition by an operation which is different from that for numbers. We now write

$$\mathbf{AB} + \mathbf{BC} = \mathbf{AC}$$

and this symbolic statement tells us the rule or law by which displacements and in general vectors are added. We shall in chapter 2 state this law formally as the triangle law of vector addition, and use this law as the starting-point of the development of vector analysis.

We shall see later that there are other physical quantities apart from displacement which have the combined number–direction property and which are added by the triangle law. We call such quantities by the general name of vectors. Thus we can say displacement is a vector.

In continuing our discussion on displacements we shall use 'vector' for 'displacement' whenever we want to emphasize the applicability of any statement to vectors in general. Referring to Fig. 1.5, since we have in number-pair forms **PQ** = (1, 2), **UV** = (3, −1), **MN** = (4, 1) and since the relation between the number-pairs is

$$(1, 2)+(3, -1) = (4, 1),$$

we can say **PQ**+**UV** = **MN**.

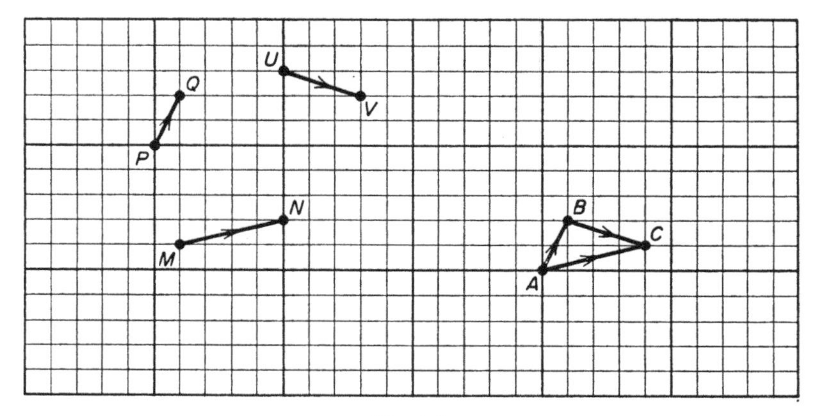

Fig. 1.5

The geometrical significance of this vector equation is that **PQ**+**UV** = **MN** if a triangle *ABC* exists where **PQ** = **AB**, **UV** = **BC**, **MN** = **AC**.

In order to appreciate further developments in our study of vectors we must now discuss symbols more thoroughly.

Symbols

When we use the symbols 3 and 7 in the normal way we understand them to represent the size of two numbers. The nature of our number system may be brought out by algebraic generalizations. For example, if we introduce the symbols x and y to denote two numbers then we can write

$$x+y = y+x.$$

However, is the statement $x+y = y+x$ always true if the symbols x and y do not stand for numbers? This question must be asked since in the study of displacements we have used symbols such as **AB**, **BC**, + and have connected them up as **AB**+**BC**. Thus since the symbols

AB, BC do not stand for numbers but for combined number–directions, can we conclude that $\mathbf{AB}+\mathbf{BC} = \mathbf{BC}+\mathbf{AB}$? Before answering the question we must remember that in the statement $x+y = y+x$ we have another symbol, namely $+$. The statement $x+y$ denotes the result of adding x to y and when x and y are numbers we clearly understand how we are to proceed and how we are to write down the result as a number symbol. The $+$ sign is the name of the operation. It tells us what we are to do but it does not tell us how. How we are to add depends on what we are adding. As we have seen when we add vectors we must do it in a special way.

Now to get back to our question. Is $x+y = y+x$ true when x, y are not numbers and when $+$ means we are to add x, y according to the operation defined for x, y?

Suppose A and B are two points (Fig. 1.6). Here we are using the symbols A, B to stand for two points.

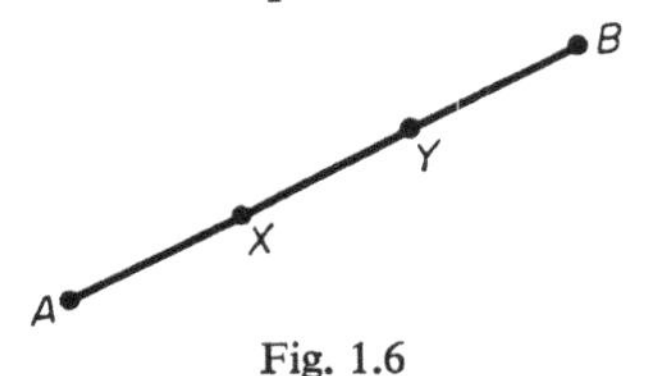

Fig. 1.6

Suppose we invent an algebra to enable us to 'add' two points. Let us define the addition of two points A and B as the point of trisection of AB nearer to the first of the points mentioned, i.e. the point X.

Therefore $A+B = X$.

Using the same definition, if we add point B to point A we obtain the point Y, the point of trisection of BA nearer to B.

Therefore $B+A = Y$.

Thus the addition of A to B and B to A gives two different points.

Thus with this algebraic system

$$A+B \neq B+A.$$

So we see that knowing $x+y = y+x$ is true for numbers we must not assume the statement is true for x and y when they are not numbers. In particular this means that if **AB, BC** are vectors we cannot conclude that $\mathbf{AB}+\mathbf{BC}$ and $\mathbf{BC}+\mathbf{AB}$ are the same.

Since the addition of vectors is an operation different from the addition of numbers the algebra developed for vectors will be a new algebra and the results for number algebra must not be assumed to apply for vector algebra. To make the point clear if x, y, z are symbols standing for numbers and **a, b, c** are symbols standing for vectors and if we know that

$$x+y = y+z,$$
$$(x+y)+z = x+(y+z),$$
$$xy = yx,$$
$$m(x+y) = mx+my \quad (m \text{ a constant number}),$$

we must not assume that the above are true when x, y, z are replaced by **a, b, c** respectively.

However, we now return to a further study of successive displacements.

Order of adding successive displacements

Consider the vectors **AB** = (4, 5), **BC** = (2, −2). Suppose we want to find **AB**+**BC** and **BC**+**AB** geometrically. Take any point O. Let **OP** = **AB** and **PQ** = **BC** (Fig. 1.7).

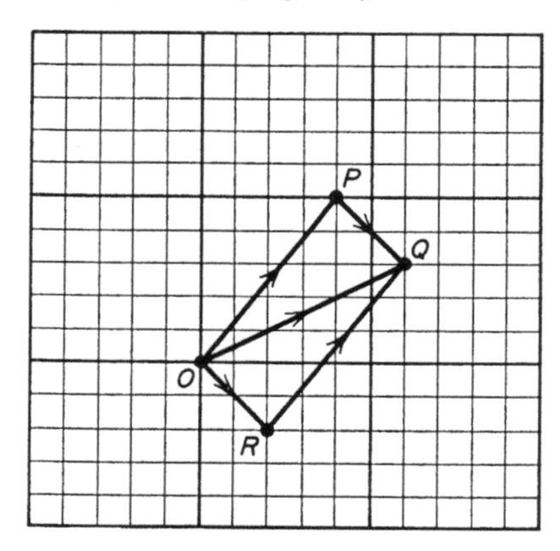

Fig. 1.7

Then $\quad$ **AB**+**BC** = **OP**+**PQ** = **OQ** = (6, 3).

Let **OR** = **BC**. Join **RQ**. We see that **RQ** = (4, 5) = **AB**.

Then $\quad$ **BC**+**AB** = **OR**+**RQ** = **OQ** = (6, 3).

From this we conclude that (**AB**+**BC**) and (**BC**+**AB**) are the same, which implies that the order of adding the vectors is immaterial. This is therefore the answer to one of our questions of the previous section.

We can, without drawing, show the equality by expressing the vectors in number-pairs:

$$\mathbf{AB} = (4, 5),$$

$$\mathbf{BC} = (2, -2),$$

$$\therefore \quad \mathbf{AB}+\mathbf{BC} = (4, 5)+(2, -2) = (6, 3),$$

and

$$\mathbf{BC}+\mathbf{AB} = (2, -2)+(4, 5) = (6, 3),$$

$$\therefore \quad \mathbf{AB}+\mathbf{BC} = \mathbf{BC}+\mathbf{AB}.$$

Now suppose we wish to add the three displacements **AB**, **MN**, **PQ**. The question immediately arises, which two of the three vectors do we add first? This difficulty is resolved by using brackets and giving them the same significance when grouping vectors as in number algebra.

Thus (**AB**+**MN**)+**PQ** means that we must obtain the sum of the displacements **AB**, **MN** first and then add this result to **PQ**.

Another question now arises. Does the way in which we group the vectors affect the final result?

Let **AB** = (4, 8), **MN** = (12, 8), **PQ** = (−20, −4). To obtain (**AB**+**MN**)+**PQ** geometrically take any point O and let **OC** = **AB**, **CD** = **MN** (Fig. 1.8).

Then

$$(\mathbf{AB}+\mathbf{MN}) = (\mathbf{OC}+\mathbf{CD}) = \mathbf{OD}.$$

Now let

$$\mathbf{DE} = \mathbf{PQ}.$$

Then

$$\mathbf{OD}+\mathbf{PQ} = \mathbf{OD}+\mathbf{DE} = \mathbf{OE} = (-4, 12),$$

$$\therefore \quad (\mathbf{AB}+\mathbf{MN})+\mathbf{PQ} = \mathbf{OE} = (-4, 12).$$

To obtain **AB**+(**MN**+**PQ**) geometrically let **OF** = **MN**, **FG** = **PQ**.

Then

$$(\mathbf{MN}+\mathbf{PQ}) = \mathbf{OF}+\mathbf{FG} = \mathbf{OG},$$

Now we see from the figure

$$\mathbf{GE} = (4, 8) = \mathbf{AB},$$

and

$$\mathbf{OG}+\mathbf{GE} = \mathbf{OE},$$

$$\therefore \quad \mathbf{GE}+\mathbf{OG} = \mathbf{OE}$$

(since we have seen the order of addition of two vectors is immaterial).

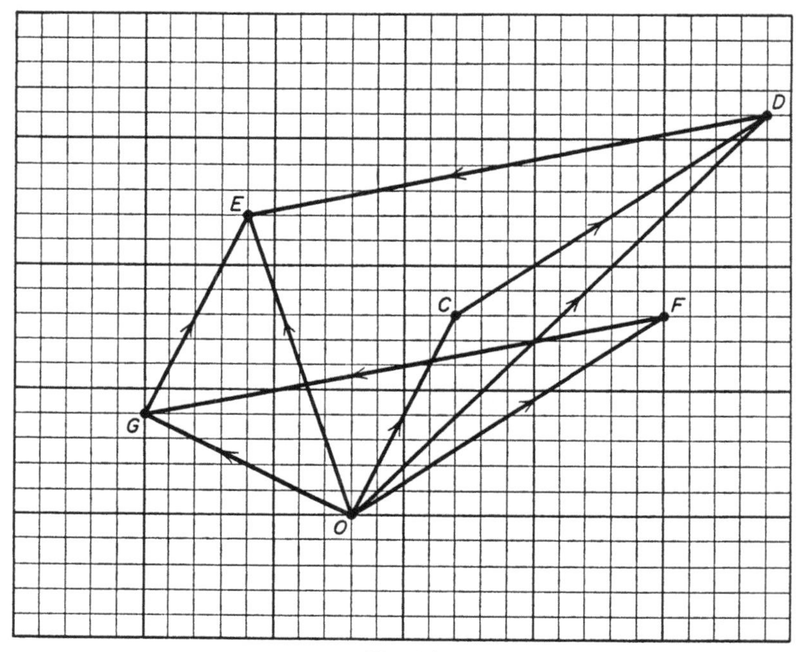

Fig. 1.8

$$\therefore \quad \mathbf{AB}+(\mathbf{MN}+\mathbf{PQ}) = \mathbf{OE} = (-4, 12),$$

$$\therefore \quad (\mathbf{AB}+\mathbf{MN})+\mathbf{PQ} = \mathbf{AB}+(\mathbf{MN}+\mathbf{PQ}).$$

Again we can obtain this result by expressing the vectors in number-pairs

$$\begin{aligned}(\mathbf{AB}+\mathbf{MN})+\mathbf{PQ} &= [(4, 8)+(12, 8)]+(-20, -4)\\ &= (16, 16)+(-20, -4)\\ &= (-4, 12).\\ \mathbf{AB}+(\mathbf{MN}+\mathbf{PQ}) &= (4, 8)+[(12, 8)+(-20, -4)]\\ &= (4, 8)+(-8, 4)\\ &= (-4, 12).\end{aligned}$$

$$\therefore \quad (\mathbf{AB}+\mathbf{MN})+\mathbf{PQ} = \mathbf{AB}+(\mathbf{MN}+\mathbf{PQ}).$$

Thus we conclude that the sum of three vectors is unaffected by the order of grouping.

The zero and inverse vectors

Suppose we start at A and carry out the following displacements **AB**, **BC**, **CA** (Fig. 1.9). We see that we end at the starting-point A, that is, the sum of the displacements, namely **AB**+**BC**+**CA** is a zero displacement.

Expressing this sum in number-pair form we have

$$\mathbf{AB}+\mathbf{BC}+\mathbf{CA} = (2, 3)+(5, 2)+(-7, -5)$$
$$= (0, 0).$$

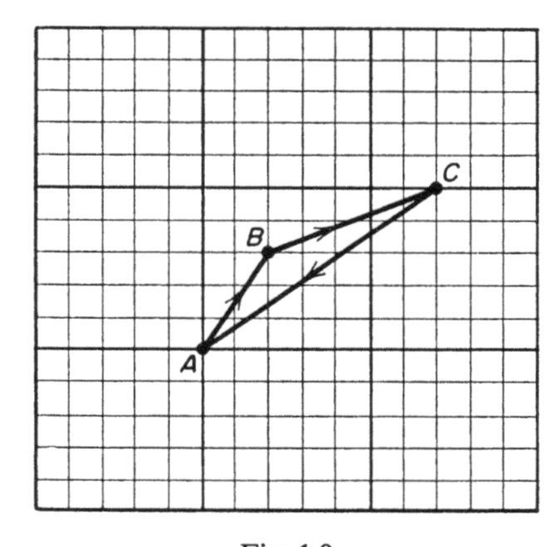

Fig. 1.9

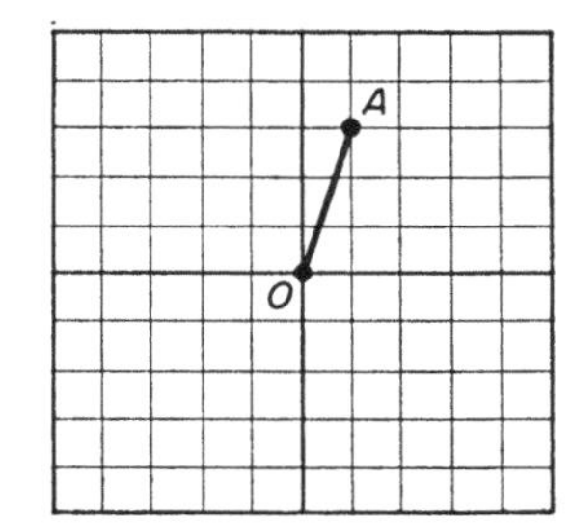

Fig. 1.10

Thus we can interpret the number-pair (0, 0) to imply that we are back at where we started, and also to stand for zero displacement. The number-pair (0, 0) is more generally called the zero vector or the null vector.

Suppose we make a displacement **OA** = (1, 3) (Fig. 1.10). The displacement **AO** necessary to bring us back to the original position denoted by the zero vector (0, 0) is given by the number-pair $(-1, -3)$.

We define $(-1, -3)$ as the inverse of (1, 3) and **AO** as the inverse of **OA**.

Vectors

We shall end this chapter by restating some of the facts we have learnt about vectors through the study of displacements. Also we shall give various definitions.

Representation of vectors

We have seen that a vector has the following essential features: (1) a magnitude or length given by a positive number; (2) a direction in space; (3) a sense.

The only exception to the above is the zero or null vector, which has zero magnitude and whose direction and sense are not defined.

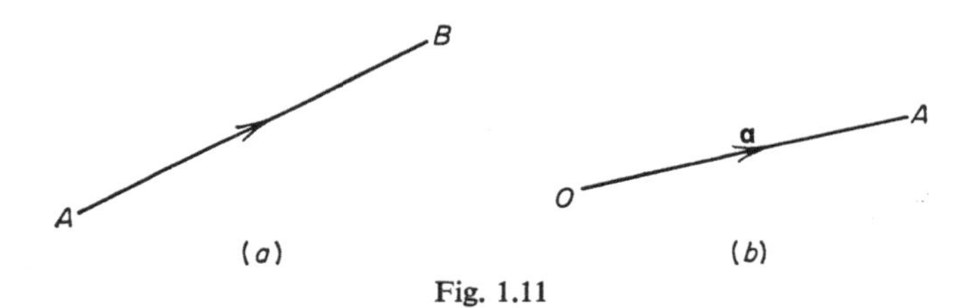

Fig. 1.11

Suppose A and B are two points in space (Fig. 1.11 *a*). Then the directed line segment **AB** represents a vector of length or magnitude AB, the direction of the vector being that of the line AB and the sense of the vector being the direction of travel from A to B. On diagrams the sense of direction can be indicated by an arrow.

Suppose the point B approaches the point A. When B reaches A we have the vector **AA** whose magnitude is zero and whose direction and sense are not specified. Thus **AA** represents the zero vector which is denoted by **0**.

The vector **BA** represents a vector in the opposite sense to the vector **AB** but whose magnitude and direction are the same.

Often a vector is denoted by a single small letter if there is no ambiguity about its direction.

The vector **OA** (Fig. 1.11 *b*) can be denoted thus:

In printing: **a**.

In writing: $\bar{a}$, $\underline{a}$, $\underset{\sim}{a}$.

Magnitude of vectors

The magnitude or modulus of a vector **AB**, denoted by $|\mathbf{AB}|$, is the length of the straight line AB.

Thus $|\mathbf{AB}| = AB$ and $|\mathbf{a}| = a$, where length $OA = a$.

A vector whose modulus is 1 is defined as a unit vector. We shall denote the unit vector having the same direction and sense as **a** by $\hat{\mathbf{a}}$.

Like and unlike vectors

Definition. *Vectors having the same direction and sense are said to be like and those which have the same direction but opposite sense are said to be unlike.*

Equal vectors

Definition. *Two vectors are said to be equal if they have the same magnitude, direction and sense.*

Thus the statement $\mathbf{AB} = \mathbf{PQ}$ means, providing the vectors are not zero vectors;

(1) the two vectors are equal in magnitude, i.e. $AB = PQ$;

(2) the two vectors have the same direction, i.e. they are parallel;

(3) the sense of direction from A to B is the same as the sense of direction from P to Q.

It must be realized that the converse of (1) is not true, i.e. if the moduli of 2 vectors are equal it does not follow that the vectors are equal since their directions may not be the same. Thus if $AB = PQ$ it does not necessarily follow that $\mathbf{AB} = \mathbf{PQ}$.

All zero vectors are equivalent to one another no matter what their directions may be.

Addition of vectors

We define the addition of two vectors by the triangle law.

Later on we shall see that before a physical quantity can be called a vector it must be shown to obey the triangle law of addition even though it is a combined number–direction quantity. It therefore follows that any physical quantity such as displacement which can be represented in magnitude, direction and sense by a directed line segment and which obeys the triangle law of addition, belongs to the class of vectors of which the directed line segment itself can be regarded as the prototype.

Free and localized vectors

We have seen that if we wish to specify the location of a vector then the term 'localized' or 'located' vector is used. Otherwise we refer to the vector as a 'free' vector. As we shall be concerned with free vectors it is important to understand thoroughly the concept of a free vector.

Consider a rigid body which is displaced a given distance in a given direction, without the body undergoing any rotation (Fig. 1.12).

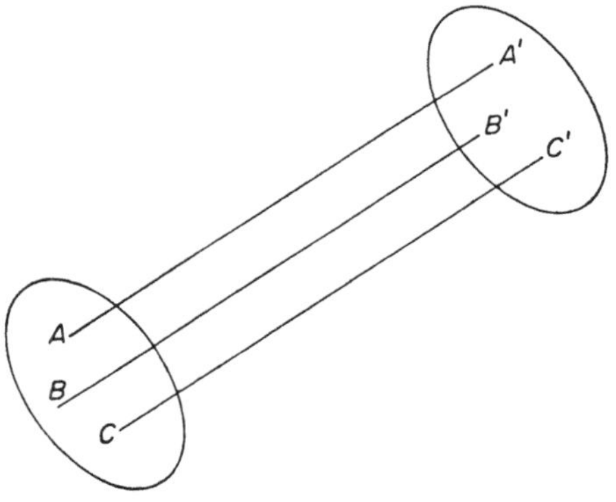

Fig. 1.12

Each point A, B, C moves to a new position A', B', C' in space. The movement of the points is given by the displacement vectors $\mathbf{AA'}$, $\mathbf{BB'}$, $\mathbf{CC'}$. Now it will be clear that any one of these displacements, such as $\mathbf{AA'}$, is sufficient to describe the displacement of the body, i.e. the vector describing the displacement of the body is independent of its location and in this sense is a free vector. It therefore follows that a free vector can be represented by many different parallel and equal directed line segments. Furthermore, we see that free vectors can be transferred from one place to another and providing their magnitude, direction and sense remain the same we consider them not only to be equal but equivalent vectors.

Sometimes it is important to know the location of a vector. We shall later show that force is a vector. If a force is acting on a rigid body its effect is dependent on its line of action. Hence we must treat the force as a localized vector or as it is sometimes called a 'line' or 'sliding' vector. Another example of a localized vector is an electric or magnetic field whose effect is fully known when it is specified with respect to some given point. In such a case the vector is sometimes called a 'tied' vector.

We shall in future use the term 'vector' to imply a 'free' vector, unless we state otherwise. Thus when we refer to the vector **AB** we are not inferring that it is located at A or that its line of action is along AB.

Exercise 1

The object of this exercise is to extend the previous work into three dimensions.

(1) See Fig. 1.13.

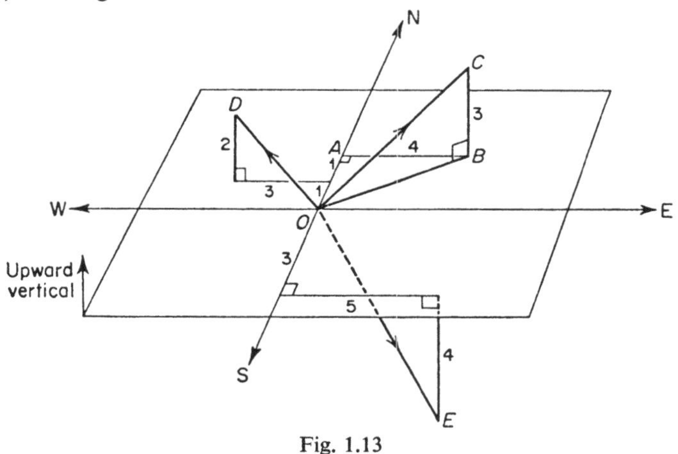

Fig. 1.13

(i) Express as a single vector **OA**+**AB**, **OB**+**BC**. Hence express **OA**+**AB**+**BC** as a single vector.

(ii) The vector **OC** can be regarded as moving 2 units N., 4 units E., 3 units vertically up and can be written as the ordered number-triple (2, 4, 3). What are the vectors **OD**, **OE** as number-triples?

(iii) **OB** is known as the projection of **OC** on the horizontal plane. What is the length of **OB**? Hence what is the length of **OC**?

If **OP** $= (a_1, a_2, a_3)$ in number-triple form deduce the length of the projection of **OP** on the horizontal plane and the length of **OP**.

The numbers a_1, a_2, a_3 are known as the components of the vector **OP**.

(2) See Fig. 1.14.

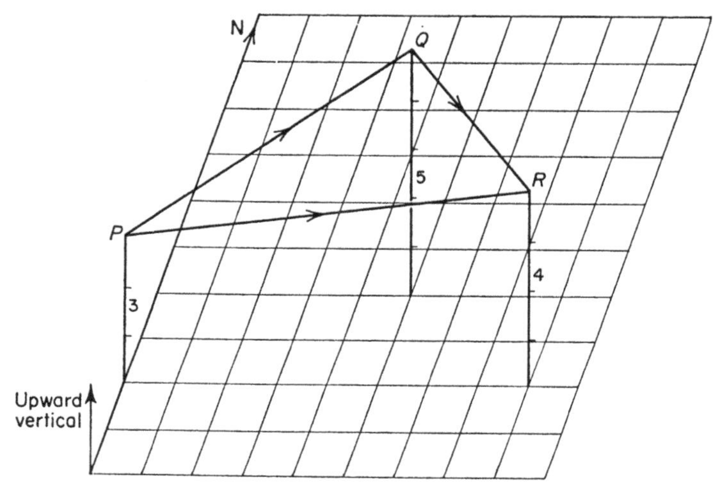

Fig. 1.14

(i) The displacement **PQ** is the result of moving 2 units N., 5 units E., 2 units vertically up and can be written as the number-triple (2, 5, 2). What are the displacements in number-triple form of **QR, PR**?

(ii) Show that the addition of corresponding components of the number-triples of **PQ, QR** results in the components of the number-triple of **PR**.

(iii) **AB** $= (2, 3, -1)$, **BC** $= (-2, 0, 4)$. What is **AC**?

(3) See Fig. 1.15.

(i) Express as number-triples **OA, OB, OC, AB, AC**.

(ii) What can we say about **OB, AC**? Hence prove that the vector sum of **OA, OB** is obtained by adding the corresponding components of the number-triples of **OA, OB**.

(iii) Substract the components of **OA** from the corresponding components of **OB**. What vector has these components?

Hence express **AB** in terms of **OA, OB**.

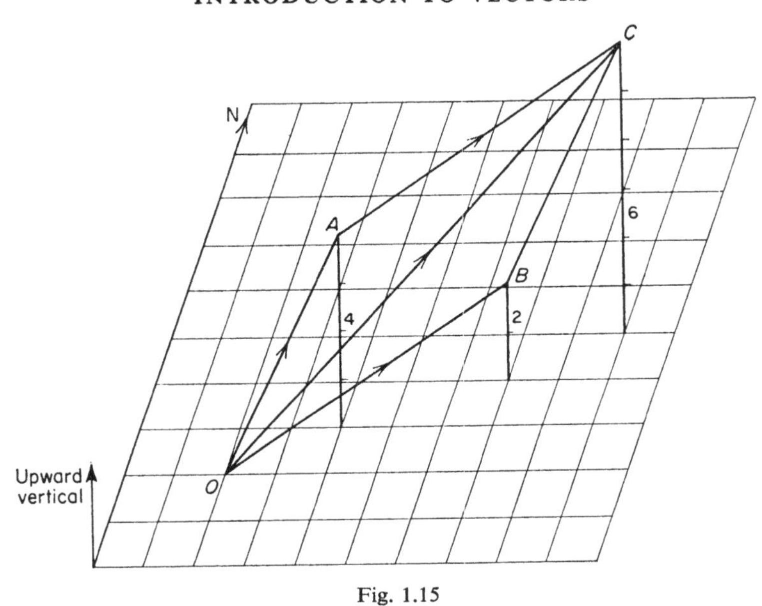

Fig. 1.15

(iv) If **OP** = (2, −4, 5), **OQ** = (7, 1, −3) what are **OP** + **OQ** and **PQ** in number-triple form?

2

ADDITION AND SUBTRACTION OF VECTORS

In this chapter the ideas already developed from the study of displacements will be redeveloped, but in a formal way, taking as our starting-point the definition of vector addition.

The addition of two vectors

Two vectors are added by either the triangle law of vector addition or the parallelogram law of vector addition. We shall define addition by each of these two laws and also show their equivalence.

(1) *The triangle law*

Suppose **a** and **b** are two vectors (Fig. 2.1).

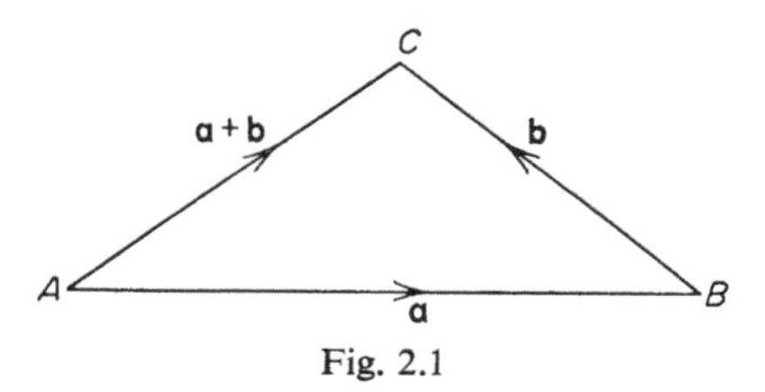

Fig. 2.1

Let **AB** represent the vector **a** and **BC** represent the vector **b**. Then we have the following definition:

***Definition* 1.** *The addition of the vector* **a** *to the vector* **b**, *written as* **a**+**b**, *is defined by*

$$\mathbf{a}+\mathbf{b} = \mathbf{AC}.$$

This sum is unique since all triangles *ABC* with **AB**, **BC** representing **a** and **b** in size and direction are congruent. Hence **AC** will be of fixed length and direction, i.e. **AC** is unique.

Also since $AC < AB+BC$ we have $|\mathbf{a}+\mathbf{b}| < |\mathbf{a}|+|\mathbf{b}|$, i.e. the modulus of the sum of two vectors is less than the sum of the moduli of the two vectors. There is an exception to this, as will be seen later.

Applying this definition to Fig. 2.2 we have

$$\mathbf{PQ}+\mathbf{QR} = \mathbf{PR},$$

and

$$\mathbf{a}+\mathbf{b} = \mathbf{c}.$$

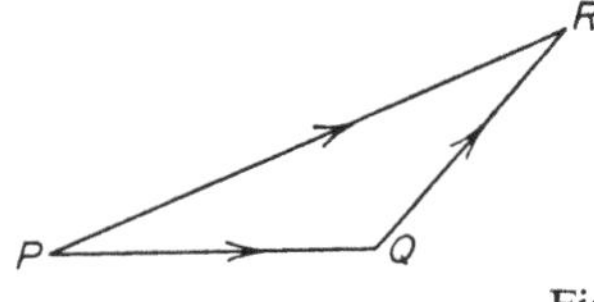

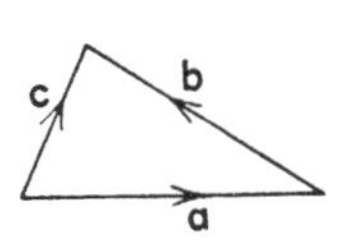

Fig. 2.2

From a geometrical point of view, $\mathbf{p}+\mathbf{q} = \mathbf{r}$ (Fig. 2.3) if a triangle OAB exists where $\mathbf{OA} = \mathbf{p}$, $\mathbf{AB} = \mathbf{q}$ and $\mathbf{OB} = \mathbf{r}$.

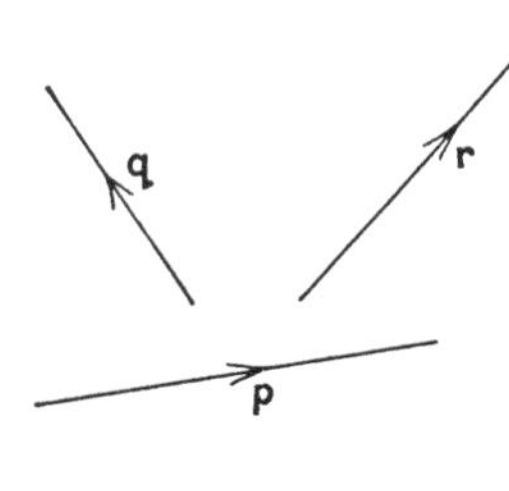

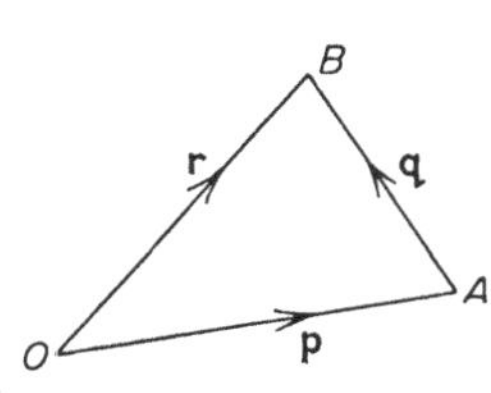

Fig. 2.3

Definition 1 agrees with the way in which two displacements are added (chapter 1) and it also represents mathematically how the vector quantities in physics and mathematics are added.

(2) *The parallelogram law*

Suppose **a** and **b** are two vectors (Fig. 2.4).

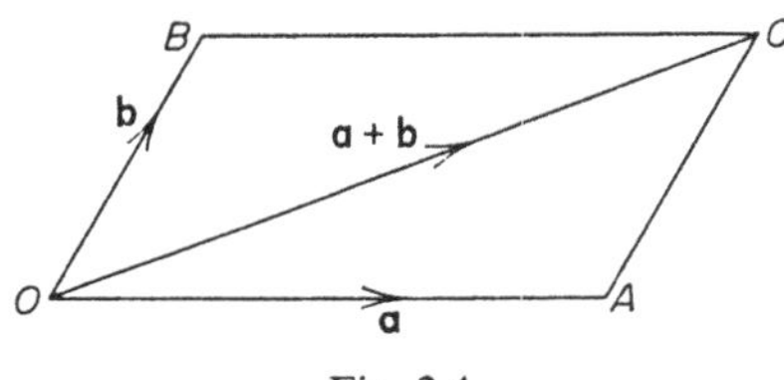

Fig. 2.4

Let **OA** and **OB** represent the vectors **a** and **b** respectively. Complete the parallelogram $OACB$. Then we have the following definition:

Definition 2. *The addition of the vector* **a** *to the vector* **b**, *written* **a**+**b**, *is defined by*

$$\mathbf{a}+\mathbf{b} = \mathbf{OC}.$$

We shall now show that this definition gives the same sum as that defined by the triangle law of addition.

By the triangle law, $$\mathbf{a}+\mathbf{AC} = \mathbf{OC}.$$

But $$\mathbf{AC} = \mathbf{OB} = \mathbf{b},$$

$$\therefore \quad \mathbf{a}+\mathbf{b} = \mathbf{OC}.$$

Thus the equivalence is shown.

We shall as a rule prefer the triangle law to the parallelogram law.

Special cases of the addition of two vectors

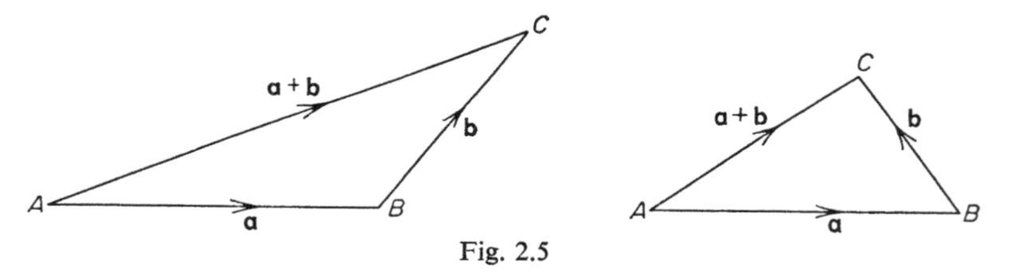

Fig. 2.5

The triangle ABC of Fig. 2.5 exists if **a**, **b** are not parallel or if neither **a** nor **b** is the zero vector **0**. If **a** and **b** are parallel or if $\mathbf{a} = \mathbf{0}$ or $\mathbf{b} = \mathbf{0}$ then the triangle degenerates to a straight line but definition 1 can still be applied. We shall now consider these cases:

(1)

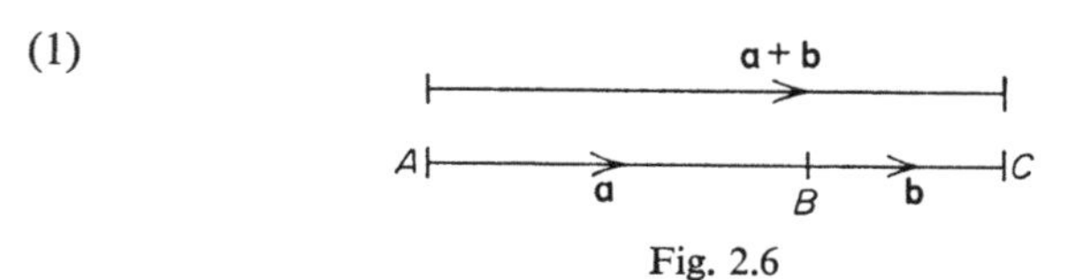

Fig. 2.6

Suppose **a** and **b** have the same direction and sense as in Fig. 2.6.

From the triangle law $$\mathbf{a}+\mathbf{b} = \mathbf{AC}.$$

Thus **AC** has the same direction and sense as **a** and **b**. Now

$$AC = AB+BC,$$

i.e. $$|\mathbf{a}+\mathbf{b}| = |\mathbf{a}|+|\mathbf{b}|.$$

Thus the modulus of the sum of two like vectors is equal to the sum of the moduli of the vectors.

(2)

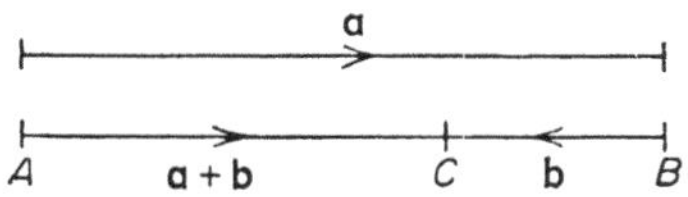

Fig. 2.7

Suppose **a** and **b** have the same direction but opposite sense and suppose that C is between A and B as in Fig. 2.7.

From the triangle law

$$\mathbf{a}+\mathbf{b} = \mathbf{AC}.$$

Thus **AC** has the same direction and sense as **a**.

Since
$$AC = AB-BC,$$

$$|\mathbf{a}+\mathbf{b}| = |\mathbf{a}|-|\mathbf{b}|.$$

Thus the modulus of the sum of two unlike unequal vectors is equal to the difference of the moduli of the vectors.

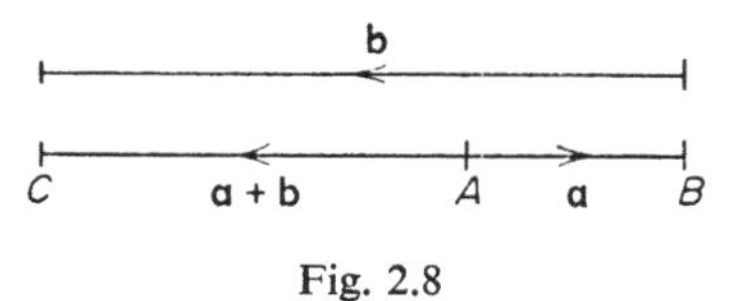

Fig. 2.8

Suppose **a** and **b** have the same direction but opposite sense and suppose C is on BA produced as in Fig. 2.8.

From the triangle law

$$\mathbf{a}+\mathbf{b} = \mathbf{AC}.$$

Thus **AC** has the same direction and sense as **b**.

Since
$$AC = BC-AB,$$

$$|\mathbf{a}+\mathbf{b}| = |\mathbf{b}|-|\mathbf{a}|.$$

Thus, as before, the modulus of the sum of two unlike unequal vectors is equal to the difference of the moduli of the vectors.

(3)

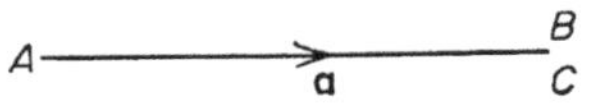

Fig. 2.9

Suppose $\mathbf{b} = \mathbf{BC} = \mathbf{0}$, that is B and C coincide as in Fig. 2.9.

From the triangle law $$\mathbf{a}+\mathbf{b} = \mathbf{AC},$$

$$\therefore \quad \mathbf{a}+\mathbf{0} = \mathbf{a}.$$

In general, there is a vector $\mathbf{0}$ such that for all $\mathbf{p}$,

$$\mathbf{p}+\mathbf{0} = \mathbf{p}.$$

Suppose $\mathbf{a} = \mathbf{AB} = \mathbf{0}$, that is B and A coincide as in Fig. 2.10.

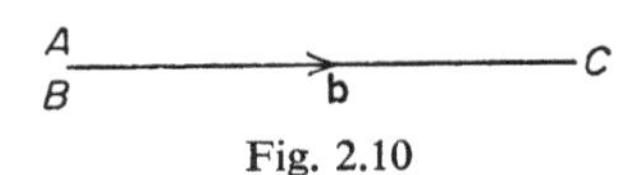

Fig. 2.10

From the triangle law $$\mathbf{a}+\mathbf{b} = \mathbf{AC},$$

$$\therefore \quad \mathbf{0}+\mathbf{b} = \mathbf{b}.$$

The statements $\quad \mathbf{a}+\mathbf{0} = \mathbf{a} \quad$ and $\quad \mathbf{0}+\mathbf{b} = \mathbf{b}$

have their analogy in the identities

$$x+0 = x \quad \text{and} \quad 0+y = y$$

from the algebra of numbers.

(4)

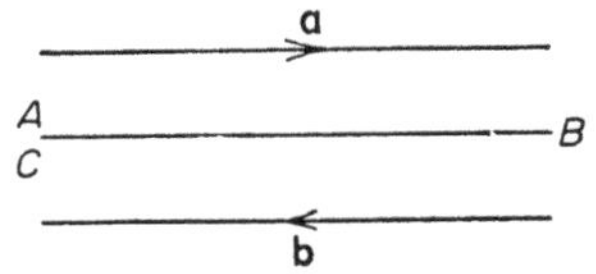

Fig. 2.11

Suppose $\mathbf{a}$ and $\mathbf{b}$ have the same direction but are opposite in sense (Fig. 2.11). Further suppose that their moduli are equal, i.e. A and C coincide.

From the triangle law $$\mathbf{a}+\mathbf{b} = \mathbf{AC},$$

$$\therefore \quad \mathbf{AB}+\mathbf{BA} = \mathbf{0}.$$

This means that the sum of two vectors of equal magnitude and direction but of opposite sense is the zero vector.

If we now define the vector $-\mathbf{AB}$ (known as the inverse or negative of $\mathbf{AB}$) as the vector having the same magnitude and direction as $\mathbf{AB}$ but of opposite sense we then have

$$-\mathbf{AB} = \mathbf{BA},$$

and
$$\mathbf{AB}+(-\mathbf{AB}) = \mathbf{0}.$$

Thus, in general, for all $\mathbf{p}$ there is an inverse or negative vector $-\mathbf{p}$ such that

$$\mathbf{p}+(-\mathbf{p}) = \mathbf{0}.$$

Since by definition
$$-\mathbf{AB} = \mathbf{BA}$$

we have
$$-(-\mathbf{BA}) = \mathbf{BA}$$

or, in general,
$$\mathbf{p} = -(-\mathbf{p}).$$

The order of adding two vectors

The Commutative Law

We now return to the question asked in chapter 1. In number algebra the result of adding a number y to a number x is the same as adding the number x to the number y, i.e. the numbers x, y obey the Commutative Law:

$$x+y = y+x.$$

Now is the result of adding a vector $\mathbf{a}$ to a vector $\mathbf{b}$ the same as adding $\mathbf{b}$ to $\mathbf{a}$?

Suppose we represent the vectors $\mathbf{a}$ and $\mathbf{b}$ by $\mathbf{AB}$ and $\mathbf{BC}$ (Fig. 2.12).

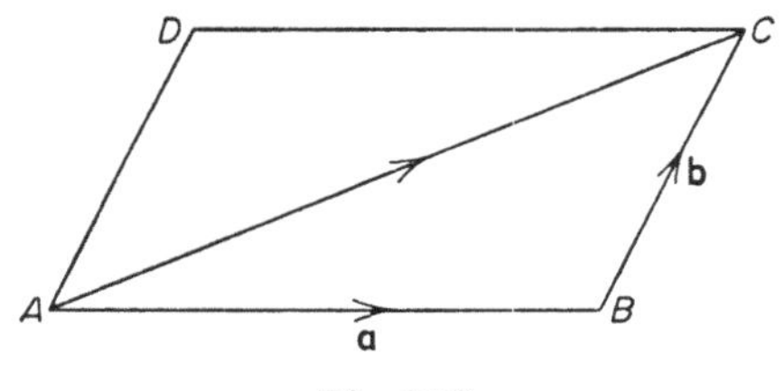

Fig. 2.12

Applying the triangle law of addition

$$\mathbf{a}+\mathbf{b} = \mathbf{AC}.$$

Now $\mathbf{a}$, $\mathbf{b}$ do not enter symmetrically into this relation, since the starting-point of $\mathbf{b}$ coincides with the end-point of $\mathbf{a}$. Thus we cannot conclude that **a** and **b** are interchangeable in the above relation. In other words, it is not obvious that

$$\mathbf{a}+\mathbf{b} = \mathbf{b}+\mathbf{a} = \mathbf{AC}.$$

To obtain $\mathbf{b}+\mathbf{a}$ the parallelogram $ABCD$ is completed and the triangle law is applied to the triangle ADC.

$$\mathbf{AD}+\mathbf{DC} = \mathbf{AC}.$$

But $$\mathbf{AD} = \mathbf{b} \quad \text{and} \quad \mathbf{DC} = \mathbf{a},$$

$$\therefore \quad \mathbf{b}+\mathbf{a} = \mathbf{AC}.$$

Thus we conclude $$\mathbf{a}+\mathbf{b} = \mathbf{b}+\mathbf{a}.$$

This has shown that vectors obey the Commutative Law when they are added by the triangle law.

Now consider the alternative method of addition, i.e. by the parallelogram law.

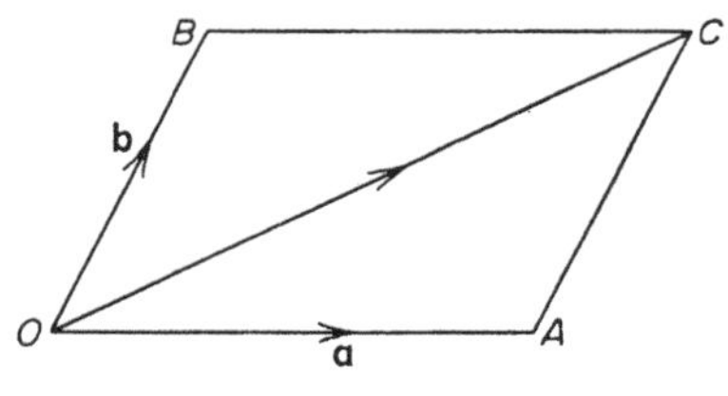

Fig. 2.13

Let **a** and **b** be represented by **OA** and **OB** (Fig. 2.13). Complete the parallelogram $OACB$. Then by the parallelogram law

$$\mathbf{a}+\mathbf{b} = \mathbf{OC}.$$

Now **a**, **b** enter symmetrically in this relation since the initial points of **a**, **b** coincide. This means that **a**, **b** are interchangeable and thus it is obvious that

$$\mathbf{a}+\mathbf{b} = \mathbf{b}+\mathbf{a}.$$

Alternatively by the parallelogram law

$$\mathbf{b}+\mathbf{a} = \mathbf{OC},$$

$$\therefore \quad \mathbf{a}+\mathbf{b} = \mathbf{b}+\mathbf{a}.$$

Thus addition defined by both the triangle and parallelogram laws leads to the same result

$$\mathbf{a}+\mathbf{b} = \mathbf{b}+\mathbf{a}.$$

This is to be expected in any case for we have shown the equivalence of the two definitions.

The order of adding several vectors

The Associative Law

We now discuss the addition of the vectors **a**, **b**, **c**. Suppose x, y, z are numbers. Consider adding x and y and then adding z to the result. This is denoted symbolically by using brackets as $(x+y)+z$. Now consider adding y and z and then adding the result to x, i.e. $x+(y+z)$.

The Associative Law for number–algebra states that

$$(x+y)+z = x+(y+z).$$

Now do vectors obey the same law? Is $(\mathbf{a}+\mathbf{b})+\mathbf{c}$ the same as $\mathbf{a}+(\mathbf{b}+\mathbf{c})$?

Suppose the result of $(\mathbf{a}+\mathbf{b})$ is $\mathbf{d}$, and the result of $(\mathbf{b}+\mathbf{c})$ is $\mathbf{e}$.

Then $$(\mathbf{a}+\mathbf{b})+\mathbf{c} = \mathbf{d}+\mathbf{c},$$

and $$\mathbf{a}+(\mathbf{b}+\mathbf{c}) = \mathbf{a}+\mathbf{e}.$$

It is by no means obvious that these last two results are the same.

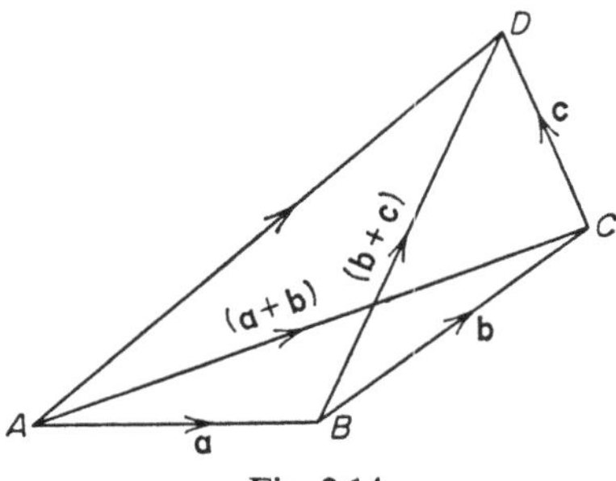

Fig. 2.14

Let **a**, **b**, **c**, not necessarily in the same plane, be represented by **AB**, **BC**, **CD** (Fig. 2.14). Then

$$\mathbf{a}+\mathbf{b} = \mathbf{AC},$$

and $$\mathbf{b}+\mathbf{c} = \mathbf{BD},$$

$$\therefore \quad (\mathbf{a}+\mathbf{b})+\mathbf{c} = \mathbf{AC}+\mathbf{c} = \mathbf{AD},$$

and $$\mathbf{a}+(\mathbf{b}+\mathbf{c}) = \mathbf{a}+\mathbf{BD} = \mathbf{AD},$$

$$\therefore \quad (\mathbf{a}+\mathbf{b})+\mathbf{c} = \mathbf{a}+(\mathbf{b}+\mathbf{c}).$$

Therefore the addition of three vectors obeys the Associative Law. This means that the order of adding is unimportant and we may omit the brackets, writing the vector sum as $\mathbf{a}+\mathbf{b}+\mathbf{c}$.

Extending this idea to several vectors $\mathbf{a}+\mathbf{b}+\mathbf{c}+\mathbf{d}+\mathbf{e}+\dots$ we conclude that the sum of several vectors is independent of the order in which we add them.

Subtraction of vectors

Definition. *The subtraction of* **b** *from* **a** *is defined as the addition of the negative or inverse of* **b** *to* **a**, *i.e.*

$$\mathbf{a}-\mathbf{b} = \mathbf{a}+(-\mathbf{b}).$$

Thus to subtract **b** from **a** reverse the direction of **b** and add to **a**. Fig. 2.15 compares geometrically the processes of addition and subtraction of two vectors.

Let $\mathbf{OA} = \mathbf{a}$, $\mathbf{AB} = \mathbf{b}$. Then $\mathbf{OB} = \mathbf{a}+\mathbf{b}$.

Now to subtract **b** from **a** draw **AC** to represent **b** but in the opposite sense, i.e. draw $-\mathbf{b}$.

Then $$\mathbf{OC} = \mathbf{a}-\mathbf{b}.$$

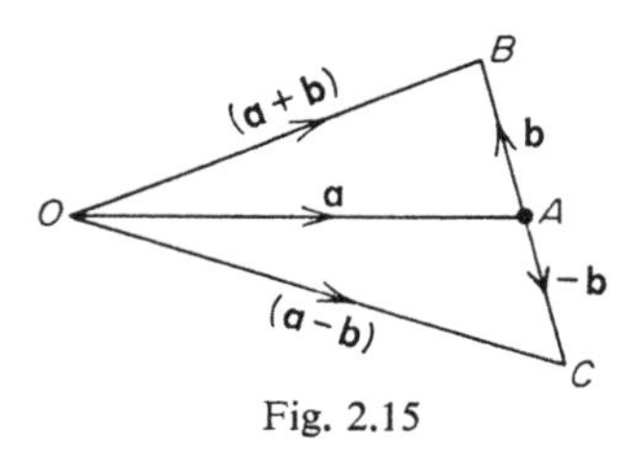

Fig. 2.15

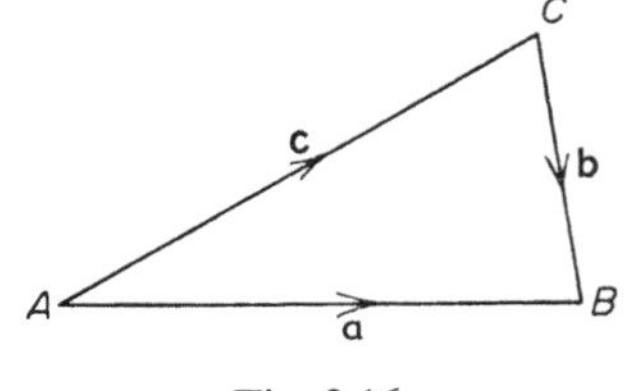

Fig. 2.16

Transposition of a vector in an equation

In Fig. 2.16, $\mathbf{AB} = \mathbf{a}$, $\mathbf{CB} = \mathbf{b}$, $\mathbf{AC} = \mathbf{c}$.

$$\left.\begin{aligned} \mathbf{AB}+\mathbf{BC} &= \mathbf{AC} \\ \therefore \quad \mathbf{a}-\mathbf{b} &= \mathbf{c}. \end{aligned}\right\} \tag{1}$$

Also $$\left.\begin{aligned} \mathbf{AB} &= \mathbf{AC}+\mathbf{CB} \\ \therefore \quad \mathbf{a} &= \mathbf{c}+\mathbf{b}. \end{aligned}\right\} \tag{2}$$

From equations (1) and (2) we see that a vector may be transposed from one side of an equation to the other side providing the rules of algebra are obeyed.

Addition by inspection

It is not always necessary to draw a diagram to add two vectors. For example

$$\mathbf{PQ}+\mathbf{QR} = \mathbf{PR},$$

and

$$\mathbf{CD}-\mathbf{FD} = \mathbf{CD}+\mathbf{DF} = \mathbf{CF}.$$

We see that when the two inner letters are the same the vector representing the sum is given by the outer letters. This can be verified by drawing the appropriate diagrams.

Examples

(1) *Prove* $\mathbf{AB}-\mathbf{CB} = \mathbf{AC}$.

From Fig. 2.17,

$$\begin{aligned}\mathbf{AB}-\mathbf{CB} &= \mathbf{AB}+(-\mathbf{CB})\\ &= \mathbf{AB}+\mathbf{BC}\\ &= \mathbf{AC}.\end{aligned}$$

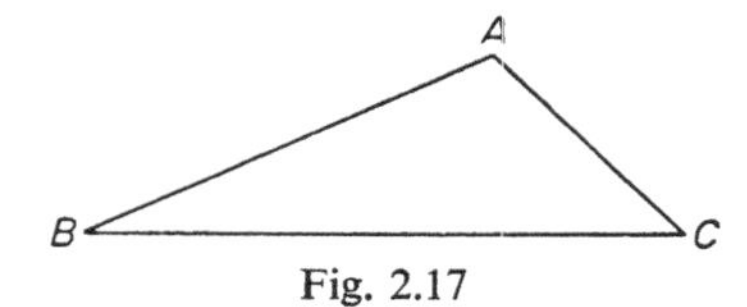

Fig. 2.17

(2) *ABCD is a quadrilateral. P and Q are the mid-points of AD and DC respectively. Show that* $\mathbf{PQ} = \mathbf{AP}+\mathbf{QC}$.

From Fig. 2.18,

$$\mathbf{PQ} = \mathbf{PD}+\mathbf{DQ}.$$

But

$$\mathbf{PD} = \mathbf{AP} \quad \text{and} \quad \mathbf{DQ} = \mathbf{QC},$$

$$\therefore \quad \mathbf{PQ} = \mathbf{AP}+\mathbf{QC}.$$

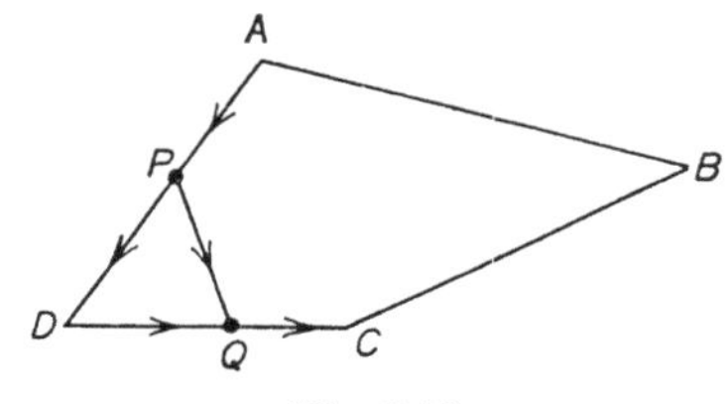

Fig. 2.18

(3) *ABCD is a quadrilateral. Show that* **BD** − **AC** = **CD** − **AB**.

Referring to Fig. 2.19 we see that **AD** can be expressed in two ways bringing in the required vectors:

$$\mathbf{AD} = \mathbf{AB}+\mathbf{BD}.$$

Also
$$\mathbf{AD} = \mathbf{AC}+\mathbf{CD},$$

$$\therefore \quad \mathbf{AB}+\mathbf{BD} = \mathbf{AC}+\mathbf{CD},$$

$$\therefore \quad \mathbf{BD}-\mathbf{AC} = \mathbf{CD}-\mathbf{AB}.$$

N.B. Instead of **AD** we can use **BC**.

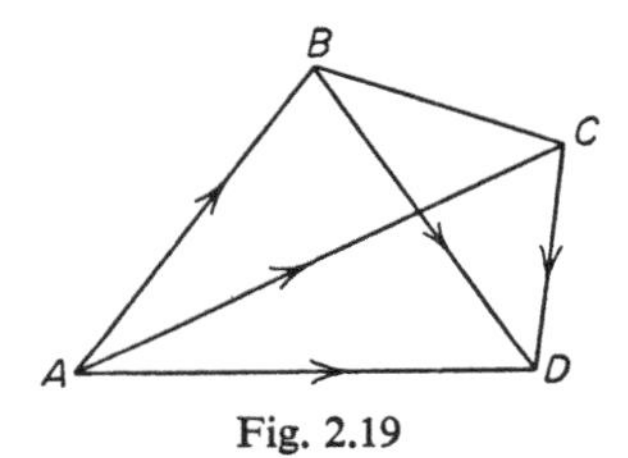

Fig. 2.19

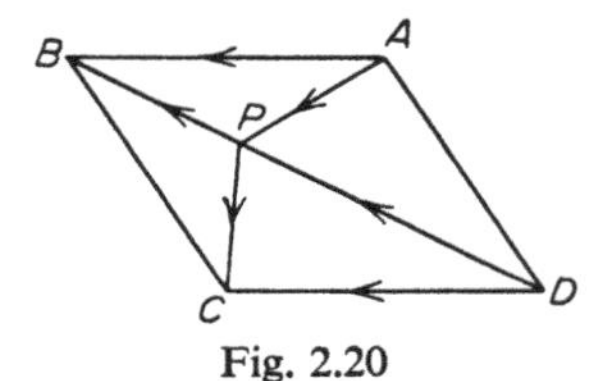

Fig. 2.20

(4) *ABCD is a quadrilateral and P is any point on BD. If*

$$\mathbf{AP}+\mathbf{PB}+\mathbf{PD} = \mathbf{PC}$$

prove that ABCD is a parallelogram.

We aim to get **AB** = **DC** (see Fig. 2.20).

$$\mathbf{AP}+\mathbf{PB}+\mathbf{PD} = \mathbf{PC} \text{ (given)},$$

$$\therefore \quad \mathbf{AP}+\mathbf{PB} = \mathbf{PC}-\mathbf{PD}$$

$$= \mathbf{PC}+\mathbf{DP},$$

$$\therefore \quad \mathbf{AB} = \mathbf{DC}.$$

Similarly,
$$\mathbf{AD} = \mathbf{BC}.$$

∴ *AB* is parallel to *DC* and *AD* is parallel to *BC*.

∴ *ABCD* is a parallelogram.

Note (*a*) In fact **AB** = **DC** is sufficient to prove *ABCD* is a parallelogram since this means that *AB* is parallel and equal to *DC* which leads to *ABCD* being a parallelogram by the well-known theorem in elementary geometry.

(*b*) The geometrical significance of **PQ** = **RS** is that *PQSR* is a parallelogram.

(5) *ABCDEF is a hexagon. If* **AB** = **a**, **BC** = **b**, **CD** = **c**, **DE** = **d**, *and* **EF** = **e**, *show that* **AF** = **a**+**b**+**c**+**d**+**e**.

From Fig. 2.21,

$$\begin{aligned} \mathbf{AC} &= \mathbf{AB}+\mathbf{BC} \\ &= \mathbf{a}+\mathbf{b}, \\ \mathbf{AD} &= \mathbf{AC}+\mathbf{CD} \\ &= \mathbf{a}+\mathbf{b}+\mathbf{c}, \\ \mathbf{AE} &= \mathbf{AD}+\mathbf{DE} \\ &= \mathbf{a}+\mathbf{b}+\mathbf{c}+\mathbf{d}, \\ \mathbf{AF} &= \mathbf{AE}+\mathbf{EF} \\ &= \mathbf{a}+\mathbf{b}+\mathbf{c}+\mathbf{d}+\mathbf{e}. \end{aligned}$$

This example shows how several vectors can be added by means of repeated applications of the triangle law. We shall now state this formally.

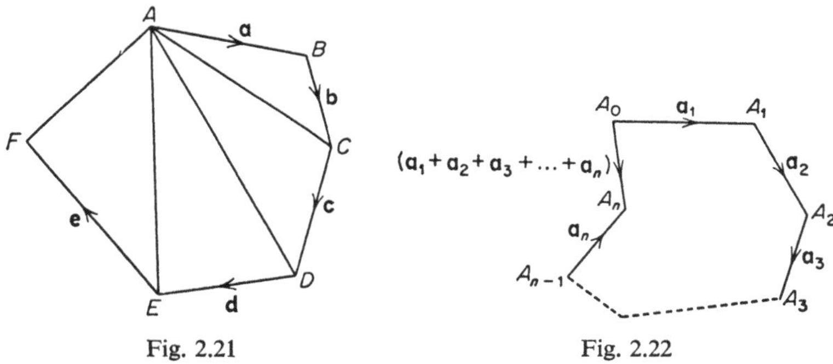

Fig. 2.21

Fig. 2.22

Addition of several vectors

Suppose vectors $\mathbf{a}_1$, $\mathbf{a}_2$, $\mathbf{a}_3$, ..., $\mathbf{a}_n$ are represented by

$$\mathbf{A}_0\mathbf{A}_1,\ \mathbf{A}_1\mathbf{A}_2,\ \mathbf{A}_2\mathbf{A}_3,\ \ldots,\ \mathbf{A}_{n-1}\mathbf{A}_n$$

(Fig. 2.22). It is clear that A_0, A_1, A_2, ..., A_n are not necessarily points in the same plane.

The addition of the vectors is given by the law

$$\mathbf{a}_1+\mathbf{a}_2+\mathbf{a}_3+\ldots+\mathbf{a}_n = \mathbf{A}_0\mathbf{A}_n.$$

The vector $\mathbf{A}_0\mathbf{A}_n$ is sometimes known as the resultant of the vectors and $A_0A_1A_2A_3\ldots A_n$ as the vector polygon.

Components of a vector

If the resultant of vectors **a** and **b** is **c**, i.e. $\mathbf{a}+\mathbf{b} = \mathbf{c}$, then the vectors **a** and **b** are known as the components in their particular directions of the vector **c**.

Suppose **AB** is a given vector and we require its components in two given directions.

Draw straight lines through A and B in the given directions (Fig. 2.23). Let the straight lines meet at C. Then since by construction,

$$\mathbf{AC}+\mathbf{CB} = \mathbf{AB},$$

the required components are **AC** and **CB.**

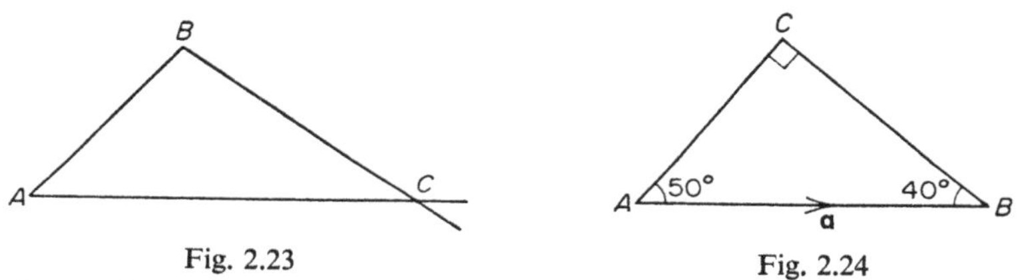

Fig. 2.23

Fig. 2.24

Example

Obtain the moduli of the components of a vector **a** *which are inclined at* 50° *and* 40° *with it.*

Let $$\mathbf{AB} = \mathbf{a} \quad \text{(Fig. 2.24).}$$

$$\therefore \quad |\mathbf{AB}| = |\mathbf{a}|.$$

Since $\mathbf{AC}+\mathbf{CB} = \mathbf{AB}$, **AC** and **CB** are the components.

$$|\mathbf{AC}| = |\mathbf{a}| \cos 50°,$$

and $$|\mathbf{CB}| = |\mathbf{a}| \cos 40°.$$

Note. This example is a special case arising from the two components being at right angles. We have already used this idea in writing vectors in ordered number-pair form, e.g. if $\mathbf{a} = (3, 2)$ the 3 and 2 can be regarded as the components of **a** in two particular directions. We shall develop this further in chapter 5.

Vector equations of a triangle and quadrilateral

The equations $$\mathbf{AB}+\mathbf{BC}+\mathbf{CA} = \mathbf{0}$$

and $$\mathbf{AB}+\mathbf{BC}+\mathbf{CD}+\mathbf{DA} = \mathbf{0},$$

are the vector equations of a triangle ABC and a quadrilateral $ABCD$ respectively (Fig. 2.25).

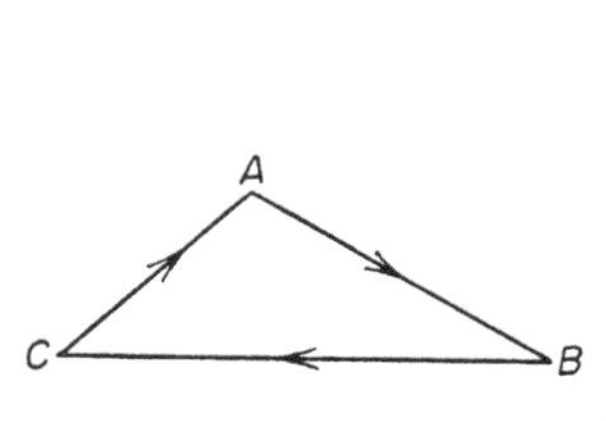

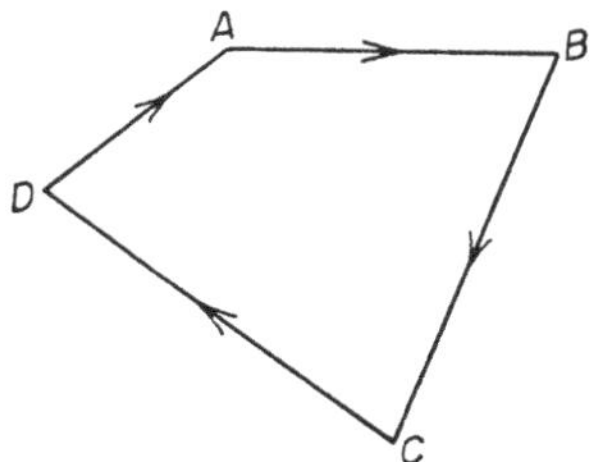

Fig. 2.25

Their proofs are left as an exercise (see Ex. 2, Question 1). These equations should be remembered since they are often useful as a starting-point in the solution of problems involving a triangle or quadrilateral (see Ex. 2, Question 8).

Similar vector equations hold for polygons of five or more sides. Thus in general the sum of the vectors forming the sides of a polygon taken in order is zero.

Exercise 2

(1) Prove: (i) $\mathbf{AB}+\mathbf{BC}+\mathbf{CA} = \mathbf{0}$, and (ii) $\mathbf{AB}+\mathbf{BC}+\mathbf{CD}+\mathbf{DA} = \mathbf{0}$.

(2) If $\mathbf{BC} = \mathbf{BA}+\mathbf{PQ}-\mathbf{CB}$ show that $\mathbf{AB}$ is parallel to $\mathbf{PQ}$.

(3) Prove that P coincides with R if $\mathbf{PQ}+\mathbf{PS} = \mathbf{RQ}+\mathbf{RS}$.

(4) O is any point within a triangle ABC. Prove that

$$\mathbf{OA}+\mathbf{CO} = \mathbf{CB}+\mathbf{BA}.$$

(5) **ABCD** is a parallelogram. If $\mathbf{AB} = \mathbf{a}$ and $\mathbf{BC} = \mathbf{b}$ show that $\mathbf{AC} = \mathbf{a}+\mathbf{b}$ and $\mathbf{BD} = \mathbf{b}-\mathbf{a}$.

(6) Show that the moduli of the components of a vector $\mathbf{r}$ which are inclined at 30° and 60° to it are

$$\frac{\sqrt{3}}{2}|\mathbf{r}| \quad \text{and} \quad \tfrac{1}{2}|\mathbf{r}|$$

respectively.

(7) If $\mathbf{AO}+\mathbf{OB} = \mathbf{BO}+\mathbf{OC}$ prove that A, B and C are collinear (i.e. they lie in a straight line).

(8) $ABCD$ is a quadrilateral with $\mathbf{AB} = \mathbf{DC}$. Prove that $ABCD$ is a parallelogram.

(9) Prove that a quadrilateral whose diagonals bisect each other is a parallelogram.

3

MULTIPLICATION AND DIVISION OF A VECTOR BY A NUMBER

Multiplication of a vector by a number

In number–algebra the repeated addition of x, for example $x+x+x+x+x$, is contracted to $5x$. We can extend this idea to the repeated addition of a vector. The vector $4\mathbf{p}$ represents $\mathbf{p}+\mathbf{p}+\mathbf{p}+\mathbf{p}$. Referring to Fig. 3.1,

$$\begin{aligned}\mathbf{AC} &= \mathbf{AB}+\mathbf{BC}\\ &= \mathbf{a}+\mathbf{a}+\mathbf{a}+\mathbf{a}+\mathbf{b}+\mathbf{b}+\mathbf{b}\\ &= 4\mathbf{a}+3\mathbf{b}.\end{aligned}$$

In the same way we can shorten $-\mathbf{a}-\mathbf{a}-\mathbf{a}-\mathbf{a}-\mathbf{a}$ to $(-5)\,\mathbf{a}$.

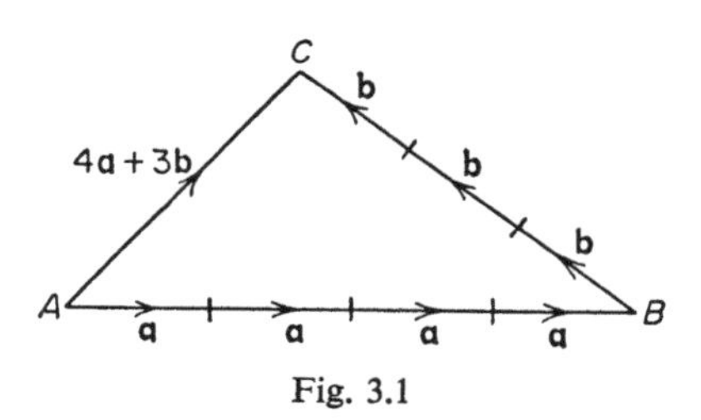

Fig. 3.1

Note. As in number–algebra we write $5x$ rather than $x.5$ so we write $7\mathbf{a}$ rather than $\mathbf{a}.7$

Referring to Fig. 3.2, the vector $5\mathbf{a}$ represents a vector whose magnitude is 5 times that of $\mathbf{a}$ and whose direction and sense are those of $\mathbf{a}$. The vector $(-4)\,\mathbf{a}$ represents a vector whose magnitude is 4 times that of $\mathbf{a}$ and whose direction is that of $\mathbf{a}$ but whose sense is opposite to that of $\mathbf{a}$.

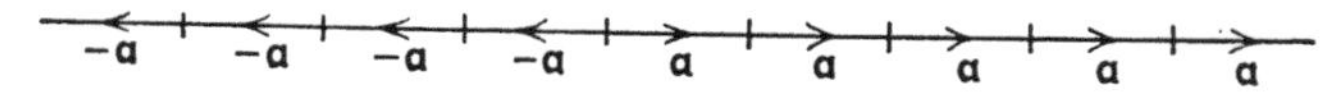

Fig. 3.2

Extending this idea if m is a positive integer then $m\mathbf{a}$ and $(-m)\mathbf{a}$ will represent vectors with magnitudes m times that of $\mathbf{a}$, with direction the same as that of $\mathbf{a}$, and in the case of $m\mathbf{a}$ with the same sense as $\mathbf{a}$ but in the case of $(-m)\mathbf{a}$ with the opposite sense as that of $\mathbf{a}$.

Now in order to further extend this idea to give a meaning to $\sqrt{3}\mathbf{a}$ or $\frac{3}{4}\mathbf{a}$ which cannot be shown geometrically by successive additions of $\mathbf{a}$ we make the following general definition.

Definition. *If m is any real number (positive or negative) and $\mathbf{a}$ a vector then $m\mathbf{a}$ is defined as the vector whose magnitude is $|m|$ times that of $\mathbf{a}$, whose direction is the same as that of $\mathbf{a}$ and with the same sense as $\mathbf{a}$ if m is positive but with opposite sense if m is negative.*

This definition does not apply when $\mathbf{a}$ is the zero vector or m is zero, in which case $m\mathbf{a}$ is the zero vector whose direction and sense are not defined. Thus

$$0\mathbf{a} = m\mathbf{0} = \mathbf{0}.$$

We have already defined a unit vector $\hat{\mathbf{a}}$ as the vector of modulus 1 and with the same sense and direction as that of $\mathbf{a}$. It therefore follows that

$$\mathbf{a} = |\mathbf{a}|\hat{\mathbf{a}} = a\hat{\mathbf{a}}$$

and

$$m\mathbf{a} = m|\mathbf{a}|\hat{\mathbf{a}} = ma\hat{\mathbf{a}}.$$

Division of a vector by a number

The definition of $m\mathbf{a}$ covers any real number. So if $m = 1/n$ where m is any real number then

$$m\mathbf{a} = (1/n)\mathbf{a} = \mathbf{a}/n.$$

We define the division of the vector $\mathbf{a}$ by the number n as the multiplication of $\mathbf{a}$ by $1/n$ resulting in a vector whose:

(1) magnitude is $1/|n|$ that of $\mathbf{a}$;

(2) direction is the same as that of $\mathbf{a}$;

(3) sense is the same as that of $\mathbf{a}$ if n is positive and opposite to that of $\mathbf{a}$ if n is negative.

This definition excludes the case of $\mathbf{a}$ being the zero vector.

We point out at this stage that if $\mathbf{a} = k\mathbf{b}$ then it is not correct to write $\mathbf{a}/\mathbf{b} = k$ since no division of vectors as an operation inverse to multiplication is possible. We can write $\mathbf{a}/k = \mathbf{b}$, however. We shall later show that we can attach a meaning to multiplication of two vectors.

Parallel vectors

In Fig. 3.3, $\mathbf{a} = 3\mathbf{b}$. This represents two parallel vectors of the same sense, the modulus of **a** being three times that of **b**, i.e. $|\mathbf{a}| = 3|\mathbf{b}|$. In the same way $\mathbf{p} = -5\mathbf{q}$ represents two parallel vectors of opposite senses, the modulus of **p** being five times that of **q**.

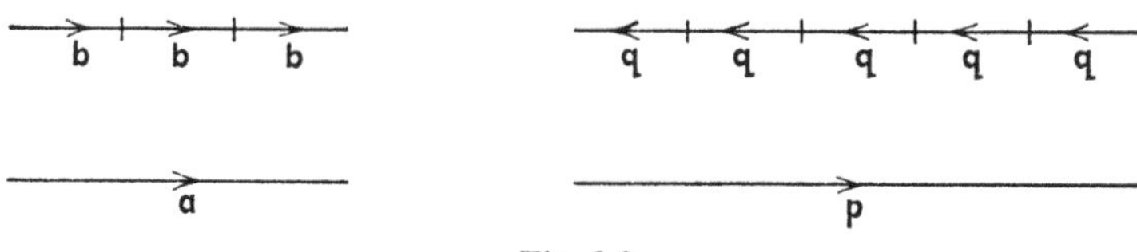

Fig. 3.3

In general $\mathbf{a} = m\mathbf{b}$ ($m \neq 0$, **a**, **b** not zero vectors) means that **a**, **b** are parallel vectors of the same sense if m is positive, but of opposite sense if m is negative, the modulus of **a** being $|m|$ times that of **b**.

Concersely, if **p** and **q** are two parallel vectors then **p** can be expressed as a multiple of **q**, i.e.

$$\mathbf{p} = k\mathbf{q}.$$

The value of k is given by $\pm|\mathbf{p}|/|\mathbf{q}|$, the plus sign to be taken if **p** and **q** have the same sense and the minus sign if they have opposite senses.

The laws of vector-number multiplication

In the algebra of numbers we have the following laws:

(1) *The Commutative Law:* $xy = yx$ which states the product of two numbers x, y, is independent of their order.

(2) *The Associative Law:* $x(yz) = (xy)z$ which states the order of multiplication of the numbers x, y, z does not matter. This law enables us to omit brackets in the product xyz.

(3) *The Distributive Law:* $(x+y)z = xz+yz$ which states the product of the sum of two numbers x, y and the number z is equal to the sum of the products xz and yz.

We have shown that numbers and vectors obey the Commutative and Associative Laws pertaining to addition. Now we shall show that vectors obey the same laws of number-multiplication. In terms of vectors $\mathbf{a}$, $\mathbf{b}$ and numbers p, q, m, n these laws are

$$m\mathbf{a} = \mathbf{a}m \qquad \text{(Commutative Law)},$$

$$p(q\mathbf{a}) = (pq)\mathbf{a} \qquad \text{(Associative Law)},$$

$$\left.\begin{aligned}(p+q)\mathbf{a} &= p\mathbf{a}+q\mathbf{a}\\ n(\mathbf{a}+\mathbf{b}) &= n\mathbf{a}+n\mathbf{b}\end{aligned}\right\} \qquad \text{(Distributive Law)}.$$

The Commutative Law

$m\mathbf{a} = \mathbf{a}m$

By definition the vector whose magnitude is $|m|$ times that of $\mathbf{a}$ and whose direction is the same as that of $\mathbf{a}$, the sense of direction being determined by the sign of m, is written as $m\mathbf{a}$. However apart from the question of preference and analogy with number–algebra (5 times x written $5x$, not $x5$) there is no reason why it cannot be written as $\mathbf{a}m$. Thus by definition $m\mathbf{a} = \mathbf{a}m$.

The Associative Law

$p(q\mathbf{a}) = (pq)\mathbf{a}$

There are really five separate cases for the complete proof. We shall consider only two of them and leave the other three as an exercise.

(i) $p = 0$ or $q = 0$ or $\mathbf{a} = \mathbf{0}$.

In this case each side of $p(q\mathbf{a}) = (pq)\mathbf{a}$ is the zero vector and so the law is true for either $p = 0$ or $q = 0$ or $\mathbf{a} = \mathbf{0}$.

(ii) $p > 0$, $q > 0$, $\mathbf{a} \neq \mathbf{0}$.

$p(q\mathbf{a})$ is a vector of modulus $|p|\,|q\mathbf{a}|$, i.e. $|p|\,|q|\,|\mathbf{a}|$ and with the same direction and sense as $q\mathbf{a}$, i.e. with the same direction and sense as $\mathbf{a}$.

$(pq)\mathbf{a}$ is a vector of modulus $|pq|\,|\mathbf{a}|$, i.e. $|p|\,|q|\,|\mathbf{a}|$ and with the same direction and sense as $\mathbf{a}$.

Thus
$$p(q\mathbf{a}) = (pq)\mathbf{a}.$$

Exercise

Prove $p(q\mathbf{a}) = (pq)\mathbf{a}$ $(\mathbf{a} \neq \mathbf{0})$ for the following cases

(i) $p < 0$, $q < 0$;

(ii) $p > 0$, $q < 0$;

(iii) $p < 0$, $q > 0$.

The Distributive Laws

(1) $(p+q)\mathbf{a} = p\mathbf{a}+q\mathbf{a}$.

Again there are several cases and we shall prove three of them.

(i) $p = 0$, or $q = 0$, or $\mathbf{a} = \mathbf{0}$.

It immediately follows that $(p+q)\mathbf{a} = p\mathbf{a}+q\mathbf{a}$.

(ii) $p > 0, q > 0$.

Since p, q are both positive, $(p+q)$ is positive and hence $(p+q)\,\mathbf{a}$ has the same direction and sense as $\mathbf{a}$.

Modulus of $(p+q)\mathbf{a} = |p+q|\,|\mathbf{a}| = (p+q)\,|\mathbf{a}|$, since $p > 0, q > 0$.

Also since p, q are both positive, $p\mathbf{a}$ and $q\mathbf{a}$ have the same direction and sense as $\mathbf{a}$. Therefore the vector sum $p\mathbf{a}+q\mathbf{a}$ has the same direction and sense as $\mathbf{a}$.

Modulus of $(p\mathbf{a}+q\mathbf{a}) = |p\mathbf{a}|+|q\mathbf{a}|$ since we have shown the modulus of the sum of two like vectors is equal to the sum of their moduli.

$\therefore$ modulus of

$$(p\mathbf{a}+q\mathbf{a}) = |p|\,|\mathbf{a}|+|q|\,|\mathbf{a}| = (|p|+|q|)\,|\mathbf{a}| = (p+q)\,|\mathbf{a}|,$$

since $p > 0, q > 0$.

Hence $(p+q)\,\mathbf{a}$ and $(p\mathbf{a}+q\mathbf{a})$ both have the same modulus, direction and sense and therefore $(p+q)\mathbf{a} = p\mathbf{a}+q\mathbf{a}$.

(iii) $p < 0, q < 0, \mathbf{a} \neq \mathbf{0}$.

Since p, q are negative we can write $p = -m, q = -n$, where m, n are positive numbers.

From case (ii) since $m > 0, n > 0$,

$$(m+n)\,\mathbf{a} = m\mathbf{a}+n\mathbf{a},$$

$$\therefore\quad (-p-q)\,\mathbf{a} = -p\mathbf{a}-q\mathbf{a},$$

$$\therefore\quad (p+q)\,\mathbf{a} = p\mathbf{a}+q\mathbf{a}.$$

Exercise

Draw diagrams to illustrate $p\mathbf{a}+q\mathbf{a}$ $(\mathbf{a}\neq\mathbf{0})$ for

(i) $p < 0, q < 0$;

(ii) $p > 0, q < 0, |p| > |q|$;

(iii) $p > 0, q < 0, |p| < |q|$;

(iv) $p < 0, q > 0, |p| < |q|$;

(v) $p < 0, q > 0, |p| > |q|$;

(vi) $p = -q$.

Hence prove $(p+q)\,\mathbf{a} = p\mathbf{a}+q\mathbf{a}$ for each case. Also prove (ii)–(vi) using the method of case (iii).

(2) $n(\mathbf{a}+\mathbf{b}) = n\mathbf{a}+n\mathbf{b}$

This time there are three cases to be considered:

(i) $n = 0$ or $\mathbf{a} = \mathbf{0}$ or $\mathbf{b} = \mathbf{0}$.

It immediately follows that $n(\mathbf{a}+\mathbf{b}) = n\mathbf{a}+n\mathbf{b}$.

(ii) $n > 0$ ($\mathbf{a} \neq \mathbf{0}, \mathbf{b} \neq \mathbf{0}$).

Let OAC be a triangle with $\mathbf{OA} = \mathbf{a}$, $\mathbf{AC} = \mathbf{b}$ (Fig. 3.4).

$$\therefore \quad \mathbf{OC} = \mathbf{a}+\mathbf{b}.$$

Produce OA to B so that

$$\mathbf{OB} = n\mathbf{a} \quad (n>1).$$

From B draw BD parallel to AC, meeting OC produced at D.

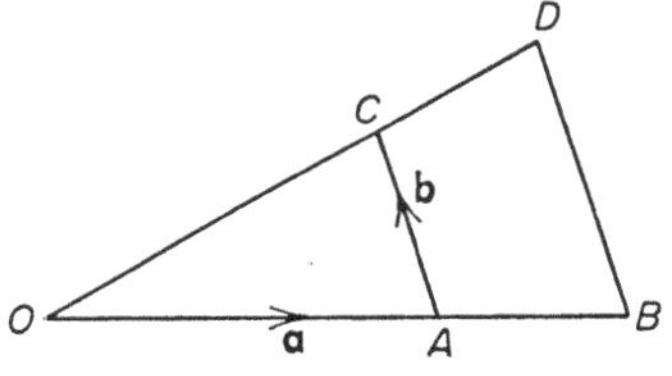

Fig. 3.4

Since triangles OAC, OBD are similar

$$\mathbf{BD} = n\mathbf{AC} = n\mathbf{b},$$

and

$$\mathbf{OD} = n\mathbf{OC} = n(\mathbf{a}+\mathbf{b}).$$

Also

$$\mathbf{OD} = \mathbf{OB}+\mathbf{BD} = n\mathbf{a}+n\mathbf{b},$$

$$\therefore \quad n(\mathbf{a}+\mathbf{b}) = n\mathbf{a}+n\mathbf{b}.$$

The proof is the same when $0 < n < 1$ but B is then on OA.

(iii) $n < 0$ ($\mathbf{a} \neq \mathbf{0}, \mathbf{b} \neq \mathbf{0}$).

Let $n = -m$ where $m > 0$.

Since $m > 0$ we can use the result of case (ii) and write

$$m(\mathbf{a}+\mathbf{b}) = m\mathbf{a}+m\mathbf{b},$$

$$\therefore \quad -n(\mathbf{a}+\mathbf{b}) = -n\mathbf{a}-n\mathbf{b},$$

$$\therefore \quad n(\mathbf{a}+\mathbf{b}) = n\mathbf{a}+n\mathbf{b}.$$

Summary

We have shown that the product of a vector and several numbers is independent of the order of multiplying. Also the product of a sum of numbers and a vector, or a sum of vectors and a number, is the same as the sum of the separate products obtained by the normal operation of removing brackets as in number–algebra.

Examples

(1) *If* $\mathbf{a}+2\mathbf{b} = \mathbf{c}$ *and* $\mathbf{a}-3\mathbf{b} = 2\mathbf{c}$ *show that* $\mathbf{a}$ *has the same sense as* $\mathbf{c}$ *and the opposite sense as* $\mathbf{b}$.

$$\mathbf{a}+2\mathbf{b} = \mathbf{c}, \qquad \text{(i)}$$

$$\mathbf{a}-3\mathbf{b} = 2\mathbf{c}. \qquad \text{(ii)}$$

Multiplying (i) by 3 and (ii) by 2 and adding we get

$$5\mathbf{a} = 7\mathbf{c},$$

$\therefore$ $\mathbf{a}$ and $\mathbf{c}$ have the same sense.

Multiplying (i) by 2 and subtracting (ii) we get

$$\mathbf{a}+7\mathbf{b} = \mathbf{0},$$

$$\therefore \quad \mathbf{a} = -7\mathbf{b},$$

$\therefore$ $\mathbf{a}$ and $\mathbf{b}$ have opposite senses.

(2) *Prove that the opposite sides of a parallelogram are equal.*

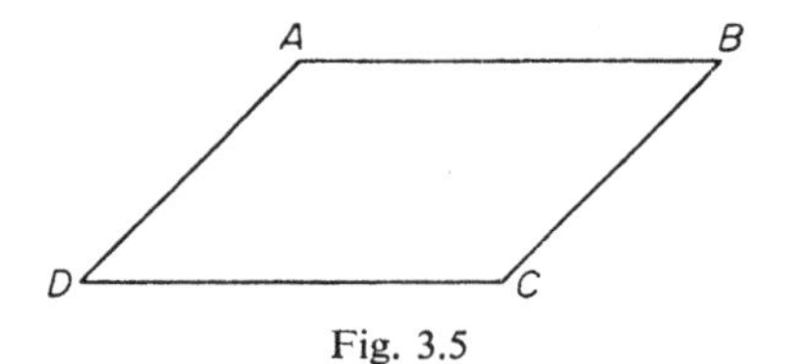

Fig. 3.5

Since AB and DC are parallel (Fig. 3.5),

$$\mathbf{AB} = x\mathbf{DC} \quad \text{where } x \text{ is a number.}$$

Similarly, $$\mathbf{BC} = y\mathbf{AD} \quad \text{where } y \text{ is a number.}$$

Now $$\mathbf{AB}+\mathbf{BC}+\mathbf{CD}+\mathbf{DA} = \mathbf{0},$$

$$\therefore \quad x\mathbf{DC}+y\mathbf{AD}+\mathbf{CD}+\mathbf{DA} = \mathbf{0},$$

$$\therefore \quad x\mathbf{DC}+y\mathbf{AD}-\mathbf{DC}-\mathbf{AD} = \mathbf{0},$$

$$\therefore \quad (x-1)\,\mathbf{DC}+(y-1)\,\mathbf{AD} = \mathbf{0},$$

$$\therefore \quad (x-1)\,\mathbf{DC} = -(y-1)\,\mathbf{AD}.$$

This implies that a vector parallel to **DC** is equal to a vector parallel to **AD** which is impossible unless both sides of the equation shown at bottom of page 38 are zero vectors.

$\therefore$ for the equation to hold true both $(x-1)$ and $(y-1)$ must be zero, since **DC, AD** are not zero vectors.

$$\therefore \quad x = 1 \quad \text{and} \quad y = 1,$$

$$\therefore \quad \mathbf{AB} = \mathbf{DC} \quad \text{and} \quad \mathbf{BC} = \mathbf{AD},$$

$$\therefore \quad AB = DC \quad \text{and} \quad BC = AD.$$

The vector median property of a triangle

$\mathbf{AB}+\mathbf{AC} = 2\mathbf{AO}$ where O is the mid-point of BC.

Proof

Let O be the mid-point of the side BC of the triangle ABC (Fig. 3.6):

$$\mathbf{AB} = \mathbf{AO}+\mathbf{OB},$$

$$\mathbf{AC} = \mathbf{AO}+\mathbf{OC},$$

$$\therefore \quad \mathbf{AB}+\mathbf{AC} = 2\mathbf{AO} \text{ (since } \mathbf{OB} = -\mathbf{OC}).$$

This important result should be remembered.

An alternative way of deriving it is to complete the parallelogram $ABDC$ and to use the parallelogram law of vector addition.

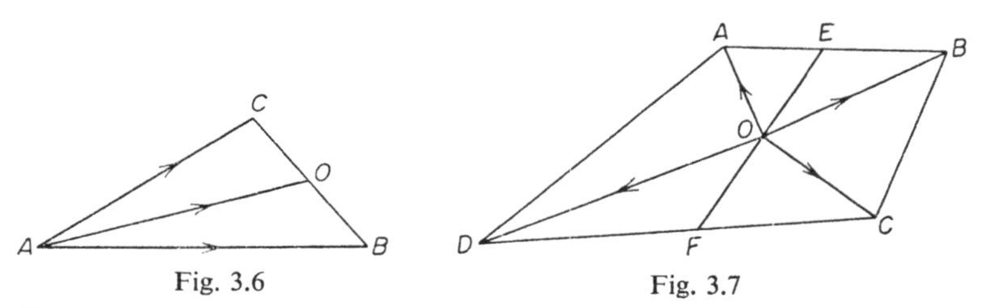

Fig. 3.6 Fig. 3.7

Examples

(1) *$ABCD$ is a quadrilateral and O is any point in its plane. Show that if* $\mathbf{OA}+\mathbf{OB}+\mathbf{OC}+\mathbf{OD} = \mathbf{0}$ *then O is the point of intersection of the lines joining the mid-points of the opposite sides of $ABCD$.*

Let E and F be the mid-points of AB and DC respectively (Fig. 3.7).

Then

$$\mathbf{OA}+\mathbf{OB} = 2\mathbf{OE},$$

and

$$\mathbf{OC}+\mathbf{OD} = 2\mathbf{OF},$$

$$\therefore \quad \mathbf{OA}+\mathbf{OB}+\mathbf{OC}+\mathbf{OD} = 2(\mathbf{OE}+\mathbf{OF}).$$

But $$\mathbf{OA}+\mathbf{OB}+\mathbf{OC}+\mathbf{OD} = \mathbf{0} \quad \text{(given)},$$

$$\therefore \quad 2(\mathbf{OE}+\mathbf{OF}) = \mathbf{0},$$

$$\therefore \quad \mathbf{OE} = -\mathbf{OF},$$

$$= \mathbf{FO}.$$

Therefore EOF is a straight line and O is the mid-point of EF. Similarly, O is the mid-point of GH where G and H are the mid-points of BC and AD respectively.

Therefore O is the point of intersection of the lines joining the mid-points of opposite sides of the quadrilateral.

(2) *$ABCDEF$ is a regular hexagon. If* $\mathbf{AB} = \mathbf{a}$ *and* $\mathbf{BC} = \mathbf{b}$, *show that* $\mathbf{AE} = 2\mathbf{b}-\mathbf{a}$ *and* $\mathbf{AF} = \mathbf{b}-\mathbf{a}$.

Referring to Fig. 3.8 since AD is parallel to BC and is twice BC

$$\mathbf{AD} = 2\mathbf{BC} = 2\mathbf{b},$$

$$\mathbf{DE} = -\mathbf{AB} = -\mathbf{a}.$$

Now $$\mathbf{AE} = \mathbf{AD}+\mathbf{DE},$$

$$\therefore \quad \mathbf{AE} = 2\mathbf{b}-\mathbf{a}.$$

Also $$\mathbf{AF} = \mathbf{AE}+\mathbf{EF},$$

$$= 2\mathbf{b}-\mathbf{a}-\mathbf{b} \quad (\text{since } \mathbf{EF} = -\mathbf{b}),$$

$$\therefore \quad \mathbf{AF} = \mathbf{b}-\mathbf{a}.$$

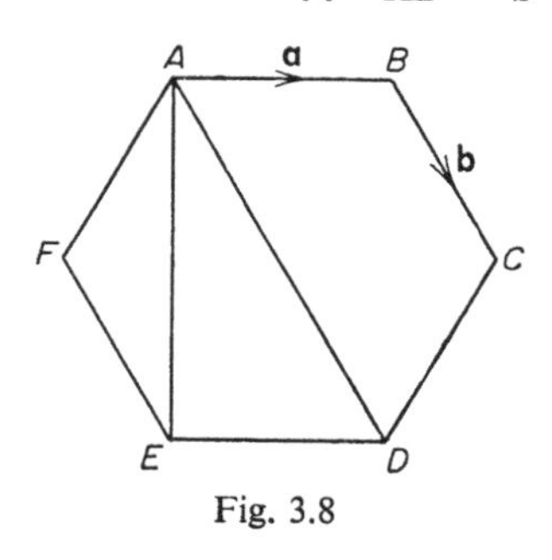

Fig. 3.8

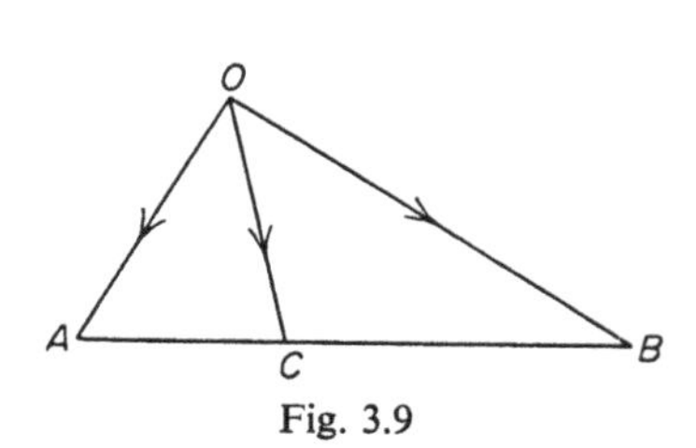

Fig. 3.9

(3) *Prove that* $m\mathbf{OA}+n\mathbf{OB} = (m+n)\,\mathbf{OC}$ *where C is a point dividing AB internally in the ratio of $n{:}m$ i.e. $AC{:}CB = n{:}m$ (m, n positive).*

From Fig. 3.9,

$$m\mathbf{OA}+n\mathbf{OB} = m(\mathbf{OC}+\mathbf{CA})+n(\mathbf{OC}+\mathbf{CB})$$

$$= (m+n)\,\mathbf{OC}+n\mathbf{CB}+m\mathbf{CA}.$$

But $$\frac{AC}{CB} = \frac{n}{m}$$

$$\therefore \quad m\mathbf{AC} = n\mathbf{CB},$$

$$\therefore \quad n\mathbf{CB} - m\mathbf{AC} = \mathbf{0},$$

$$\therefore \quad n\mathbf{CB} + m\mathbf{CA} = \mathbf{0},$$

$$\therefore \quad m\mathbf{OA} + n\mathbf{OB} = (m+n)\,\mathbf{OC}.$$

This result should be remembered.

(4) *ABCD is a parallelogram. E is the mid-point of AB. F is a point on DE such that DE = 3FE. Prove that A, F and C are collinear and that F is a point of trisection of AC.*

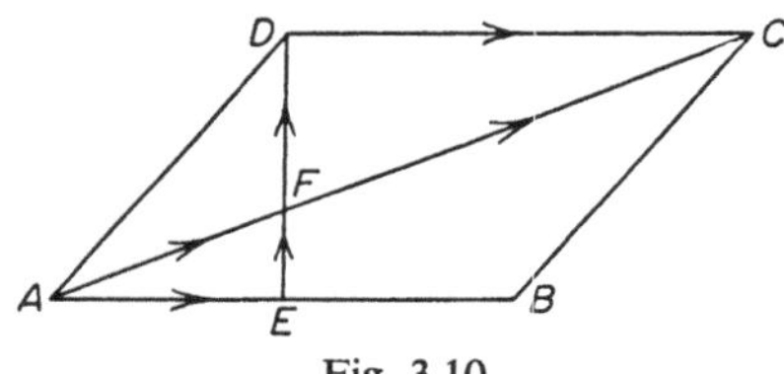

Fig. 3.10

Since $DE = 3FE$, $DF = 2FE$ (Fig. 3.10).

$$\mathbf{AF} = \mathbf{AE} + \mathbf{EF},$$

$$\mathbf{FC} = \mathbf{FD} + \mathbf{DC}.$$

But $$\mathbf{FD} = 2\mathbf{EF} \quad \text{and} \quad \mathbf{DC} = 2\mathbf{AE},$$

$$\therefore \quad \mathbf{FC} = 2\mathbf{AE} + 2\mathbf{EF},$$

$$\therefore \quad \mathbf{FC} = 2\mathbf{AF},$$

$\therefore$ A, F, C are collinear and $AC = 3AF$, i.e. F is a point of trisection of AC nearer to A.

Exercise 3

(1) Prove by completing the parallelogram $OADB$ that

$$\mathbf{OA} + \mathbf{OB} = 2\mathbf{OC},$$

where C is the mid-point of AB.

(2) Prove A, B and C are collinear if $2\mathbf{OA} - 3\mathbf{OB} + \mathbf{OC} = \mathbf{0}$.

(3) Prove that the line joining the mid-points of two sides of a triangle is parallel to the third side and equal to a half of it.

(4) A, B, C and D are any points. Prove

$$\mathbf{OA}+\mathbf{OB}+\mathbf{OC} = 3\mathbf{OD}+\mathbf{DA}+\mathbf{DB}+\mathbf{DC}.$$

(5) $ABCD$ is a skew quadrilateral and E, F, G and H are the mid-points of AB, BC, CD and DA respectively. Prove $EFGH$ is a parallelogram.

(6) $ABCD$ is a quadrilateral with E and F the mid-points of AB and DC respectively. Show that $\mathbf{AD}+\mathbf{BC} = 2\mathbf{EF}$. Further if X and Y are the mid-points of AC and BD respectively show that

$$\mathbf{AB}+\mathbf{AD}+\mathbf{CB}+\mathbf{CD} = 4\mathbf{XY}.$$

(7) ABC is a triangle with G a point on the median AD such that $AG:GD = 2:1$. Prove that $\mathbf{BA}+\mathbf{BC} = 3\mathbf{BG}$.

(8) $ABCDEF$ is a regular hexagon. If $\mathbf{AB} = \mathbf{a}$ and $\mathbf{BC} = \mathbf{b}$ show that $\mathbf{CD} = \mathbf{b}-\mathbf{a}$, $\mathbf{DE} = -\mathbf{a}$, $\mathbf{EF} = -\mathbf{b}$ and $\mathbf{FA} = \mathbf{a}-\mathbf{b}$.

(9) Prove that the diagonals of a parallelogram bisect one another.

(*Hint.* Let O be the mid-point of one of the diagonals.)

(10) ABC is a triangle and D any point in BC. If

$$\mathbf{AD}+\mathbf{DB}+\mathbf{DC} = \mathbf{DE}$$

show that $ABEC$ is a parallelogram and hence E is a fixed point.

(11) If O is the circumcentre of a triangle ABC and H the orthocentre prove that

$$\mathbf{OA}+\mathbf{OB}+\mathbf{OC} = \mathbf{OH} \quad \text{and} \quad \mathbf{HA}+\mathbf{HB}+\mathbf{HC} = 2\mathbf{HO}.$$

(*Hint.* Use geometrical fact $AH = 2OD$,

where D is mid-point of BC.)

(12) O is a point within triangle ABC such that

$$\mathbf{OA}+\mathbf{OB}+\mathbf{OC} = \mathbf{0}.$$

Prove that O is the point of intersection of the medians.

4

POSITION VECTORS AND CENTROIDS

Definition. *If P is any point and O an origin then the position vector of P relative to O is the vector* **OP**.

We shall use **a**, **b**, **p**, **r**, ... to denote the position vectors of the points A, B, P, R, ... relative to the origin O.

The use of position vectors enables vectors to be interpreted in algebraic terms resulting in a conciseness of expression.

The vector AB in terms of position vectors

Let the position vectors of A and B relative to the origin O be **a** and **b** (Fig. 4.1).

$$\mathbf{OA}+\mathbf{AB} = \mathbf{OB},$$

$$\therefore \quad \mathbf{AB} = \mathbf{OB}-\mathbf{OA},$$

$$\therefore \quad \mathbf{AB} = \mathbf{b}-\mathbf{a}.$$

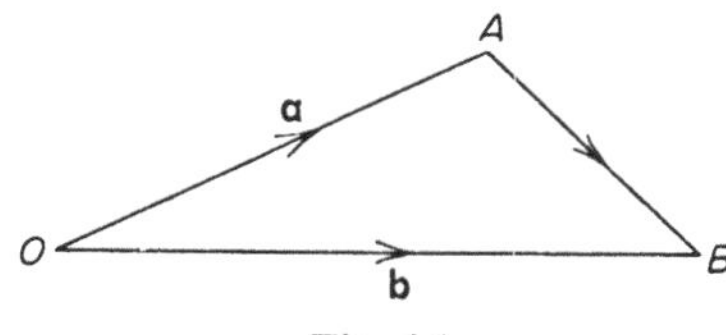

Fig. 4.1

This is an important result to remember, since it enables us to write a vector **AB** in terms of the position vectors of the points A and B.

Examples

(1) *Prove* $\mathbf{AB}+\mathbf{BC}+\mathbf{CA} = \mathbf{0}$.

Let the position vectors of A, B and C relative to an origin O be **a**, **b** and **c**.

Then $\quad \mathbf{AB}+\mathbf{BC}+\mathbf{CA} = \mathbf{b}-\mathbf{a}+\mathbf{c}-\mathbf{b}+\mathbf{a}-\mathbf{c} = \mathbf{0}.$

This is a result already known and proved without the use of position vectors in Ex. 2, Question 1.

(2) *If* $\mathbf{AB}-\mathbf{BC}-\mathbf{DC}+\mathbf{AD} = \mathbf{0}$ *prove ABCD is a parallelogram.*

Let the position vectors of A, B, C and D relative to an origin O be $\mathbf{a}$, $\mathbf{b}$, $\mathbf{c}$ and $\mathbf{d}$.

$$\mathbf{AB}-\mathbf{BC}-\mathbf{DC}+\mathbf{AD} = \mathbf{0},$$

$$\therefore \quad \mathbf{AB}+\mathbf{CB}+\mathbf{CD}+\mathbf{AD} = \mathbf{0},$$

$$\therefore \quad (\mathbf{b}-\mathbf{a})+(\mathbf{b}-\mathbf{c})+(\mathbf{d}-\mathbf{c})+(\mathbf{d}-\mathbf{a}) = \mathbf{0},$$

$$\therefore \quad 2(\mathbf{b}-\mathbf{a})-2(\mathbf{c}-\mathbf{d}) = \mathbf{0},$$

$$\therefore \quad \mathbf{b}-\mathbf{a} = \mathbf{c}-\mathbf{d},$$

$$\therefore \quad \mathbf{AB} = \mathbf{DC}.$$

Similarly, $\mathbf{BC} = \mathbf{AD}.$

Therefore $ABCD$ is a parallelogram.

Position vector of the point dividing a given straight line in a given ratio

A point can divide a straight line in a given ratio either internally or externally.

In Fig. 4.2 P divides AB internally in the ratio $m:n$. Taking the ratio $m:n$ to be positive we can write $\mathbf{AP} = (m/n)\,\mathbf{PB}$.

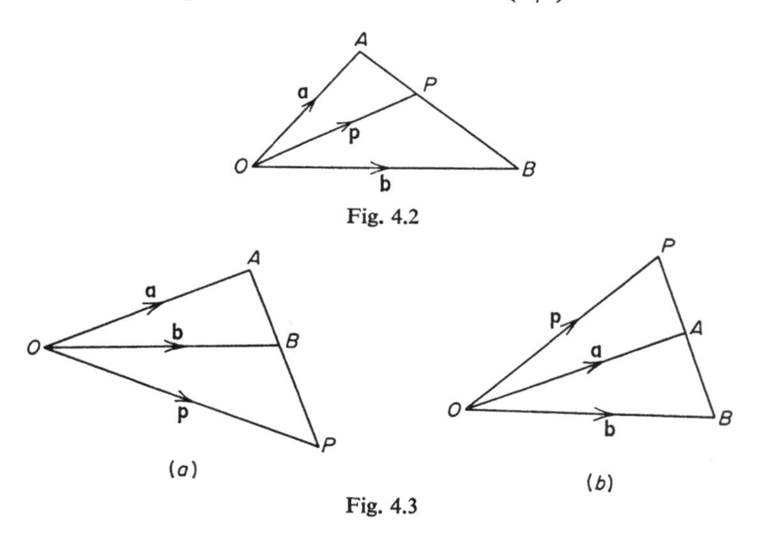

Fig. 4.2

Fig. 4.3

In Fig. 4.3 (*a*) P divides AB externally and in Fig. 4.3 (*b*) P divides BA externally. Now if we take the ratio $m:n$ to be negative for the cases of external division we can still write **AP** $=$ (m/n) **PB**. We note that in these cases if P is nearer to B then $m:n$ is between -1 and $-\infty$ and if P is nearer to A then $m:n$ is between 0 and -1.

Thus in all cases of division of a straight line into a given ratio by a point we have **AP** $=$ (m/n) **PB**.

$$\therefore \quad n(\mathbf{p}-\mathbf{a}) = m(\mathbf{b}-\mathbf{p}),$$

where **a**, **b**, **p** are the position vectors of A, B, P relative to an origin O.

$$\therefore \quad n\mathbf{p}-n\mathbf{a} = m\mathbf{b}-m\mathbf{p},$$

$$\therefore \quad (m+n)\,\mathbf{p} = n\mathbf{a}+m\mathbf{b},$$

$$\therefore \quad \mathbf{p} = \frac{n\mathbf{a}+m\mathbf{b}}{m+n}. \qquad \text{(i)}$$

Thus (i) gives the required position vector whether the line is divided internally or externally. In the internal case $m:n$ is taken to be positive and in the external case $m:n$ is taken to be negative when substituting in (i).

A particular case of (i) occurs when $m = n = 1$, i.e. P is the mid-point of AB. We then have,

$$\text{position vector of mid-point of } AB = \frac{\mathbf{a}+\mathbf{b}}{2}. \qquad \text{(ii)}$$

The results (i) and (ii) should be remembered.

The result (i) can be written as

$$(a) \quad n\mathbf{OA}+m\mathbf{OB} = (m+n)\,\mathbf{OP} \text{ (see Ex. 3, p. 40)},$$

or

$$(b) \quad -(m+n)\,\mathbf{p}+n\mathbf{a}+m\mathbf{b} = \mathbf{0}.$$

Writing $l = -(m+n)$ we have

$$l\mathbf{p}+n\mathbf{a}+m\mathbf{b} = \mathbf{0}.$$

Thus we see that if **p**, **a**, **b** are the position vectors of three distinct collinear points there are numbers l, m, n, different from zero, such that

$$l\mathbf{p}+n\mathbf{a}+m\mathbf{b} = \mathbf{0} \quad \text{and} \quad l+m+n = 0.$$

The converse of this is true as now will be shown.

Condition for three points to be collinear

If $p\mathbf{a}+q\mathbf{b}+r\mathbf{c} = \mathbf{0}$ where **a**, **b**, **c** are the position vectors relative to an origin of the points A, B, C and $p+q+r = 0$ then A, B, C are collinear.

First proof. We have

$$p+q+r = 0 \text{ and } p\mathbf{a}+q\mathbf{b}+r\mathbf{c} = \mathbf{0}.$$

$$\therefore \quad p\mathbf{a}-(p+r)\,\mathbf{b}+r\mathbf{c} = \mathbf{0},$$

$$\therefore \quad \mathbf{b} = \frac{p\mathbf{a}+r\mathbf{c}}{p+r}.$$

Therefore B is the point which divides the straight line AC internally or externally in the ratio $r:p$ according as whether $r:p$ is positive or negative.

Therefore A, B, C are collinear.

Second proof. Let the origin be O. Then

$$p\mathbf{OA}+q\mathbf{OB}+r\mathbf{OC} = \mathbf{0}.$$

$$\therefore \quad p(\mathbf{OB}+\mathbf{BA})+q(\mathbf{OB})+r(\mathbf{OB}+\mathbf{BC}) = \mathbf{0},$$

$$\therefore \quad (p+q+r)\,\mathbf{OB}+p\mathbf{BA}+r\mathbf{BC} = \mathbf{0}.$$

But

$$p+q+r = 0,$$

$$\therefore \quad p\mathbf{BA}+r\mathbf{BC} = \mathbf{0},$$

$$\therefore \quad p\mathbf{BA} = r\mathbf{CB}.$$

If $p:r$ is positive we have Fig. 4.4 (*a*).

If $p:r$ is negative we have Fig. 4.4 (*b*) or Fig. 4.4 (*c*).

Therefore A, B, C are collinear.

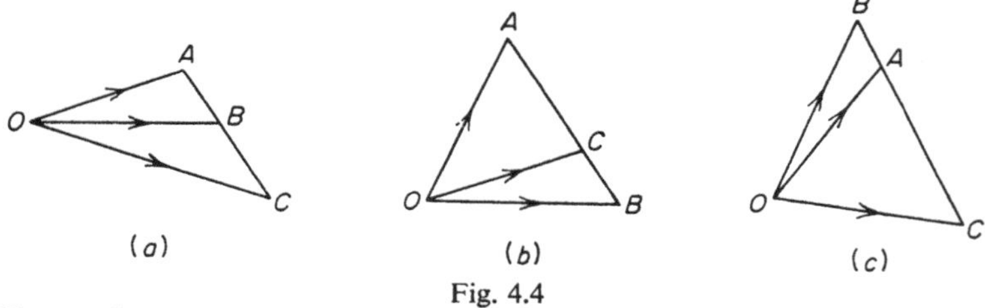

Fig. 4.4

Examples

(1) *ABC is a triangle with D the mid-point of BC and E a point on AC such that $AE:EC = 2:1$. Prove that the sum of the vectors* **BA**, **CA**, 2**BC** *is parallel to* **DE**.

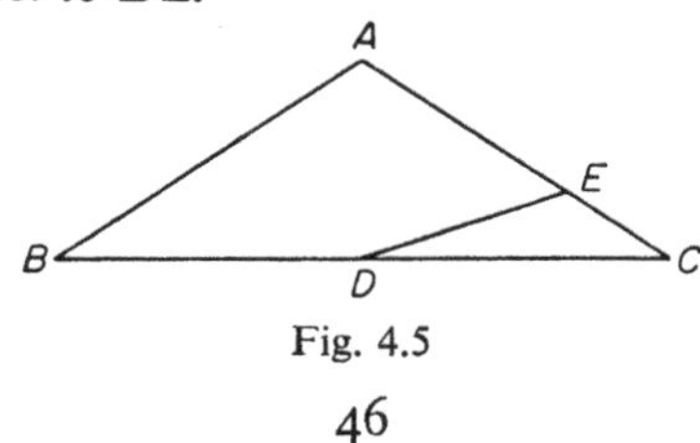

Fig. 4.5

Let $\mathbf{a}$, $\mathbf{b}$, $\mathbf{c}$, $\mathbf{d}$ and $\mathbf{e}$ be the position vectors relative to an origin O of the points A, B, C, D and E (Fig. 4.5).

$$\mathbf{d} = \frac{\mathbf{b}+\mathbf{c}}{2},$$

$$\mathbf{e} = \frac{\mathbf{a}+2\mathbf{c}}{3}.$$

$$\begin{aligned}\mathbf{DE} &= \mathbf{e}-\mathbf{d}\\ &= \frac{\mathbf{a}+2\mathbf{c}}{3}-\frac{\mathbf{b}+\mathbf{c}}{2}\\ &= \frac{2\mathbf{a}-3\mathbf{b}+\mathbf{c}}{6}.\end{aligned}$$

$$\begin{aligned}\mathbf{BA}+\mathbf{CA}+2\mathbf{BC} &= (\mathbf{a}-\mathbf{b})+(\mathbf{a}-\mathbf{c})+2(\mathbf{c}-\mathbf{b})\\ &= 2\mathbf{a}-3\mathbf{b}+\mathbf{c}\\ &= 6\mathbf{DE}.\end{aligned}$$

Therefore $\mathbf{BA}+\mathbf{CA}+2\mathbf{BC}$ is parallel to $\mathbf{DE}$.

(2) *Show that if* $7\mathbf{AB}-2\mathbf{AC}-5\mathbf{AD} = \mathbf{0}$ *the points* B, C *and* D *are collinear.*

Using the condition for collinearity since

$$7\mathbf{AB}-2\mathbf{AC}-5\mathbf{AD} = \mathbf{0}$$

and sum of coefficients $= 7-2-5$

$$= 0,$$

then B, C and D are collinear.

Centroid of a number of points

Definition. *The centroid or mean centre of n points* $A_1, A_2, \ldots, A_n$ *with position vectors* $\mathbf{a}_1, \mathbf{a}_2, \ldots, \mathbf{a}_n$ *respectively is the point with position vector*

$$\frac{\mathbf{a}_1+\mathbf{a}_2+\ldots+\mathbf{a}_n}{n}.$$

If $m_1, m_2, \ldots, m_n$ *are real numbers the point with position vector*

$$\frac{m_1\mathbf{a}_1+m_2\mathbf{a}_2+\ldots+m_n\mathbf{a}_n}{m_1+m_2+\ldots+m_n}$$

is defined as the centroid or weighted mean centre of the given points with associated numbers or weights $m_1, m_2, \ldots, m_n$.

There are two important theorems concerning centroids; the first showing that the position of the centroid is independent of the origin taken for the position vectors, and the second showing how the centroid of a number of points is obtained from a consideration of the centroids of systems forming the points.

Theorem I. *The centroid is independent of the origin.*

Proof. Consider points A, B, C, ... with associated numbers p, q, r, ... Let the position vector of point A relative to O be $\mathbf{a}$ and relative to O_1 be $\mathbf{a}_1$ (Fig. 4.6). Let position vector of O_1 relative to O be $\mathbf{h}$.

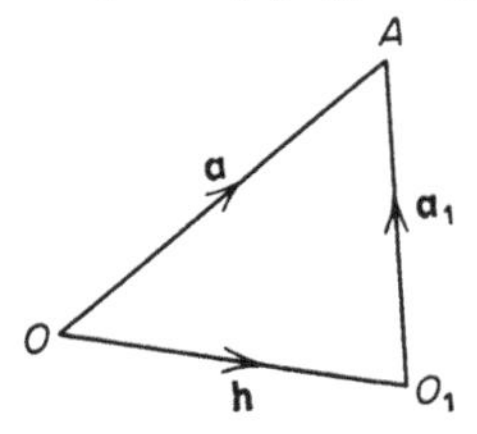

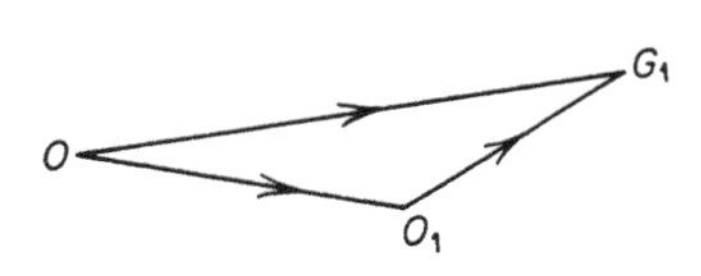

Fig. 4.6

Then
$$\mathbf{OO}_1+\mathbf{O}_1\mathbf{A} = \mathbf{OA},$$
$$\therefore \quad \mathbf{h}+\mathbf{a}_1 = \mathbf{a},$$
$$\therefore \quad \mathbf{a}_1 = \mathbf{a}-\mathbf{h}.$$

Similarly, if $\mathbf{b}$, $\mathbf{c}$, ... are position vectors of the points B, C, ... relative to the origin O then $(\mathbf{b}-\mathbf{h})$, $(\mathbf{c}-\mathbf{h})$, ... are their position vectors relative to O_1.

Let G and G_1 be the centroids of the points with O and O_1 as origins respectively.

$$\therefore \quad \mathbf{OG} = \frac{p\mathbf{a}+q\mathbf{b}+r\mathbf{c}+\dots}{p+q+r+\dots},$$

and
$$\mathbf{O}_1\mathbf{G}_1 = \frac{p\mathbf{a}_1+q\mathbf{b}_1+r\mathbf{c}_1+\dots}{p+q+r+\dots}$$
$$= \frac{p(\mathbf{a}-\mathbf{h})+q(\mathbf{b}-\mathbf{h})+r(\mathbf{c}-\mathbf{h})+\dots}{p+q+r+\dots}$$
$$= \frac{(p\mathbf{a}+q\mathbf{b}+r\mathbf{c}+\dots)-\mathbf{h}(p+q+r+\dots)}{p+q+r+\dots}$$
$$= \frac{p\mathbf{a}+q\mathbf{b}+r\mathbf{c}+\dots}{p+q+r+\dots}-\mathbf{h}.$$

Now
$$\mathbf{OG_1} = \mathbf{OO_1}+\mathbf{O_1G_1},$$
$$\therefore\quad \mathbf{OG_1} = \mathbf{h}+\frac{p\mathbf{a}+q\mathbf{b}+r\mathbf{c}+\ldots}{p+q+r+\ldots}-\mathbf{h}$$
$$= \frac{p\mathbf{a}+q\mathbf{b}+r\mathbf{c}+\ldots}{p+q+r+\ldots},$$
$$\therefore\quad \mathbf{OG_1} = \mathbf{OG}.$$

Therefore G_1 coincides with G.
Therefore centroid is independent of the origin.

Theorem II. *If H is the centroid of a system of points $A, B, C, \ldots$ with associated numbers $p, q, r, \ldots$ and H' is the centroid of a second system of points $A', B'\ C' \ldots$ with associated numbers $p', q'\ r', \ldots$ then the centroid of all the points is the centroid of the two points H and H' with associated numbers $(p+q+r+\ldots)$ and $(p'+q'+r'+\ldots)$ respectively.*

Proof. Let the origin be O and G the centroid of all the points.

$$\mathbf{OH} = \frac{p\mathbf{a}+q\mathbf{b}+r\mathbf{c}+\ldots}{p+q+r+\ldots} = \frac{\Sigma(p\mathbf{a})}{\Sigma p},$$
$$\mathbf{OH'} = \frac{p'\mathbf{a}'+q'\mathbf{b}'+r'\mathbf{c}'+\ldots}{p'+q'+r'+\ldots} = \frac{\Sigma(p'\mathbf{a}')}{\Sigma p'},$$
$$\mathbf{OG} = \frac{(p\mathbf{a}+q\mathbf{b}+r\mathbf{c}+\ldots)+(p'\mathbf{a}'+q'\mathbf{b}'+r'\mathbf{c}'+\ldots)}{(p+q+r+\ldots)+(p'+q'+r'+\ldots)}.$$
$$\therefore\quad \mathbf{OG} = \frac{\Sigma(p\mathbf{a})+\Sigma(p'\mathbf{a}')}{\Sigma p+\Sigma p'}$$
$$= \frac{(\Sigma p)\,\mathbf{OH}+(\Sigma p')\,\mathbf{OH'}}{\Sigma p+\Sigma p'}.$$

But this represents the position vector of the centroid of points H and H' associated with the numbers Σp and $\Sigma p'$ respectively. Thus the theorem is proved.

The theorem can be extended for more than two systems, that is,

$$\mathbf{OG} = \frac{(\Sigma p)\,\mathbf{OH}+(\Sigma p')\,\mathbf{OH'}+(\Sigma p'')\,\mathbf{OH''}+\ldots}{\Sigma p+\Sigma p'+\Sigma p''+\ldots}.$$

Examples

(1) *Show that the centroid of the points A and B with associated numbers p and q respectively is the point C on AB such that*

$$AC:CB = q:p.$$

Let **a** and **b** be the position vectors relative to a fixed origin of the points A and B respectively.

Position vector of centroid $= \dfrac{p\mathbf{a}+q\mathbf{b}}{p+q}$.

Position vector of point $C = \dfrac{p\mathbf{a}+q\mathbf{b}}{p+q}$.

Therefore centroid and the point C coincide.

Note. When $p = q = 1$, the centroid is the mid-point of AB.

(2) *Prove that the centroid of the vertices of a triangle is a point of trisection of a median and hence deduce that the medians are concurrent.*

Let D be the mid-point of BC and G the centroid of A, B and C (Fig. 4.7). Let **a**, **b** and **c** be the position vectors of A, B and C relative to a fixed origin.

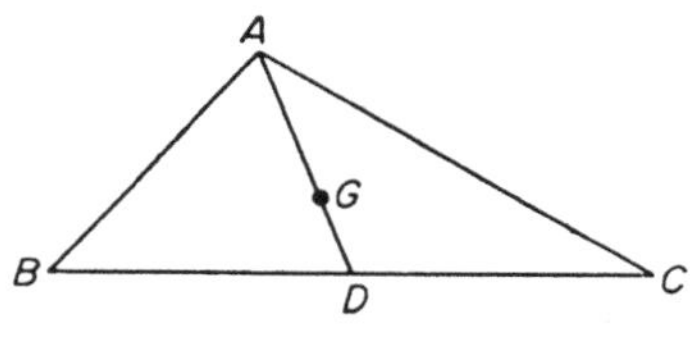

Fig. 4.7

Position vector of $G = \dfrac{\mathbf{a}+\mathbf{b}+\mathbf{c}}{3}$.

Position vector of $D = \dfrac{\mathbf{b}+\mathbf{c}}{2}$.

Position vector of centroid of B and $C = \dfrac{\mathbf{b}+\mathbf{c}}{2}$.

Therefore D is the centroid of B and C.

Therefore centroid of A, B and C must lie on AD, i.e. G is on AD.

Suppose
$$AG:GD = m:n.$$

Therefore position vector of $G = \dfrac{n\mathbf{a}+m\dfrac{\mathbf{b}+\mathbf{c}}{2}}{m+n}$,

$$\therefore \quad \frac{n\mathbf{a}+m\dfrac{\mathbf{b}+\mathbf{c}}{2}}{m+n} = \frac{\mathbf{a}+\mathbf{b}+\mathbf{c}}{3},$$

$$\therefore \quad m:n = 2:1.$$

Therefore centroid is a point of trisection of the median AD. Since the position vector of the centroid is symmetrical in **a, b** and **c** the centroid is a point of trisection of all the medians and thus all the medians are concurrent. The point of intersection of the medians, i.e. the centroid is one third of the way up each median from the base.

(3) *ABCD is a tetrahedron. Show that the lines joining the vertices of the tetrahedron to the centroids of opposite faces intersect in a point dividing these lines in the ratio* 3:1. *Also show that this point is the centroid of the vertices of the tetrahedron.*

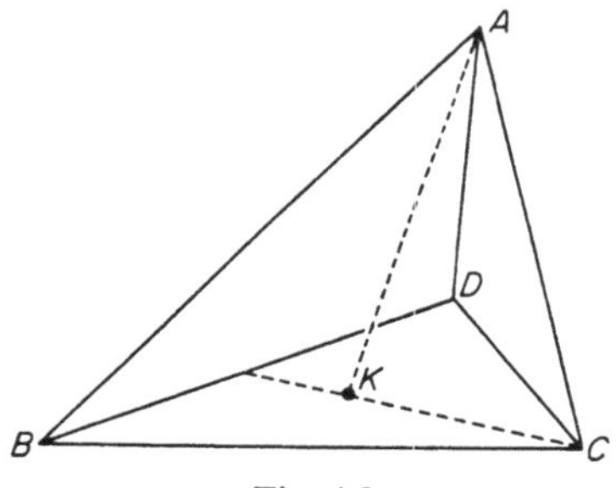

Fig. 4.8

Let the position vectors of A, B, C and D relative to a fixed origin be **a, b, c** and **d** (Fig. 4.8). Let centroid of face BCD be K.

Position vector of centroid of B, C and $D = \dfrac{\mathbf{b}+\mathbf{c}+\mathbf{d}}{3}$.

Centroid of vertices A, B, C and D must lie on AK.

But position vector of centroid of A, B, C and $D = \dfrac{\mathbf{a}+\mathbf{b}+\mathbf{c}+\mathbf{d}}{4}$

$$= \frac{1.\mathbf{a}+3\dfrac{\mathbf{b}+\mathbf{c}+\mathbf{d}}{3}}{4}.$$

But this is the position vector of a point G on AK such that $AG:GK = 3:1$.

By symmetry G must also lie on the corresponding lines through B, C and D.

Therefore the lines joining the vertices to the centroids of the opposite faces are concurrent at a point of quadrisection of each line, this point also being the centroid of the four vertices.

(4) *Prove that if G is the centroid of n points $A_1, A_2, \ldots, A_n$ and G' is the centroid of n points $B_1, B_2, \ldots, B_n$ then*

$$\mathbf{A_1B_1}+\mathbf{A_2B_2}+\ldots+\mathbf{A_nB_n} = n\mathbf{GG'}.$$

Let the position vectors of $A_1, A_2, \ldots, A_n$ and $B_1, B_2, \ldots, B_n$ relative to a fixed origin be

$$\mathbf{a}_1, \mathbf{a}_2, \ldots, \mathbf{a}_n \quad \text{and} \quad \mathbf{b}_1, \mathbf{b}_2, \ldots, \mathbf{b}_n \quad \text{respectively.}$$

Then

$$\begin{aligned}\mathbf{A_1B_1}+\mathbf{A_2B_2}+\ldots+\mathbf{A_nB_n} &= (\mathbf{b}_1-\mathbf{a}_1)+(\mathbf{b}_2-\mathbf{a}_2)+\ldots+(\mathbf{b}_n-\mathbf{a}_n)\\ &= (\mathbf{b}_1+\mathbf{b}_2+\ldots+\mathbf{b}_n)-(\mathbf{a}_1+\mathbf{a}_2+\ldots+\mathbf{a}_n).\end{aligned}$$

Now position vector of $G = \dfrac{\mathbf{a}_1+\mathbf{a}_2+\ldots+\mathbf{a}_n}{n} = \mathbf{g}$, and position vector of $G' = \dfrac{\mathbf{b}_1+\mathbf{b}_2+\ldots+\mathbf{b}_n}{n} = \mathbf{g}'$.

$$\begin{aligned}\therefore\quad \mathbf{A_1B_1}+\mathbf{A_2B_2}+\ldots+\mathbf{A}_n\mathbf{B}_n &= n(\mathbf{g}'-\mathbf{g})\\ &= n\mathbf{GG'}.\end{aligned}$$

Exercise 4

(1) Prove that the position vector of the point Q which divides AB externally such that $AQ:QB = m:n$ ($m:n$ positive) is $\dfrac{n\mathbf{a}-m\mathbf{b}}{n-m}$, where $\mathbf{a}$ and $\mathbf{b}$ are the position vectors of A and B relative to a fixed origin.

(2) The position vectors of the points P and Q are $\mathbf{p}$ and $\mathbf{q}$ respectively. PQ is divided internally at R and externally at S so that $PR:RQ = PS:QS = m:1$. Show that $\mathbf{RS} = \dfrac{2m(\mathbf{p}-\mathbf{q})}{1-m^2}$.

(3) The medians of a triangle ABC intersect at G. Prove that $\mathbf{GA}+\mathbf{GB}+\mathbf{GC} = \mathbf{0}$.

(4) $OABC$ is a tetrahedron with $\mathbf{OA} = \mathbf{a}$, $\mathbf{OB} = \mathbf{b}$ and $\mathbf{OC} = \mathbf{c}$. P and Q are the mid-points of OA and BC respectively. Find in terms of $\mathbf{a}$, $\mathbf{b}$ and $\mathbf{c}$ the position vector of the mid-point of PQ relative to

O as the origin, and hence deduce that the joins of the mid-points of opposite edges of a tetrahedron are concurrent and bisect each other.

(5) G and G' are the mid-points of PQ and $P'Q'$ respectively. Prove that $\mathbf{PP'}+\mathbf{QQ'} = 2\mathbf{GG'}$.

(6) Prove that the diagonals of a parallelepiped are concurrent and bisect one another.

(7) The position vectors of four points P, Q, R and S are $\mathbf{a}$, $\mathbf{a}+\mathbf{p}$, $\mathbf{a}+\mathbf{q}$, and $\mathbf{a}+\mathbf{p}+\mathbf{q}$ respectively. Prove that $PQSR$ is a parallelogram.

(8) The position vectors of three points are $\mathbf{p}$, $\mathbf{q}$ and $5\mathbf{p}-4\mathbf{q}$. Show that the points are collinear.

(9) Show that the centroid of two points A and B associated with the numbers 3 and 1 respectively is the point of quadrisection of AB nearer to A.

(10) $ABCD$ is a quadrilateral with P and Q the mid-points of AB and DC respectively. Prove that the centroid of A, B, C and D is the mid-point of PQ. Hence deduce that the straight lines joining the mid-points of opposite sides and the straight line joining the mid-points of the diagonals are concurrent.

(11) $ABCD$ is a parallelogram. E is on AD such that

$AE:ED = 1:n-1$. BE meets the diagonal AC at P. If

$$AP:AC = 1:x$$

show that $\mathbf{AB}+n\mathbf{AE} = x\mathbf{AP}$. Hence show that BE divides AC in the ratio of $1:n$.

(12) Show that the centroid of the points A, B and C with associated numbers a, b and c respectively is the incentre of the triangle ABC with sides BC, CA and AB of length a, b and c respectively.

(13) Prove that $p_1\mathbf{A_1B_1}+p_2\mathbf{A_2B_2}+\ldots+p_n\mathbf{A_nB_n} = N\mathbf{GG'}$ where $N = p_1+p_2+\ldots+p_n$, G is the centroid of $A_1, A_2, \ldots, A_n$ with associated numbers $p_1, p_2, \ldots, p_n$ respectively, and G' is the centroid of $B_1, B_2, \ldots, B_n$ with associated numbers $p_1, p_2, \ldots, p_n$ respectively.

5

PROJECTION AND COMPONENTS OF A VECTOR

Before we discuss what is meant by the projection of a vector we must define the angle between two vectors.

Angle between two vectors

Definition. *The angle between two vectors* **a**, **b** *is the angle* AOB *where* **OA** = **a**, **OB** = **b**.

Referring to Fig. 5.1 the angle between the vectors **a**, **b** is denoted by θ. The usual convention is used, i.e. angles measured in an anti-clockwise sense are positive.

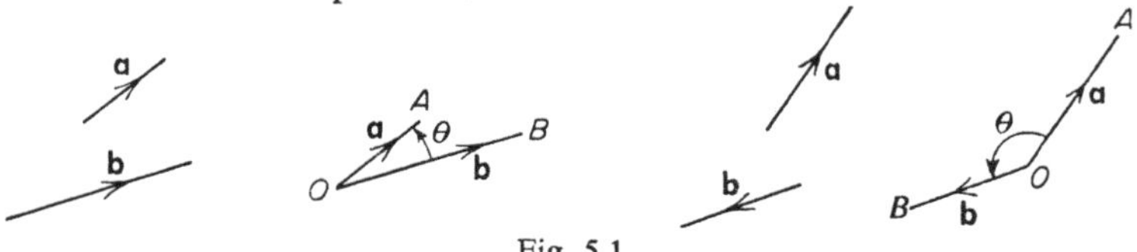

Fig. 5.1

Projection of a vector upon a vector

Definition. *The projection of a vector* **b** *upon a vector* **a** *is defined as the number* $|\mathbf{b}| \cos\theta$ *where* θ *is the angle between the vectors* **a**, **b**.

For convenience we shall denote the projection of a vector **b** upon another vector by $p(\mathbf{b})$. The geometrical significance of the definition is seen from Figs. 5.2 and 5.3.

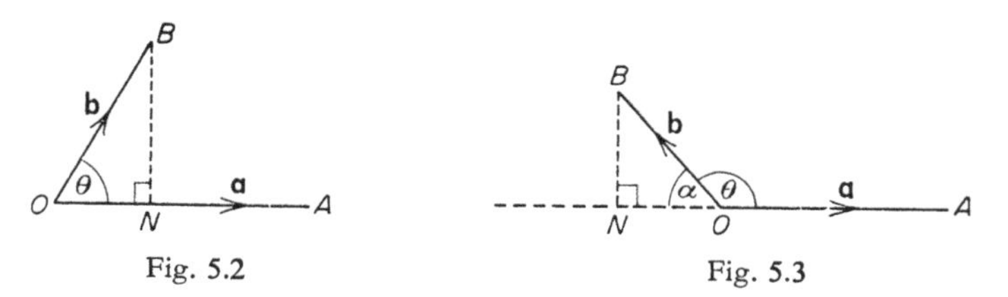

Fig. 5.2

Fig. 5.3

Let **OA** = **a** and **OB** = **b**. From B draw BN perpendicular to OA or AO produced.

In Fig. 5.2

$$p(\mathbf{b}) \text{ on } \mathbf{a} = |\mathbf{b}|\cos\theta = OB\cos\theta = ON.$$

In Fig. 5.3

$$p(\mathbf{b}) \text{ on } \mathbf{a} = |\mathbf{b}|\cos\theta = OB\cos(180-\alpha) = -OB\cos\alpha = -ON.$$

From these it is seen that the projection is either a positive or negative number depending whether θ is acute or obtuse.

The projection of a vector **b** upon a vector **a** is also known as the resolute or resolved part of the vector **b** upon the vector **a**.

We now obtain an important theorem on the projection of the sum of several vectors.

The angle between **b** and **p** is acute in Fig. 5.4 and obtuse in Fig. 5.5. Hence the projections of **b** on **p** are opposite in signs in the two figures.

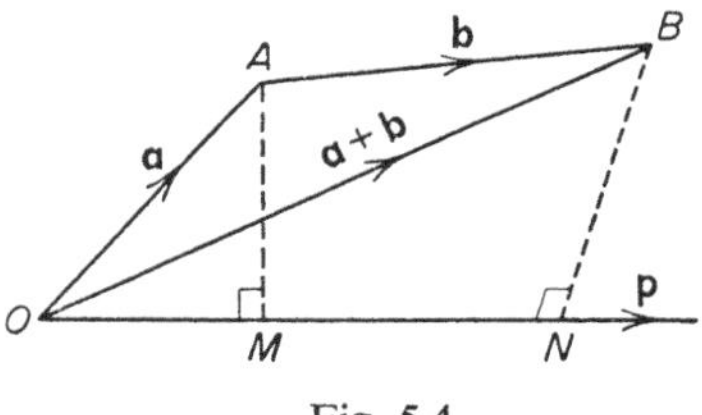

Fig. 5.4

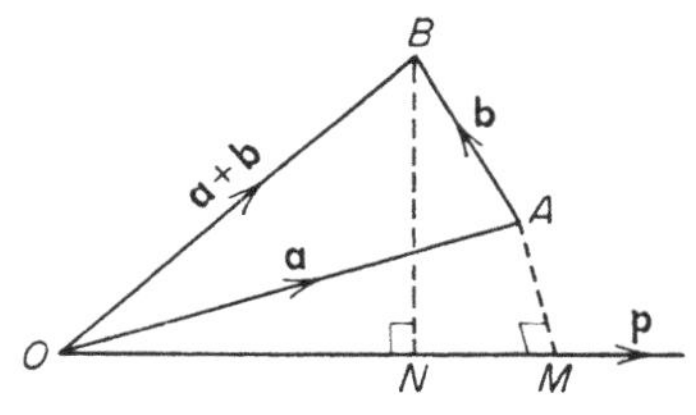

Fig. 5.5

From A, B draw perpendiculars AM, BN on to the vector **p**, the vectors not necessarily being in the same plane.

We shall consider the projections of **a**, **b** and $\mathbf{a}+\mathbf{b}$ upon **p** in each case.

In Fig. 5.4, $\quad p(\mathbf{a}) = OM, \quad p(\mathbf{b}) = MN, \quad p(\mathbf{a}+\mathbf{b}) = ON.$

But $\quad ON = OM+MN,$

$$\therefore \quad p(\mathbf{a}+\mathbf{b}) = p(\mathbf{a})+p(\mathbf{b}).$$

In Fig. 5.5, $\quad p(\mathbf{a}) = OM, \quad p(\mathbf{b}) = -MN, \quad p(\mathbf{a}+\mathbf{b}) = ON.$

But $\quad OM = ON+NM,$

$$\therefore \quad p(\mathbf{a}) = p(\mathbf{a}+\mathbf{b})-p(\mathbf{b}),$$

$$\therefore \quad p(\mathbf{a}+\mathbf{b}) = p(\mathbf{a})+p(\mathbf{b}).$$

In the same way by extending the argument we can show that

$$p(\mathbf{a}+\mathbf{b}+\mathbf{c}\ldots) = p(\mathbf{a})+p(\mathbf{b})+p(\mathbf{c})+\ldots.$$

Thus the projection of a sum of vectors on a given vector is equal to the sum of the projections of the separate vectors on the given vector.

Projection of a vector upon a plane

In Fig. 5.6 we require the projection of the vector **AB** on the plane XY.

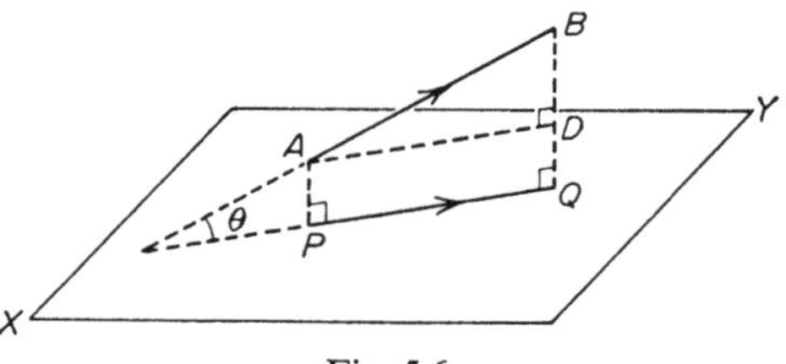

Fig. 5.6

Definition. *The projection of a vector* **AB** *on a plane is the vector* **PQ** *where P, Q are the feet of the perpendiculars from A, B respectively to the plane.*

Since the projection **PQ** is a vector we denote the projection of a vector **a** upon a plane by the bold type **P(a)**.

If θ is the angle between **AB** and the plane then

$$|\mathbf{PQ}| = AD = |\mathbf{AB}| \cos\theta,$$

where AD is perpendicular to BQ.

We now obtain a theorem about the projection of the sum of several vectors upon a plane.

In Fig. 5.7 **p, q, r** are the projections of the vectors **a, b, (a+b)** on the plane XY, i.e. $\mathbf{P(a)} = \mathbf{p}$, $\mathbf{P(b)} = \mathbf{q}$, $\mathbf{P(a+b)} = \mathbf{r}$.

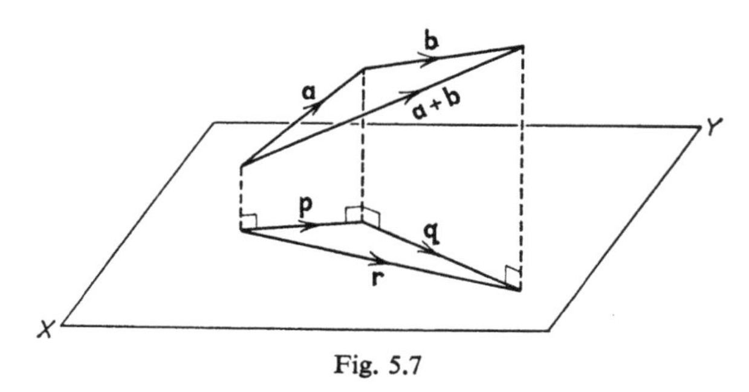

Fig. 5.7

Now $$\mathbf{r} = \mathbf{p}+\mathbf{q},$$

$$\therefore \quad \mathbf{P}(\mathbf{a}+\mathbf{b}) = \mathbf{P}(\mathbf{a})+\mathbf{P}(\mathbf{b}).$$

In the same way we can deduce that

$$\mathbf{P}(\mathbf{a}+\mathbf{b}+\mathbf{c}...) = \mathbf{P}(\mathbf{a})+\mathbf{P}(\mathbf{b})+\mathbf{P}(\mathbf{c})+....$$

Thus the projection of a sum of vectors on a given plane is equal to the sum of the projections of the separate vectors on the plane.

Components of a vector

The position of a point P can be completely specified by the vector **OP** if we agree upon a fixed origin O. We have previously used the term position vector of P relative to the origin O for the vector **OP**. Furthermore, if we have agreed axes OX, OY the point P is completely specified in position by its co-ordinates, namely OA, AP (Fig. 5.8).

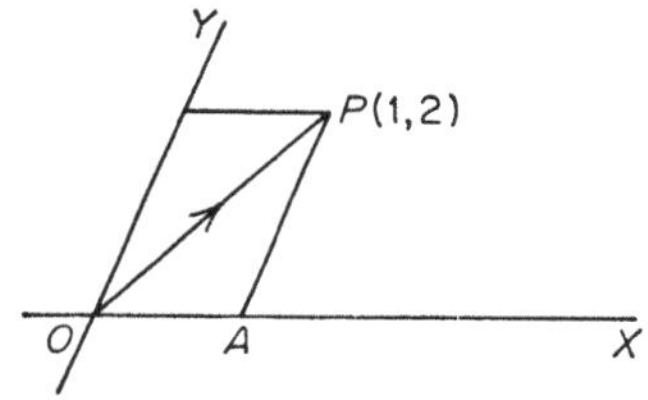

Fig. 5.8

Now since **OA**+**AP** = **OP**, **OA** and **AP** are the components parallel to the axes of the vector **OP**. In particular if P is the point (1, 2) then $OA = 1$ and $AP = 2$ and thus the co-ordinates of a point may be defined as the components of the position vector. This means that when we label a point $P(1, 2)$ we may think of the separate numbers 1, 2 as co-ordinates of the point P, or we may think of the number-pair (1, 2) as the position vector of P with the numbers 1, 2 referring to the components of **OP**.

We shall now show generally that a vector can be uniquely expressed in terms of two components in two given directions or three components in three given directions not parallel to the same plane.

The process of expressing a vector into components is known as the decomposition or resolution of the vector.

Decomposition of a vector in two directions

Suppose $\mathbf{r}$ is a given vector and $\hat{\mathbf{u}}$, $\hat{\mathbf{v}}$ are two unit vectors in any two directions. Take any point O as origin and let $\mathbf{OP} = \mathbf{r}$. Draw a triangle OPA with OA, AP parallel to the directions of $\hat{\mathbf{u}}$, $\hat{\mathbf{v}}$ respectively (Fig. 5.9).

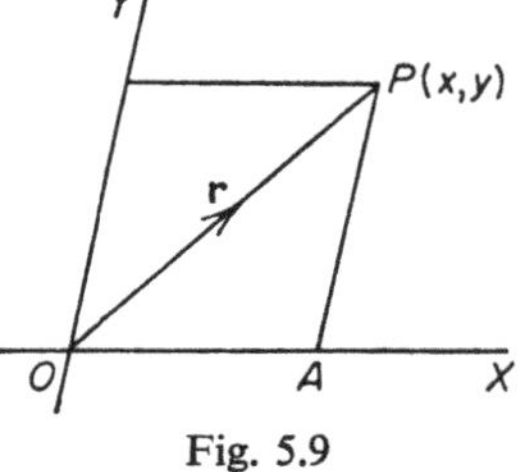

Fig. 5.9

Then $$\mathbf{OP} = \mathbf{OA}+\mathbf{AP}.$$

This resolution into the vector components $\mathbf{OA}$, $\mathbf{AP}$ is unique since only one triangle with sides parallel to the given directions can be constructed with $\mathbf{OP}$ as side.

Through O draw OX, OY with the same direction and sense as $\hat{\mathbf{u}}$, $\hat{\mathbf{v}}$. Then taking OX, OY as axes let P have the co-ordinates (x, y). We now have

$$\mathbf{OA} = x\hat{\mathbf{u}}, \quad \mathbf{AP} = y\hat{\mathbf{v}},$$

and hence we can write $\qquad \mathbf{r} = x\hat{\mathbf{u}}+y\hat{\mathbf{v}}.$

Here $x\hat{\mathbf{u}}$, $y\hat{\mathbf{v}}$ are known as the vector components of $\mathbf{r}$ and the numbers x, y are known as the components of $\mathbf{r}$ for the given directions.

As we have seen it is convenient to denote the vector $x\hat{\mathbf{u}}+y\hat{\mathbf{v}}$ by the number-pair (x, y). Thus the vector $(3, -4)$ denotes the vector with components $(3, -4)$ for the given directions, i.e. the vector $3\hat{\mathbf{u}}-4\hat{\mathbf{v}}$.

Addition of vectors

Consider now several vectors $\mathbf{r}_1, \mathbf{r}_2, \mathbf{r}_3, \ldots$.

In component form they are

$$\mathbf{r}_1 = x_1\hat{\mathbf{u}}+y_1\hat{\mathbf{v}} = (x_1, y_1),$$
$$\mathbf{r}_2 = x_2\hat{\mathbf{u}}+y_2\hat{\mathbf{v}} = (x_2, y_2),$$
$$\mathbf{r}_3 = x_3\hat{\mathbf{u}}+y_3\hat{\mathbf{v}} = (x_3, y_3).$$

Adding we have

$$\mathbf{r}_1+\mathbf{r}_2+\mathbf{r}_3+\ldots = (x_1+x_2+x_3+\ldots)\,\hat{\mathbf{u}}+(y_1+y_2+y_3+\ldots)\,\hat{\mathbf{v}}.$$

Thus the vector $(\mathbf{r}_1+\mathbf{r}_2+\mathbf{r}_3+\ldots)$ has components $(x_1+x_2+x_3+\ldots)$, $(y_1+y_2+y_3+\ldots)$. In general we see that the components of the sum of several vectors is obtained by adding up the corresponding components of the separate vectors.

Special case

The case of rectangular axes, i.e. axes at right angles is very important. In this case it is usual to indicate the unit vectors $\hat{\mathbf{u}}$, $\hat{\mathbf{v}}$ by **i**, **j** respectively, i.e. **i**, **j** are the unit vectors with the same direction and sense as the positive OX, OY axes.

In this case the length of **OP** is easily calculated by using Pythagoras's Theorem (Fig. 5.10):

$$OP = \sqrt{(OA^2+AP^2)} = \sqrt{(x^2+y^2)}.$$

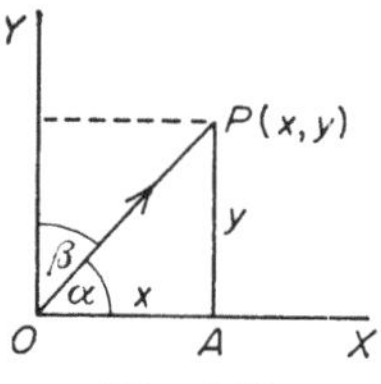

Fig. 5.10

Also if α, β are the angles that **OP** make with the positive OX, OY axes respectively, we have

$$\cos\alpha = \frac{x}{|\mathbf{OP}|}, \quad \cos\beta = \frac{y}{|\mathbf{OP}|}.$$

These are known as the direction cosines of **OP**.

Examples

(1) *If* $\mathbf{a} = 7\mathbf{i}+4\mathbf{j}$ *find* $|\mathbf{a}|$ *and the angle* **a** *makes with the x axis.*

$$\mathbf{a} = 7\mathbf{i}+4\mathbf{j},$$
$$\therefore\quad |\mathbf{a}|^2 = 7^2+4^2 = 65,$$
$$\therefore\quad |\mathbf{a}| = \sqrt{65}.$$

Angle is $\tan^{-1} y/x$, i.e. $\tan^{-1}\frac{4}{7}$.

(2) *If* $\mathbf{a} = 3\mathbf{i}+4\mathbf{j}$, $\mathbf{b} = 2\mathbf{i}-3\mathbf{j}$ *and* $\mathbf{c} = -\mathbf{i}+\mathbf{j}$ *show that* $\mathbf{a}+3\mathbf{b}+5\mathbf{c}$ *is parallel to the x axis.*

$$\begin{aligned}\mathbf{a}+3\mathbf{b}+5\mathbf{c} &= (3\mathbf{i}+4\mathbf{j})+3(2\mathbf{i}-3\mathbf{j})+5(-\mathbf{i}+\mathbf{j})\\ &= (3+6-5)\,\mathbf{i}+(4-9+5)\mathbf{j}\\ &= 4\mathbf{i}.\end{aligned}$$

Therefore $\mathbf{a}+3\mathbf{b}+5\mathbf{c}$ is parallel to the x axis.

Decomposition of a vector into three non-coplanar directions

We shall now show how any vector $\mathbf{r}$ can be expressed as the sum of three vectors which are parallel to any three vectors which are not in the same plane.

Let $\hat{\mathbf{u}}$, $\hat{\mathbf{v}}$, $\hat{\mathbf{w}}$ be unit vectors in the three given non-coplanar directions. Take any point O as origin and let $\mathbf{OP} = \mathbf{r}$. Draw a parallelepiped on $\mathbf{OP}$ as diagonal and with edges OA, OB, OC parallel to $\hat{\mathbf{u}}$, $\hat{\mathbf{v}}$, $\hat{\mathbf{w}}$ respectively. Referring to Fig. 5.11 we have

$$\mathbf{OP} = \mathbf{OA}+\mathbf{AD}+\mathbf{DP}.$$

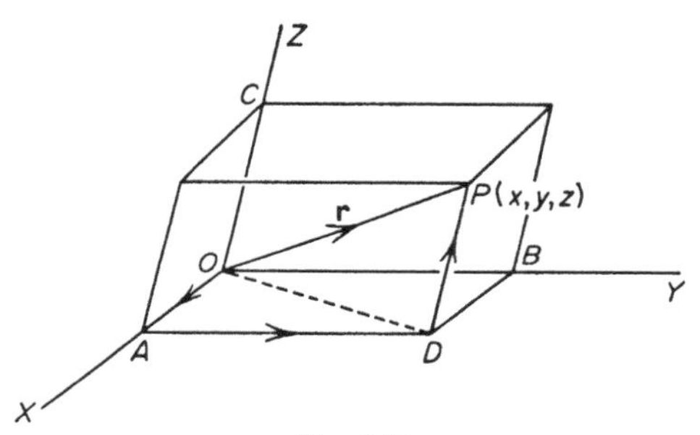

Fig. 5.11

Then $\mathbf{OP}$ has been expressed in terms of the vector components $\mathbf{OA}$, $\mathbf{AD}$, $\mathbf{DP}$. This resolution of $\mathbf{r}$ is unique because only one parallelepiped can be drawn on $\mathbf{OP}$ as diagonal and with edges parallel to the given directions.

Through O draw OX, OY, OZ with the same direction and sense as $\hat{\mathbf{u}}$, $\hat{\mathbf{v}}$, $\hat{\mathbf{w}}$. Then taking OX, OY, OZ as axes let P have the co-ordinates (x, y, z). We now can write

$$\mathbf{OA} = x\hat{\mathbf{u}}, \quad \mathbf{AD} = y\hat{\mathbf{v}}, \quad \mathbf{DP} = z\hat{\mathbf{w}},$$

$$\therefore \quad \mathbf{r} = x\hat{\mathbf{u}}+y\hat{\mathbf{v}}+z\hat{\mathbf{w}}.$$

As before $x\hat{\mathbf{u}}$, $y\hat{\mathbf{v}}$, $z\hat{\mathbf{w}}$ are known as the vector components of $\mathbf{r}$ and the numbers x, y, z are known as the components of $\mathbf{r}$ for the given directions. Also as before we can denote the vector $x\hat{\mathbf{u}}+y\hat{\mathbf{v}}+z\hat{\mathbf{w}}$ by the number-triple (x, y, z).

Addition of vectors

Given several vectors $\mathbf{r}_1$, $\mathbf{r}_2$, $\mathbf{r}_3$, ... we can obtain the vector sum by expressing each in component form and adding.

$$\mathbf{r}_1 = x_1\hat{\mathbf{u}}+y_1\hat{\mathbf{v}}+z_1\hat{\mathbf{w}},$$

$$\mathbf{r}_2 = x_2\hat{\mathbf{u}}+y_2\hat{\mathbf{v}}+z_2\hat{\mathbf{w}},$$

$$\mathbf{r}_3 = x_3\hat{\mathbf{u}}+y_3\hat{\mathbf{v}}+z_3\hat{\mathbf{w}},$$

$$\therefore \quad (\mathbf{r}_1+\mathbf{r}_2+\mathbf{r}_3+\ldots) = (x_1+x_2+x_3+\ldots)\hat{\mathbf{u}}+(y_1+y_2+y_3+\ldots)\hat{\mathbf{v}} + (z_1+z_2+z_3+\ldots)\hat{\mathbf{w}}.$$

Thus the components of the sum of a number of vectors is obtained by adding up the corresponding components of the separate vectors.

Multiplication by a number

Suppose we have a vector $\mathbf{r} = x\hat{\mathbf{u}}+y\hat{\mathbf{v}}+z\hat{\mathbf{w}}$ and we require the components of the vector $m\mathbf{r}$ where m is any real number. We have

$$m\mathbf{r} = m(x\hat{\mathbf{u}}+y\hat{\mathbf{v}}+z\hat{\mathbf{w}}) = mx\hat{\mathbf{u}}+my\hat{\mathbf{v}}+mz\hat{\mathbf{w}}.$$

Hence each component of $m\mathbf{r}$ is obtained by multiplying the corresponding component of $\mathbf{r}$ by m.

Special case

The most important and common case of resolution of vectors is that in which the three directions form a right-handed rectangular Cartesian co-ordinate frame. In this case the axes OX, OY, OZ are mutually perpendicular and the resulting parallelepiped is a cuboid (Fig. 5.12).

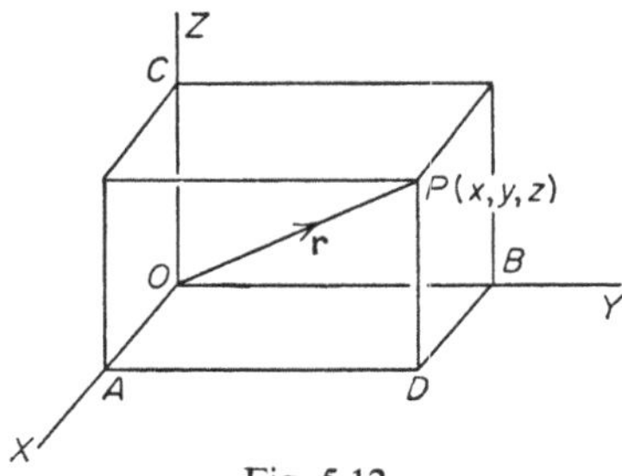

Fig. 5.12

If **i**, **j**, **k** are the unit vectors for this system then $\mathbf{i} = \hat{\mathbf{u}}, \mathbf{j} = \hat{\mathbf{v}}, \mathbf{k} = \hat{\mathbf{w}}$ and we write the vector **r** in component form as $\mathbf{r} = x\mathbf{i}+y\mathbf{j}+z\mathbf{k}$.

In this case the components x, y, z are also known as the projections, resolutes or resolved parts of **r** upon **i**, **j**, **k** respectively.

The sum of several vectors is given by

$$(\mathbf{r}_1+\mathbf{r}_2+\mathbf{r}_3+\ldots) = (x_1+x_2+x_3+\ldots)\,\mathbf{i}+(y_1+y_2+y_3+\ldots)\,\mathbf{j} \\ +(z_1+z_2+z_3+\ldots)\,\mathbf{k}.$$

The length of OP can be calculated by successive applications of Pythagoras's Theorem.

$$\begin{aligned} OP^2 &= OD^2+DP^2 \text{ (Fig. 5.12)} \\ &= OA^2+AD^2+DP^2 \\ &= x^2+y^2+z^2, \\ \therefore \quad OP &= \sqrt{(x^2+y^2+z^2)}. \end{aligned}$$

Thus the length of a vector is the square root of the sum of the squares of its rectangular components.

We make the following general observations:

(1) In all the above resolutions the number x is either positive or negative according as **OA** has the same sense as **i**, or the opposite sense. The same distinction of signs apply to the numbers y, z.

(2) The decomposition of a vector into three components is more useful than the decomposition into two components, for by the former decomposition we can deal with several vectors in space but the latter is only of use when the vectors are all coplanar.

(3) A vector may be resolved into three or more components in directions which are coplanar but this resolution is not unique because a polygon is not determined by its angles and one side.

In what follows the axes OX, OY, OZ are mutually perpendicular.

Direction cosines

Let $\mathbf{OP} = \mathbf{r}$ make angles of α, β and γ with the OX, OY and OZ axes respectively.

$$\therefore \quad x = r\cos\alpha, \quad y = r\cos\beta, \quad z = r\cos\gamma.$$

Cos α, cos β and cos γ are known as the direction cosines of **OP** and are usually denoted by l, m and n respectively.

$$\therefore \quad l = x/r, \quad m = y/r \quad \text{and} \quad n = z/r.$$

Three useful facts involving direction cosines will now be obtained.

(1) $l^2+m^2+n^2 = 1$. $\qquad \mathbf{r} = x\mathbf{i}+y\mathbf{j}+z\mathbf{k}$,

$$\therefore \quad r^2 = x^2+y^2+z^2,$$

$$\therefore \quad 1 = \frac{x^2}{r^2}+\frac{y^2}{r^2}+\frac{z^2}{r^2}.$$

$$\therefore \quad l^2+m^2+n^2 = 1.$$

(2) *The coefficients of* **i**, **j** *and* **k** *of a unit vector are its direction cosines.*

Let $\hat{\mathbf{r}}$ be the unit vector.

$$\therefore \quad \hat{\mathbf{r}} = x\mathbf{i}+y\mathbf{j}+z\mathbf{k}.$$

Now $$x = |\hat{\mathbf{r}}|\, l, \quad y = |\hat{\mathbf{r}}|\, m, \quad z = |\hat{\mathbf{r}}|\, n.$$

But $$|\hat{\mathbf{r}}| = 1.$$

$$\therefore \quad x = l, \quad y = m, \quad z = n.$$

(3) *$\cos\theta = ll_1+mm_1+nn_1$ where θ is the angle between two vectors whose direction cosines are l, m, n and l_1, m_1, n_1, respectively.*

Let the position vectors of the points P and P_1 relative to the origin O of the axes be $\mathbf{r}$ and $\mathbf{r}_1$, and the angle between **OP** and $\mathbf{OP}_1$ be θ.

$$\mathbf{r} = x\mathbf{i}+y\mathbf{j}+z\mathbf{k} = r(l\mathbf{i}+m\mathbf{j}+n\mathbf{k}),$$

$$\mathbf{r}_1 = x_1\mathbf{i}+y_1\mathbf{j}+z_1\mathbf{k} = r_1(l_1\mathbf{i}+m_1\mathbf{j}+n_1\mathbf{k}).$$

$$\mathbf{PP}_1 = \mathbf{r}_1-\mathbf{r}$$

$$= (r_1l_1-rl)\,\mathbf{i}+(r_1m_1-rm)\,\mathbf{j}+(r_1n_1-rn)\,\mathbf{k}.$$

Applying the cosine rule:

$$|\mathbf{PP}_1|^2 = |\mathbf{OP}|^2+|\mathbf{OP}_1|^2-2\,|\mathbf{OP}|\,|\mathbf{OP}_1|\cos\theta,$$

$$\therefore \quad \cos\theta = \frac{r^2+r_1^2-(r_1l_1-rl)^2-(r_1m_1-rm)^2-(r_1n_1-rn)^2}{2rr_1}$$

$$= \frac{r^2+r_1^2-r^2(l^2+m^2+n^2)+2rr_1(ll_1+mm_1+nn_1)-r_1^2(l_1^2+m_1^2+n_1^2)}{2rr_1}.$$

Since $$l^2+m^2+n^2 = l_1^2+m_1^2+n_1^2 = 1,$$

$$\cos\theta = ll_1+mm_1+nn_1.$$

Modulus and direction cosines of $\mathbf{PP_1}$

Let the position vectors of the points P and P_1 relative to the origin O of the axes be $\mathbf{r}$ and $\mathbf{r_1}$. Then

$$\mathbf{r} = x\mathbf{i}+y\mathbf{j}+z\mathbf{k},$$

$$\mathbf{r_1} = x_1\mathbf{i}+y_1\mathbf{j}+z_1\mathbf{k}.$$

$$\mathbf{PP_1} = \mathbf{r_1}-\mathbf{r}$$

$$= (x_1-x)\,\mathbf{i}+(y_1-y)\,\mathbf{j}+(z_1-z)\,\mathbf{k},$$

$$\therefore \quad |\mathbf{PP_1}| = \sqrt{\{(x_1-x)^2+(y_1-y)^2+(z_1-z)^2\}}.$$

The direction cosines of $\mathbf{PP_1}$ are

$$\frac{x_1-x}{|\mathbf{PP_1}|}, \quad \frac{y_1-y}{|\mathbf{PP_1}|}, \quad \frac{z_1-z}{|\mathbf{PP_1}|}.$$

Angle between the vectors PQ and RS

Suppose we require the angle between the free vectors **PQ** and **RS** (Fig. 5.13). Consider the two corresponding localized vectors **OA** and **OB** passing through the origin O of the axes. Then the angle between **OA** and **OB** is the required angle. We have shown that this angle is given by

$$\cos\theta = ll_1+mm_1+nn_1,$$

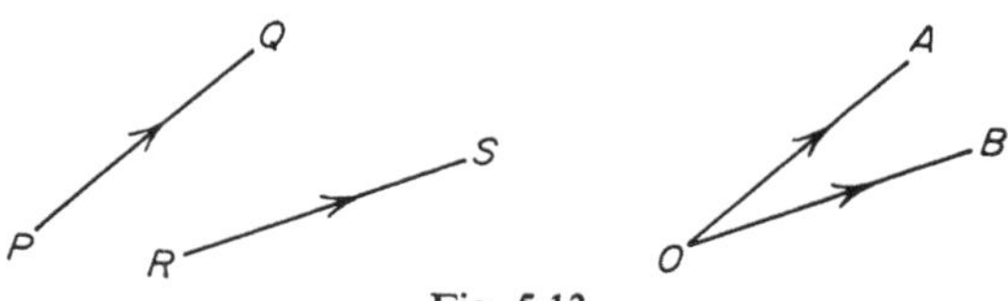

Fig. 5.13

where l, m, n and l_1, m_1, n_1, are the direction cosines of **OA** and **OB**. However, the direction cosines of the localized vectors **OA** and **OB** are the same as those of the free vectors **PQ** and **RS**. Hence the angle θ between the two vectors is given by

$$\cos\theta = ll_1+mm_1+nn_1,$$

where l, m, n and l_1, m_1, n_1 are the direction cosines of the vectors.

Examples

(1) *If the position vectors relative to the origin O of the axes of the points A and B are* $\mathbf{4i}+\mathbf{4j}-\mathbf{7k}$ *and* $\mathbf{5i}-\mathbf{2j}+\mathbf{6k}$ *respectively, find the direction cosines of* **OA, OB** *and* **AB** *and the angle between* **OA** *and* **AB**.

$$\mathbf{OA} = 4\mathbf{i}+4\mathbf{j}-7\mathbf{k} \quad \text{and} \quad \mathbf{OB} = 5\mathbf{i}-2\mathbf{j}+6\mathbf{k}.$$

$$\therefore \quad OA = \sqrt{(4^2+4^2+7^2)} \quad \text{and} \quad OB = \sqrt{(5^2+2^2+6^2)}$$
$$= 9, \qquad\qquad = \sqrt{65},$$

$$\therefore \quad \text{direction cosines are } \frac{4}{9}, \frac{4}{9}, \frac{-7}{9} \text{ and } \frac{5}{\sqrt{65}}, \frac{-2}{\sqrt{65}}, \frac{6}{\sqrt{65}}.$$

$$\mathbf{AB} = (5\mathbf{i}-2\mathbf{j}+6\mathbf{k})-(4\mathbf{i}+4\mathbf{j}-7\mathbf{k}) = \mathbf{i}-6\mathbf{j}+13\mathbf{k}.$$

$$\therefore \quad AB = \sqrt{(1^2+6^2+13^2)} = \sqrt{206},$$

$$\therefore \quad \text{direction cosines are } \frac{1}{\sqrt{206}}, \frac{-6}{\sqrt{206}}, \frac{13}{\sqrt{206}}.$$

$$\cos\theta = ll_1+mm_1+nn_1$$
$$= \frac{(4\times 1)+(4\times -6)+(-7\times 13)}{9\sqrt{206}},$$

$$\therefore \quad \text{angle is } \cos^{-1}\frac{-37}{3\sqrt{206}}.$$

(2) *Show that* $\mathbf{a} = 9\mathbf{i}+\mathbf{j}-6\mathbf{k}$ *and* $\mathbf{b} = 4\mathbf{i}-6\mathbf{j}+5\mathbf{k}$ *are at right angles to each other.*

Direction cosines of **a** are $\frac{9}{\sqrt{118}}, \frac{1}{\sqrt{118}}, \frac{-6}{\sqrt{118}}$.

Direction cosines of **b** are $\frac{4}{\sqrt{77}}, \frac{-6}{\sqrt{77}}, \frac{5}{\sqrt{77}}$.

$$\cos\theta = ll_1+mm_1+nn_1,$$

$$\therefore \quad \cos\theta = \frac{36-6-30}{\sqrt{118}.\sqrt{77}} = 0,$$

$$\therefore \quad \theta = 90^\circ.$$

Centroid of points

Let A, B, C, ... be n points (x_1, y_1, z_1), (x_2, y_2, z_2), (x_3, y_3, z_3)..., and O the origin of the axes.

Let $\mathbf{OA} = \mathbf{a}$, $\mathbf{OB} = \mathbf{b}$, $\mathbf{OC} = \mathbf{c}$, ...

$$\mathbf{a} = x_1\mathbf{i}+y_1\mathbf{j}+z_1\mathbf{k},$$

$$\mathbf{b} = x_2\mathbf{i}+y_2\mathbf{j}+z_2\mathbf{k},$$

$$\mathbf{c} = x_3\mathbf{i}+y_3\mathbf{j}+z_3\mathbf{k},$$

$$\cdots\cdots\cdots\cdots\cdots$$

If G is the centroid of these points

$$\mathbf{OG} = \frac{\mathbf{a}+\mathbf{b}+\mathbf{c}\ldots}{n}$$

$$= \frac{(x_1+x_2+x_3\ldots)\mathbf{i}+(y_1+y_2+y_3\ldots)\mathbf{j}+(z_1+z_2+z_3\ldots)\mathbf{k}}{n}$$

$$= \frac{(\Sigma x)\mathbf{i}+(\Sigma y)\mathbf{j}+(\Sigma z)\mathbf{k}}{n}.$$

Therefore G is the point $\left(\frac{\Sigma x}{n}, \frac{\Sigma y}{n}, \frac{\Sigma z}{n}\right)$.

Example

Show that the centroid of the points

$A(4, 3, 2)$, $B(5, -4, -3)$, $C(8, 3, -2)$ *and* $D(-1, 6, -5)$

is the point $G(4, 2, -2)$. *Also show that* $\mathbf{AG} = -\mathbf{j}-4\mathbf{k}$.

Let $\mathbf{a}$, $\mathbf{b}$, $\mathbf{c}$ and $\mathbf{d}$ be the position vectors of A, B, C and D respectively relative to the origin O of the axes.

$$\mathbf{a} = 4\mathbf{i}+3\mathbf{j}+2\mathbf{k},$$

$$\mathbf{b} = 5\mathbf{i}-4\mathbf{j}-3\mathbf{k},$$

$$\mathbf{c} = 8\mathbf{i}+3\mathbf{j}-2\mathbf{k},$$

$$\mathbf{d} = -\mathbf{i}+6\mathbf{j}-5\mathbf{k}.$$

$$\mathbf{OG} = \frac{\mathbf{a}+\mathbf{b}+\mathbf{c}+\mathbf{d}}{4}$$

$$= \frac{16\mathbf{i}+8\mathbf{j}-8\mathbf{k}}{4}$$

$$= 4\mathbf{i}+2\mathbf{j}-2\mathbf{k}.$$

Therefore G is the point $(4, 2, -2)$.

$$\mathbf{AG} = \mathbf{g}-\mathbf{a} \quad \text{(where } \mathbf{g} \text{ is the position vector of } G\text{)}$$
$$= (4\mathbf{i}+2\mathbf{j}-2\mathbf{k})-(4\mathbf{i}+3\mathbf{j}+2\mathbf{k})$$
$$= -\mathbf{j}-4\mathbf{k}.$$

Exercise 5

(1) Show that $|\mathbf{i}+\mathbf{j}| = \sqrt{2}$ and $|\mathbf{i}+\mathbf{j}+\mathbf{k}| = \sqrt{3}$.

(2) If $\mathbf{a} = 4\mathbf{i}+3\mathbf{j}-2\mathbf{k}$, $\mathbf{b} = 3\mathbf{i}-7\mathbf{j}+3\mathbf{k}$ and $\mathbf{c} = -2\mathbf{i}-5\mathbf{j}+\mathbf{k}$ find the values of m and n so that $\mathbf{a}+m\mathbf{b}+n\mathbf{c}$ is parallel to the x axis.

(3) $ABCD$ is a rectangle with $AB = 2AD = 2a$. E and F are the mid-points of BC and DC respectively. Show that the components in the directions AB and AD of the resultant of $\mathbf{AE}+\mathbf{AC}+\mathbf{AF}$ are $5a$ and $2\frac{1}{2}a$ respectively.

(4) OA, OB and OC are the diagonals of three adjacent faces of a cube and OD is a diagonal of the cube. Show that

$$\mathbf{OA}+\mathbf{OB}+\mathbf{OC} = 2\mathbf{OD}.$$

(5) If the position vectors of points A and B are $2\mathbf{i}-3\mathbf{j}+4\mathbf{k}$ and $3\mathbf{i}-7\mathbf{j}+12\mathbf{k}$ respectively, find the length of AB and its direction cosines.

(6) Calculate the modulus and the unit vector of the sum of the vectors $\mathbf{i}+4\mathbf{j}+2\mathbf{k}$, $3\mathbf{i}-3\mathbf{j}-2\mathbf{k}$ and $-2\mathbf{i}+2\mathbf{j}+6\mathbf{k}$.

(7) $A(2, -1, 3)$, $B(6, 3, -4)$ and $C(3, 1, 1)$ are the vertices of a triangle. Show that $AB = 3AC$ and the direction cosines of $\mathbf{BC}$ are $\frac{-3}{\sqrt{38}}, \frac{-2}{\sqrt{38}}, \frac{5}{\sqrt{38}}$.

(8) Show that the vectors $\mathbf{a} = 3\mathbf{i}-2\mathbf{j}-5\mathbf{k}$ and $\mathbf{b} = 6\mathbf{i}-\mathbf{j}+4\mathbf{k}$ are perpendicular to one another.

(9) The position vectors of the points A, B, C and D are $4\mathbf{i}+3\mathbf{j}-\mathbf{k}$, $5\mathbf{i}+2\mathbf{j}+2\mathbf{k}$, $2\mathbf{i}-2\mathbf{j}-3\mathbf{k}$ and $4\mathbf{i}-4\mathbf{j}+3\mathbf{k}$ respectively. Show that AB and CD are parallel.

(10) The position vectors of the points A, B and C are $2\mathbf{i}-\mathbf{j}+\mathbf{k}$, $3\mathbf{i}+2\mathbf{j}-\mathbf{k}$ and $6\mathbf{i}+11\mathbf{j}-7\mathbf{k}$ respectively. Show that A, B and C are collinear and that $AB:BC = 1:3$.

(11) Show that the angle between the vectors

$$4\mathbf{i}-4\mathbf{j}+7\mathbf{k} \quad \text{and} \quad -\mathbf{i}+4\mathbf{j}+8\mathbf{k}$$

is $\cos^{-1}\frac{4}{9}$.

(12) The position vectors of the points A, B and C are $8\mathbf{i}+4\mathbf{j}-3\mathbf{k}$, $6\mathbf{i}+3\mathbf{j}-4\mathbf{k}$ and $7\mathbf{i}+5\mathbf{j}-5\mathbf{k}$ respectively. Find the angle between **AB** and **BC**.

(13) Show that the centroid of the points $(2, -3, 3)$, $(6, -2, -2)$, $(-5, 1, 7)$ and $(1, -4, 4)$ is the point $(1, -2, 3)$.

(14) Show that the diagonals of a parallelepiped and the straight lines joining the mid-points of opposite edges are concurrent and bisect each other.

(15) If $\mathbf{r} = x\mathbf{i}+y\mathbf{j}+z\mathbf{k}$ show that the modulus of the sum of the vectors $\mathbf{r}_1, \mathbf{r}_2, \mathbf{r}_3, \ldots$ is $(\Sigma r^2+2\Sigma r_m r_n \cos r_{mn})^{\frac{1}{2}}$, where r_{mn} is the angle between the vectors $\mathbf{r}_m$ and $\mathbf{r}_n$.

6

APPLICATIONS IN MECHANICS

In this chapter we shall establish the vector nature of certain physical quantities and show how certain problems in Mechanics can be solved vectorially.

Vector quantities

Quantities such as displacement, velocity, acceleration, force and angular velocity are only fully specified when we know their magnitude and direction.

Let us consider velocity as an example. It is not enough to know how fast a particle is moving, that is its speed, but also in what direction it is moving. It is the combined speed–direction which determines the velocity. In dynamics we are at times concerned with the motion of a particle describing a circle with uniform speed. Although the speed is uniform its direction of motion which is in the direction of the tangent at any particular time is changing. Thus the velocity of the particle describing a circle with uniform speed is varying. So we see that for a velocity to be fully specified we must know its magnitude and direction.

Definition. *We define a vector quantity as one which has magnitude and direction and which obeys the triangle law of addition or its equivalent, the parallelogram law.*

Thus in order for a physical quantity having magnitude and direction to be a vector quantity, or briefly a vector, it is necessary to show that it obeys the triangle law of addition.

The following two examples will make this point clear. Consider displacements over the earth's surface (Fig. 6.1).

Suppose we start at point A and travel x kilometres south to B and then y kilometres east to C. Now suppose we again start at A but this time we travel y kilometres east to D and then x kilometres south to E.

Now obviously the final points C and E are not the same. This is because the longitude change depends on the latitude. However, we

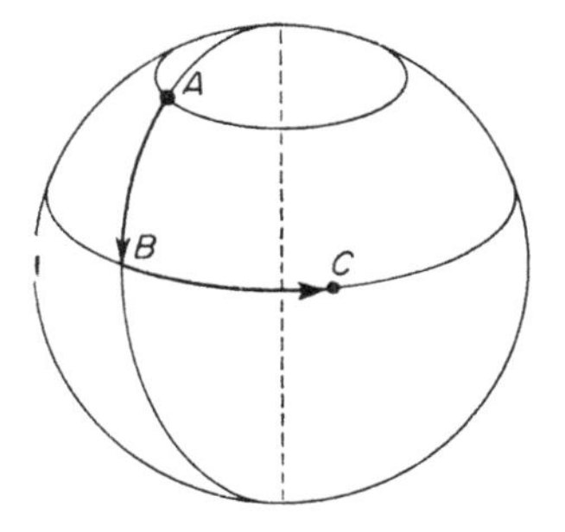

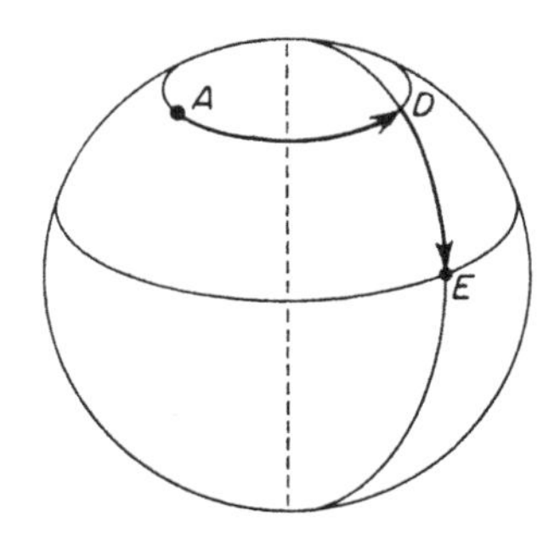

Fig. 6.1

have in each case made the same successive displacements but in the reverse order. Thus the successive displacements are not commutative. A consequence of the triangle law of addition is that the sum of vectors is commutative. We therefore conclude that displacements over the earth's surface although having magnitude and direction are not vectors.

As a second example we consider finite rotations of a rigid body about a fixed point in the body. A finite rotation has magnitude and direction since the angle turned through specifies the magnitude and the axis of rotation specifies the direction. The sense of direction is defined as the same as that moved by a right-handed corkscrew driven by the finite rotation along the axis of rotation.

Consider a book lying in the horizontal plane of the paper (Fig. 6.2*a*). We shall rotate the book in turn about two perpendicular horizontal axes XX', YY'.

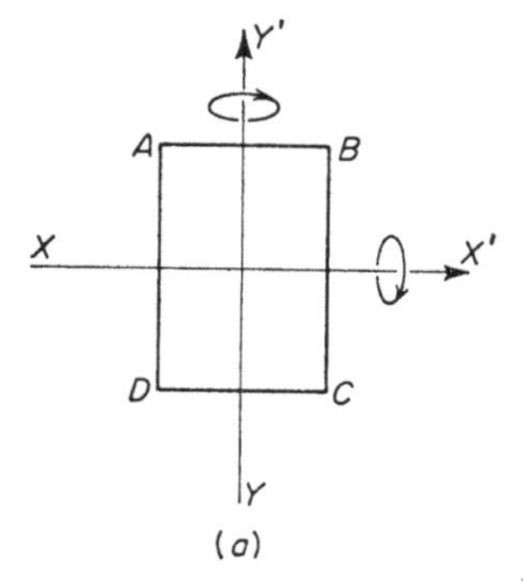

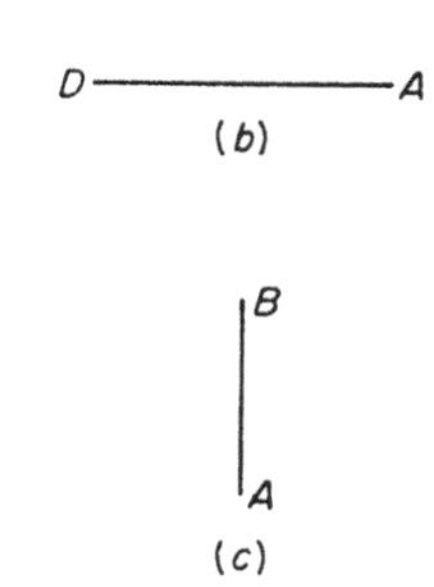

Fig. 6.2

Suppose the book is first rotated through a right angle about the axis XX', the sense of rotation being indicated by the arrow. Corners A and B will be above corners D and C, the book being in the vertical plane. Now we rotate it through another right angle about the axis YY', the sense of rotation being shown by the arrow. This will bring corners A and D above corners B and C, the plane of the book still being vertical (Fig. 6.2*b*).

We now repeat the rotations starting from the original position of the book, but in the reverse order, i.e. rotation through a right angle about the axis YY' and then about XX'. At the end of the two rotations the plane of the book is vertical and corners A and B are above corners D and C (Fig. 6.2*c*).

This clearly indicates that the order of performing finite rotations affects the final position of the book. Thus finite rotations do not commute and we conclude that a finite rotation although possessing magnitude and direction is not a vector.

We shall use the term 'directed quantity' for a physical quantity having magnitude and direction and the term 'vector quantity' or more briefly 'vector' for the same quantity when we have shown it to obey the triangle law of addition.

Representation of a vector quantity

Since we have seen that displacements have magnitude and direction and are added by the triangle law it follows that displacement is a vector. The representation of displacement by a directed line segment offers no difficulty since both displacement and the line segment have length as the common physical property.

However the magnitude of many vector quantities is not a length. By using a suitable scale a directed line segment can be used to represent any vector quantity whose magnitude is not a length, e.g. the directed line segment **AB** represents a velocity of 4 km/h in the direction of A to B if the length of AB represents 4 km/h in the chosen scale. Hence strictly speaking since we have not shown that velocity is a vector we are using the notation **AB** to imply that the velocity has the magnitude, direction and sense of **AB** but making no assumption about the nature of velocity itself.

As a consequence of the above geometrical representation of a physical quantity any vector quantity can be represented in the same way by directed line segments. Thus the algebra of vectors we have developed can also be applied to vector quantities of the physical world.

Displacement

A displacement from the point A to the point B is represented by the vector **AB**.

Two successive displacements are added vectorially.

A displacement from A to B followed by a displacement from B to C is the same as a displacement from A to C (Fig. 6.3). Thus

$$\mathbf{AB}+\mathbf{BC} = \mathbf{AC}.$$

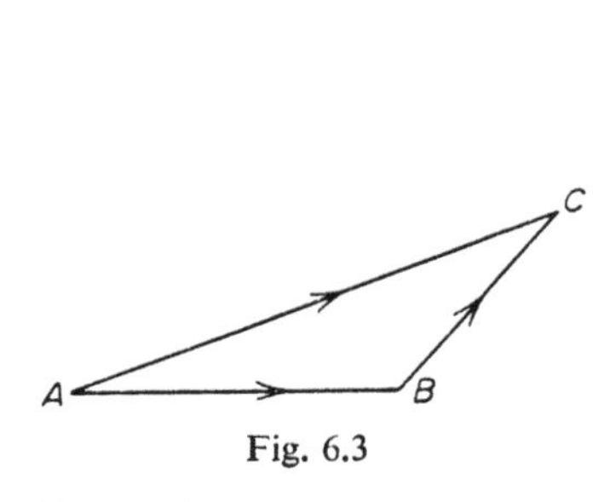

Fig. 6.3

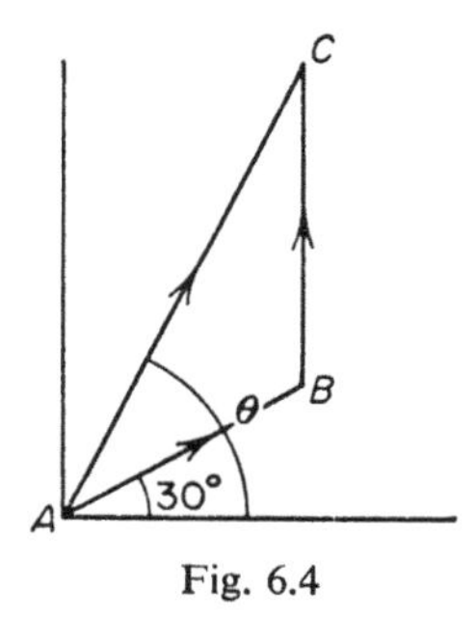

Fig. 6.4

Example

A ship starts from A, travels 4 *km in the direction* N. 60° E. *to B, and then* 5 *km* N. *to C. Find its displacement from A.*

Let $\mathbf{i}$ and $\mathbf{j}$ be unit vectors towards east and north respectively. From Fig. 6.4,

$$\mathbf{AC} = \mathbf{AB}+\mathbf{BC}$$

$$= \left(\frac{4\sqrt{3}}{2}\,\mathbf{i}+4.\frac{1}{2}\,\mathbf{j}\right)+5\mathbf{j}$$

$$= 2\sqrt{(3)}\,\mathbf{i}+7\mathbf{j},$$

$$\therefore \quad AC = \sqrt{(12+49)} = \sqrt{61}\text{ km},$$

$$\tan\theta = \frac{7}{2\sqrt{3}}.$$

Therefore displacement is $\sqrt{61}$ km at $\tan^{-1}\dfrac{7\sqrt{3}}{6}$ N. of E.

Velocity and acceleration

We assume that the reader is familiar with the idea of uniform velocity and acceleration, and with the idea that all velocities and accelerations are relative to some frame of reference.

The vector nature of velocity and acceleration when uniform will be established.

Resultant velocity and acceleration

Consider the following problem. A ship is steaming due north at 12 km/h across a current which flows due east at 5 km/h. What is the actual velocity of the ship?

Now from the context we understand that the velocity of 12 km/h due north is relative to an observer on the water, the velocity of 5 km/h due east is relative to an observer on the land, and the actual velocity required is the velocity of the ship relative to the observer on the land. It is also convenient in this type of question to say that the ship has two velocities 'at the same time' or 'two simultaneous velocities'. The actual velocity is often called the resultant velocity.

Let in general, $\mathbf{u}$ be the velocity of the ship relative to the water, $\mathbf{v}$ be the velocity of the water relative to the land and $\mathbf{V}$ be the velocity of the ship relative to the land, all velocities being uniform.

In time t the whole of the water surface is displaced vt units in the direction of $\mathbf{v}$, and so if drifting the ship will have a displacement of $\mathbf{v}t$. However, in the same time on account of its motion the ship is displaced ut units in the direction of $\mathbf{u}$, i.e. it occurs a further displacement of $\mathbf{u}t$. Since displacement is a vector the resultant displacement is $\mathbf{u}t+\mathbf{v}t$ by the triangle law of addition.

Now since the velocities $\mathbf{u}$ and $\mathbf{v}$ are uniform it is reasonable to assume that the resultant velocity $\mathbf{V}$ is uniform. Hence in time t the resultant displacement is $\mathbf{V}t$.

$$\therefore \quad \mathbf{V}t = \mathbf{u}t+\mathbf{v}t,$$

$$\therefore \quad \mathbf{V} = \mathbf{u}+\mathbf{v}.$$

From this we see that uniform velocities obey the triangle law of addition, i.e. they are added vectorially. Hence we see that velocity when uniform is a vector.

We have since $$\mathbf{V} = \mathbf{v}+\mathbf{u}$$

velocity of ship relative to land = velocity of ship relative to water + velocity of water relative to land.

In general we have

$${}_A\mathbf{v}_C = {}_A\mathbf{v}_B + {}_B\mathbf{v}_C,$$

where ${}_A\mathbf{v}_C$, ${}_A\mathbf{v}_B$, ${}_B\mathbf{v}_C$ are velocities of A relative to C, A relative to B and B relative to C respectively.

We now consider uniform accelerations. Suppose a point has a uniform acceleration $\mathbf{a}_1$, and at the same time a uniform acceleration $\mathbf{a}_2$. Then after time t the gains in velocity are $a_1 t$, $a_2 t$ units in the direction of $\mathbf{a}_1$, $\mathbf{a}_2$ respectively. Since we have shown that velocity when uniform is a vector the resultant gain in the velocity of the particle is $\mathbf{a}_1 t+\mathbf{a}_2 t$.

Again, since the accelerations are uniform it is reasonable to assume that the resultant acceleration $\mathbf{a}$ is uniform. This being so the resultant gain in velocity in time t is $\mathbf{a}t$.

$$\therefore \quad \mathbf{a}t = \mathbf{a}_1 t+\mathbf{a}_2 t,$$

$$\therefore \quad \mathbf{a} = \mathbf{a}_1+\mathbf{a}_2.$$

Hence uniform accelerations are added vectorially and we can say that acceleration when uniform is a vector.

In general we can write

$${}_A\mathbf{a}_C = {}_A\mathbf{a}_B + {}_B\mathbf{a}_C,$$

where ${}_A\mathbf{a}_C$, ${}_A\mathbf{a}_B$, ${}_B\mathbf{a}_C$ are the uniform accelerations of A relative to C, A relative to B and B relative to C respectively.

The triangle law of addition when applied to velocities and acceleration is often known as the triangle of velocities or accelerations. The vector sum of velocities or accelerations is known as the resultant velocity or acceleration and the velocities or accelerations which are combined are known as the components of the resultant.

We shall in the next chapter show that velocity and acceleration are vectors even when they are variable.

Example

A ship whose course is due south is steaming across a current due west. After 2 hours the ship has gone 36 km in the direction 15° west of south. Find the velocity of the ship and current.

Let the velocity of the ship and current be **u** km/h and **v** km/h and the resultant velocity be **V** km/h. (See Fig. 6.5.)

$$\therefore \quad \mathbf{V} = \mathbf{u}+\mathbf{v}.$$

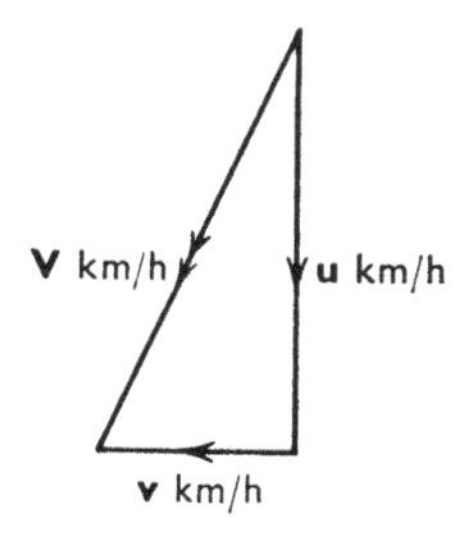

Fig. 6.5

If **i** and **j** are unit vectors in the directions due west and south

$$\mathbf{V} = 18\sin 15°\mathbf{i}+18\cos 15°\mathbf{j},$$

$$\mathbf{u} = u\mathbf{j},$$

$$\mathbf{v} = v\mathbf{i}.$$

$$\therefore \quad 18\sin 15°\mathbf{i}+18\cos 15°\mathbf{j} = u\mathbf{j}+v\mathbf{i},$$

$$\therefore \quad v = 18\sin 15° = 4{\cdot}66,$$

$$\therefore \quad u = 18\cos 15° = 17{\cdot}4.$$

Therefore velocities of ship and current are 17·4 and 4·66 km/h respectively.

Relative velocity and acceleration

Suppose points A and B are moving with uniform velocities ${}_{A}\mathbf{v}_{O}$ and ${}_{B}\mathbf{v}_{O}$ relative to a fixed origin O. Then the velocity of B relative to A denoted by ${}_{B}\mathbf{v}_{A}$ can be deduced from

$${}_{B}\mathbf{v}_{O} = {}_{B}\mathbf{v}_{A}+{}_{A}\mathbf{v}_{O},$$

from which we see that

$${}_{B}\mathbf{v}_{A} = {}_{B}\mathbf{v}_{O}-{}_{A}\mathbf{v}_{O},$$

i.e. velocity of B relative to A = vector sum of the velocities of B and the negative of A.

We give an alternative method of obtaining the relative velocity equation.

Let the velocities of two points A and B relative to a fixed point O be ${}_{A}\mathbf{v}_{O}$ and ${}_{B}\mathbf{v}_{O}$ respectively (Fig. 6.6).

Let $\mathbf{PR} = {}_{B}\mathbf{v}_{O}$ and let $\mathbf{PQ}$ and $\mathbf{QR}$ be the two components of $\mathbf{PR}$, $\mathbf{QR}$ being equal and parallel to ${}_{A}\mathbf{v}_{O}$, i.e. $\mathbf{QR} = {}_{A}\mathbf{v}_{O}$. Then $\mathbf{PQ}$ is the velocity of B relative to A.

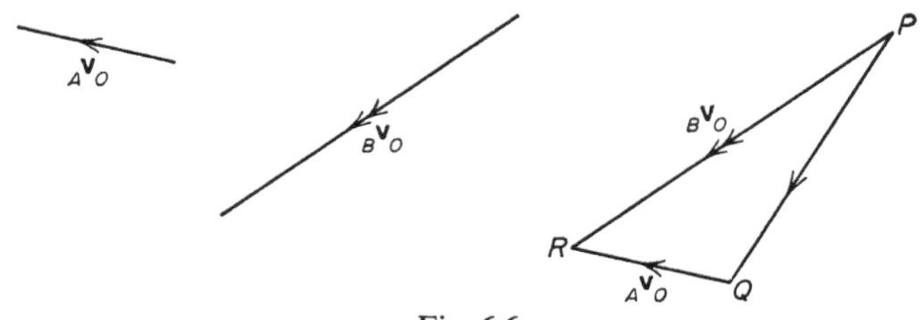

Fig. 6.6

But
$$\mathbf{PQ}+\mathbf{QR} = \mathbf{PR},$$
$$\therefore \quad \mathbf{PQ} = \mathbf{PR}-\mathbf{QR},$$
$$\therefore {}_{B}\mathbf{v}_{A} = {}_{B}\mathbf{v}_{O} - {}_{A}\mathbf{v}_{O},$$
where ${}_{B}\mathbf{v}_{A}$ is the velocity of B relative to A.

The following deductions are important:

(1) If ${}_{A}\mathbf{v}_{O} = \mathbf{0}$ we have ${}_{B}\mathbf{v}_{A} = {}_{B}\mathbf{v}_{O}$, i.e. the velocity of a point is the same relative to all fixed points.

(2) If O and B coincide we have ${}_{O}\mathbf{v}_{A} = -{}_{A}\mathbf{v}_{O}$, i.e. the velocity of A relative to O = $-$velocity of O relative to A.

By similar methods we can show that the corresponding result for uniform accelerations is

acceleration of B relative to A = acceleration of B relative to O
$-$acceleration of A relative to O.

Examples

(1) *The velocity of a particle A relative to B is* $3\mathbf{i}-4\mathbf{j}$ *and the velocity of B relative to a third particle C is* $\mathbf{i}+\mathbf{j}$. *Determine the*

magnitude and direction of the velocity of A relative to C, assuming that **i** *and* **j** *represent velocities of* 1 *m/s horizontally and vertically respectively.*

$$\underset{AC}{\mathbf{v}} = \underset{AB}{\mathbf{v}} + \underset{BC}{\mathbf{v}} \quad \text{where} \quad \underset{AB}{\mathbf{v}} = 3\mathbf{i}-4\mathbf{j} \quad \text{and} \quad \underset{BC}{\mathbf{v}} = \mathbf{i}+\mathbf{j},$$

$$\therefore \quad \underset{AC}{\mathbf{v}} = 3\mathbf{i}-4\mathbf{j}+\mathbf{i}+\mathbf{j} = 4\mathbf{i}-3\mathbf{j}.$$

Therefore magnitude $= \sqrt{(4^2+3^2)}$ m/s $=$ 5 m/s. Direction is 323° 08′ with direction of **i**.

(2) *A man on a ship steaming north-east at* 10 *km/h observes the smoke issuing from the funnel in a south-east direction. He estimates the speed of the smoke to be the same as that of the ship. What is the magnitude and direction of the velocity of the wind?*

Let **i** and **j** represent velocities of 1 km/h in the directions due east and north respectively.

Let **u** and **v** be the velocities of the ship and wind respectively, and **V** the velocity of the smoke relative to the ship.

Then

$$\mathbf{u} = \frac{10}{\sqrt{2}}\mathbf{i}+\frac{10}{\sqrt{2}}\mathbf{j},$$

$$\mathbf{v} = x\mathbf{i}+y\mathbf{j},$$

$$\mathbf{V} = \frac{10}{\sqrt{2}}\mathbf{i}-\frac{10}{\sqrt{2}}\mathbf{j}.$$

But

$$\mathbf{V} = \mathbf{v}-\mathbf{u},$$

$$\therefore \quad 5\sqrt{(2)}\mathbf{i}-5\sqrt{(2)}\mathbf{j} = (x\mathbf{i}+y\mathbf{j})-\{5\sqrt{(2)}\mathbf{i}+5\sqrt{(2)}\mathbf{j}\},$$

$$\therefore \quad 10\sqrt{(2)}\mathbf{i} = x\mathbf{i}+y\mathbf{j},$$

$$\therefore \quad x = 10\sqrt{2} \quad \text{and} \quad y = 0.$$

Therefore magnitude of velocity is $10\sqrt{2}$ km/h and direction is from the west.

Angular velocity

Angular velocity about an axis is a directed quantity. Its magnitude is the magnitude of the angular velocity, and its direction is defined as being along the axis of rotation, the sense being the same as that moved by a right-handed corkscrew driven by the angular velocity along the axis of rotation (Fig. 6.7).

In order to show that angular velocity is a vector we must show that two angular velocities are added by the triangle law or its equivalent the parallelogram law.

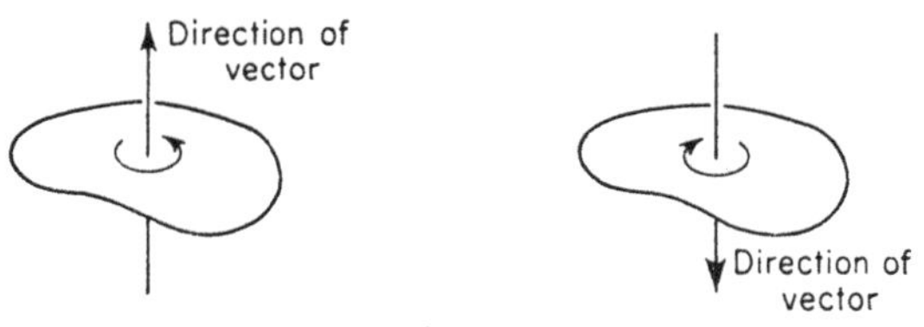

Fig. 6.7

Suppose OA is rotating about a point O with uniform angular velocity ω (Fig. 6.8). Then the velocity of the point A is $\omega . OA$ and is at right angles to OA.

Let **AB** and **AC** represent two angular velocities (Fig. 6.9).

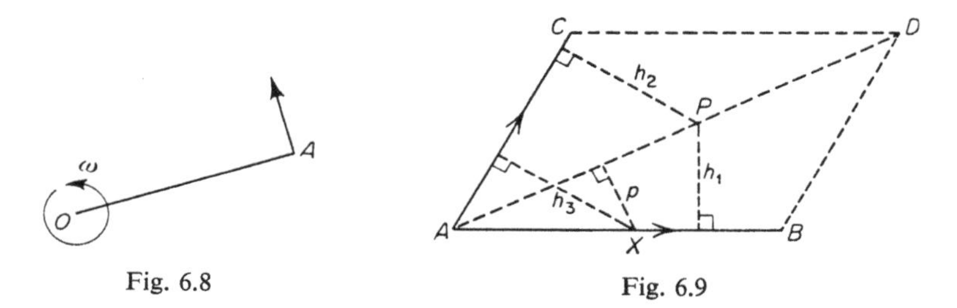

Fig. 6.8 Fig. 6.9

Complete the parallelogram $ABDC$ and take P any point on AD.

Magnitude of velocity of P due to

$$\text{angular velocity } \mathbf{AB} = h_1 . AB = 2\triangle APB.$$

Direction of this velocity is out of the paper.

Magnitude of velocity of P due to

$$\text{angular velocity } \mathbf{AC} = h_2 . AC = 2\triangle APC.$$

Direction of this velocity is into the paper.

Since velocity is a vector we have from the triangle law of addition,

$$\text{resultant velocity of } P = 2\triangle APC - 2\triangle APB, \text{ into the paper.}$$

Since $\triangle APB$ and $\triangle APC$ have the same base AP and the altitudes from B and C respectively are equal

$$\triangle APB = \triangle APC,$$

therefore resultant velocity of $P = 0$.

Thus P must lie on the axis of the resultant angular velocity, i.e. angular velocities in the direction AB and AC combine to give an angular velocity in the direction of the diagonal.

Now consider the velocity of any point X on AB. Its velocity is the resultant of the velocities due to the angular velocities **AB** and **AC**.

Magnitude of velocity of X due to $\mathbf{AB} = 0$.

Magnitude of velocity of X due to $\mathbf{AC} = h_3 . AC$

$$= 2 \triangle ACX,$$

$\therefore$ resultant velocity of $X = 2 \triangle ACX$, direction being into the paper.

But if ω is the magnitude of the resultant angular velocity about AD,

$$\text{velocity of } X \text{ due to } \omega = \omega . p = \frac{2\omega \triangle ADX}{AD},$$

where p = perpendicular from X to AD.

$$\therefore \quad 2 \triangle ACX = \frac{2\omega \triangle ADX}{AD}.$$

Now $\quad \triangle ACX = \triangle ADX$, same base AX, same parallels,

$$\therefore \quad \omega = AD.$$

Since the direction of the resultant velocity of X is into the paper the sense of the resultant angular velocity must be in the sense A to D. Thus the sum of the angular velocities **AB**, **AC** is the angular velocity **AD**, i.e.

$$\mathbf{AB} + \mathbf{AC} = \mathbf{AD}.$$

We therefore conclude that angular velocity is a vector since the parallelogram law of addition is obeyed.

Example

Find the angular velocity vector $\boldsymbol{\omega}$ *of modulus* 14 *radians per second about an axis whose direction cosines are*

$$\frac{2}{7}, \frac{-3}{7}, \frac{6}{7}.$$

The vector is $x\mathbf{i}+y\mathbf{j}+z\mathbf{k}$, where

$$x = \frac{2}{7}.14, \quad y = \frac{-3}{7}.14, \quad z = \frac{6}{7}.14.$$

$$\therefore \quad \boldsymbol{\omega} = 4\mathbf{i}-6\mathbf{j}+12\mathbf{k}.$$

Force

Newton's First Law of Motion can be stated: *Every body continues in its state of rest or of uniform motion in a straight line, except in so far as it is compelled by impressed forces to change that state.*

Now if a body is not at rest or in uniform motion relative to a frame of reference we say that it is accelerating relative to this frame of reference. Thus the first law leads us to the idea that acceleration is caused by the action of force.

The Second Law of Motion is: *The rate of change of momentum is proportional to the impressed force and takes place in the direction of the straight line in which the force is impressed.*

For a constant mass this is equivalent to the statement that acceleration is proportional to the impressed force and is in the direction of the impressed force.

Now suppose that we have two particles A and B which are moving. Furthermore, suppose they are alike in every respect including the forces acting on them except that there is a force P acting on A but not on B.

Let O be any point and ${}_{A}\mathbf{a}_{O}$ and ${}_{B}\mathbf{a}_{O}$ the accelerations of A and B relative to O respectively. Then if ${}_{A}\mathbf{a}_{B}$ is the acceleration of A relative to B we have ${}_{A}\mathbf{a}_{O} = {}_{A}\mathbf{a}_{B} + {}_{B}\mathbf{a}_{O}$.

From this we see that the additional acceleration of A relative to O due to the force P is the acceleration of A relative to B. This leads to Newton's corollary to his Second Law: *Each of the forces acting on a particle is proportional in magnitude to the additional acceleration it produces and is in the same direction as the additional acceleration.*

By defining our unit of force as that which produces unit acceleration on a particle of unit mass, our force P will have the magnitude and direction of $m\mathbf{a}$ where m is the mass of the particle and $\mathbf{a}$ is the additional acceleration produced by P.

Although $\mathbf{a}$ is a vector and m a scalar we cannot assume that $m\mathbf{a}$ is a vector quantity since $m\mathbf{a}$ is a physical quantity of a different nature from $\mathbf{a}$. If we now write $\mathbf{P} = m\mathbf{a}$ we must understand this equation to specify the magnitude and direction of $\mathbf{P}$ as being ma and that of $\mathbf{a}$ respectively and to make no assumption about the nature of force. We shall now show that force is also a vector, assuming that acceleration is a vector.

Consider three particles A, B, C each of mass m.

Let A, B and C be acted upon by equal systems of forces.

Apply an extra force $\mathbf{P}$ to B and let the acceleration of B relative to A be $\mathbf{p}$. We can write $\mathbf{P} = m\mathbf{p}$ bearing in mind that this equation gives the magnitude and direction of the directed quantity $\mathbf{P}$. Now apply additional forces $\mathbf{P}$ and $\mathbf{Q}$ to C and let the acceleration of C relative to B be $\mathbf{q}$. Then $\mathbf{Q} = m\mathbf{q}$.

Let the forces $\mathbf{P}$ and $\mathbf{Q}$ acting on C be replaced by a single force $\mathbf{R}$ and let the acceleration of C relative to A be $\mathbf{a}$. Then $\mathbf{R} = m\mathbf{a}$.

Since acceleration of C relative to A = acceleration of C relative to B + acceleration of B relative to A,

we have

$$\mathbf{a} = \mathbf{q}+\mathbf{p},$$

$$\therefore \quad \mathbf{R} = m(\mathbf{q}+\mathbf{p})$$

$$= m\mathbf{q}+m\mathbf{p},$$

$$\therefore \quad \mathbf{R} = \mathbf{Q}+\mathbf{P}.$$

From this we see that the resultant force $\mathbf{R}$ is obtained by the vector addition of the forces $\mathbf{P}$ and $\mathbf{Q}$. Thus force is a vector. We can now state that if O is an origin, $\mathbf{P}$ the vector sum of all forces acting on a particle A and $\mathbf{a}$ the acceleration of A relative to O then we have the vector equation $\mathbf{P} = m\mathbf{a}$.

Care must be taken in dealing with a system of forces. If the forces are acting on a single particle or if they are concurrent when acting on a rigid body, then they may be treated as free vectors. However, if the forces acting on a rigid body do not meet at a point then they must be treated as localized vectors, and their lines of action need to be known for the effect of the forces to be understood.

Concurrent forces

If there are several forces $\mathbf{F}_1$, $\mathbf{F}_2$, $\mathbf{F}_3$, ... acting on a point or on a single particle their net effect is the same as that of a single force equivalent to the vector sum of the forces and acting at the point or particle.

This equivalent single force is known as the resultant. If this resultant is represented by the vector **R** then

$$\mathbf{R} = \mathbf{F}_1+\mathbf{F}_2+\mathbf{F}_3+\dots.$$

The forces $\mathbf{F}_1, \mathbf{F}_2, \mathbf{F}_3, \dots$ are known as the components of the resultant **R**.

The resultant force is found by the vector polygon which has been already stated for finding the vector sum of several vectors.

Suppose $\mathbf{A}_0\mathbf{A}_1, \mathbf{A}_1\mathbf{A}_2, \mathbf{A}_2\mathbf{A}_3, \dots, \mathbf{A}_{n-1}\mathbf{A}_n$ represent in magnitude and direction the forces

$$\mathbf{F}_1, \mathbf{F}_2, \mathbf{F}_3, \dots, \mathbf{F}_n \quad \text{respectively (Fig. 6.10).}$$

If A_n does not coincide with A_0, that is the polygon is open, the resultant force is represented by the vector $\mathbf{A}_0\mathbf{A}_n$.

If A_n coincides with A_0, that is the polygon is closed, the resultant force is zero, and thus the forces are in equilibrium.

For a large number of forces the resultant is more quickly obtained by the method of resolution.

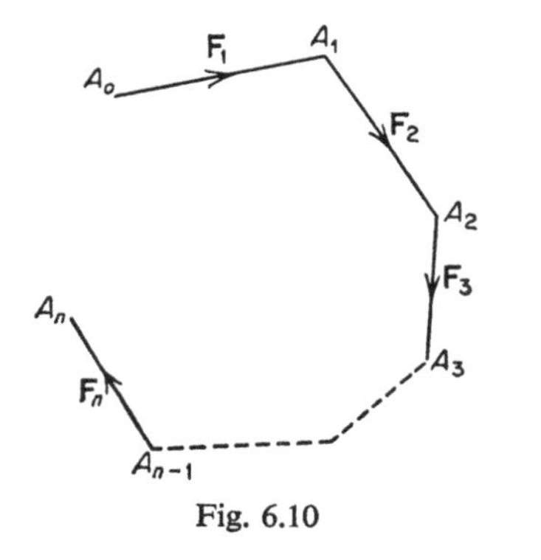

Fig. 6.10

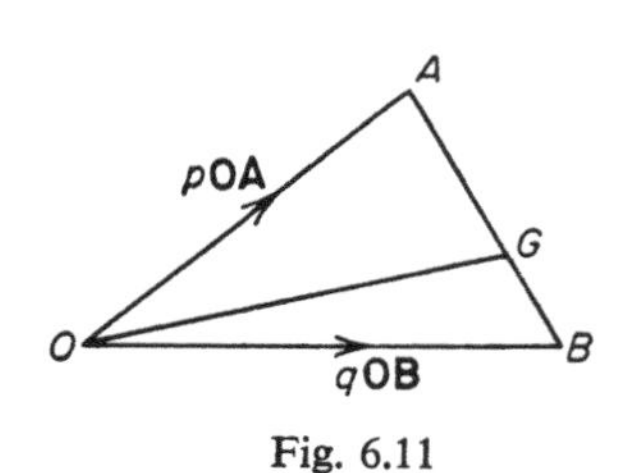

Fig. 6.11

Resultant of forces pOA and qOB

The resultant of two concurrent forces $p\mathbf{OA}$ and $q\mathbf{OB}$ is $(p+q)\mathbf{OG}$ where G is the point in AB such that $AG:GB = q:p$.

Proof. Referring to Fig. 6.11,

$$\begin{aligned} p\mathbf{OA}+q\mathbf{OB} &= p(\mathbf{OG}+\mathbf{GA})+q(\mathbf{OG}+\mathbf{GB}) \\ &= (p+q)\mathbf{OG}+p\mathbf{GA}+q\mathbf{GB}. \end{aligned}$$

But we are given $$p\mathbf{AG} = q\mathbf{GB},$$
$$\therefore \quad p\mathbf{GA}+q\mathbf{GB} = \mathbf{0},$$
$$\therefore \quad p\mathbf{OA}+q\mathbf{OB} = (p+q)\mathbf{OG}.$$

Examples

(1) *ABCDEF is a regular hexagon. Forces* **AB, AC, AD, AE, AF** *act at the vertices A. If O is the centre of the hexagon prove that the resultant is a force* **6AO.**

Method I

Let BF and CE meet AD at P and Q respectively (Fig. 6.12). From geometry P and Q are mid-points of FB and EC respectively.

$$\begin{aligned}\mathbf{AB}+\mathbf{AC}+\mathbf{AD}+\mathbf{AE}+\mathbf{AF} &= (\mathbf{AB}+\mathbf{AF})+(\mathbf{AC}+\mathbf{AE})+\mathbf{AD}\\ &= 2\mathbf{AP}+2\mathbf{AQ}+\mathbf{AD}\\ &= 2.\tfrac{1}{2}\mathbf{AO}+2.1\tfrac{1}{2}\mathbf{AO}+2\mathbf{AO}\\ &= 6\mathbf{AO}.\end{aligned}$$

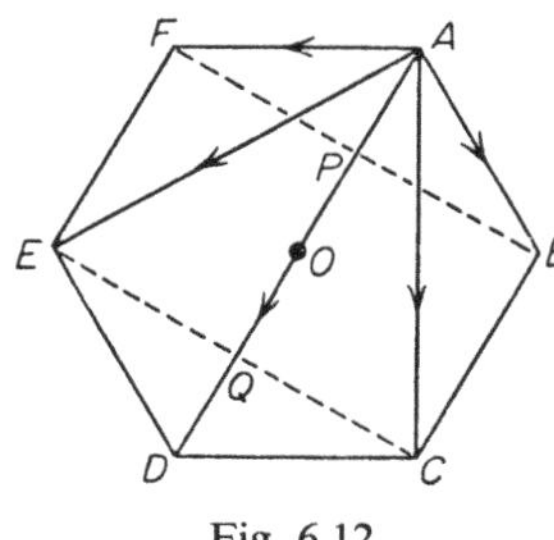

Fig. 6.12

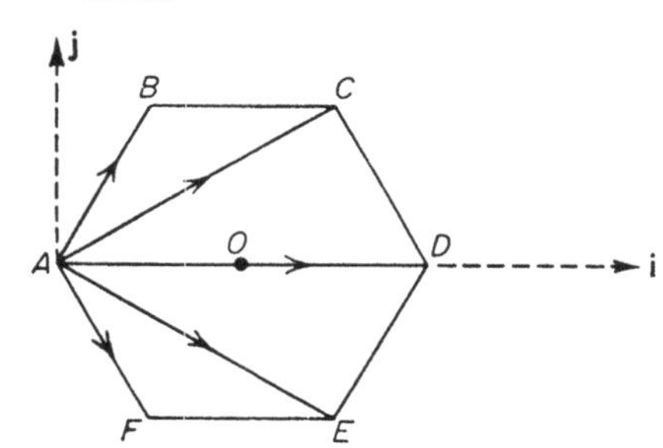

Fig. 6.13

Method II

Let **i** and **j** be unit forces in the directions AD and perpendicular to AD (Fig. 6.13). Let $AO = r$.

$$\begin{aligned}\mathbf{AB}+\mathbf{AC}+\mathbf{AD}+\mathbf{AE}+\mathbf{AF} &= (r\cos 60^\circ\mathbf{i}+r\sin 60^\circ\mathbf{j})\\ &\quad +(\sqrt{3}r\cos 30^\circ\mathbf{i}+\sqrt{3}r\sin 30^\circ\mathbf{j})\\ &\quad +(2r\mathbf{i})\\ &\quad +(\sqrt{3}r\cos 30^\circ\mathbf{i}-\sqrt{3}r\sin 30^\circ\mathbf{j})\\ &\quad +(r\cos 60^\circ\mathbf{i}-r\sin 60^\circ\mathbf{j})\\ &= (2r\cos 60^\circ+2\sqrt{3}r\cos 30^\circ+2r)\mathbf{i}\\ &= (r+3r+2r)\mathbf{i}\\ &= 6\mathbf{AO}.\end{aligned}$$

(2) *Find the resultant of the forces* **3BA, 4BC, 6CA** *which act along the sides of a triangle ABC.*

$$\mathbf{3BA}+\mathbf{4BC} = \mathbf{7BD} \quad \text{where} \quad AD:DC = 4:3. \quad \text{(See Fig. 6.14}a.)$$

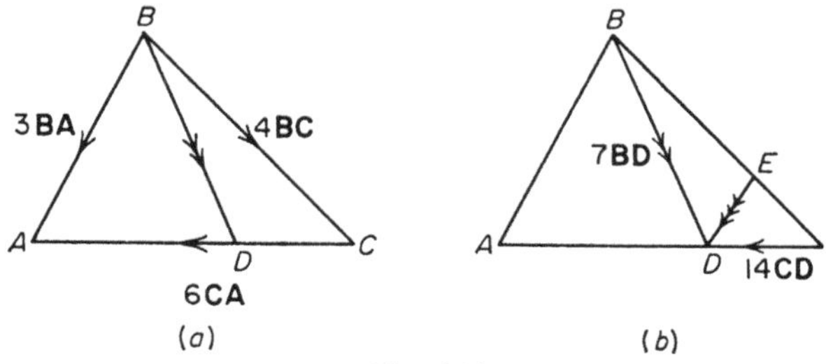

Fig. 6.14

We now have to combine **7BD** and **6CA**.

Now

$$CA:CD = 7:3,$$

$$\therefore \quad \mathbf{CA} = \tfrac{7}{3}\mathbf{CD},$$

$$\therefore \quad \mathbf{6CA} = \mathbf{14CD}.$$

We can now combine **7BD** with **14CD** since they both terminate at the same point D. (See Fig. 6.14*b*.)

$$\mathbf{7BD}+\mathbf{14CD} = \mathbf{21ED} \quad \text{where} \quad BE:EC = 2:1.$$

Therefore resultant is 21**ED** where $BE:EC = 2:1$ and $AD:DC = 4:3$.

Centre of mass

Let masses of $m_1, m_2, m_3, \ldots$ be situated at the points $A, B, C, \ldots$ whose position vectors relative to an origin O are $\mathbf{a}, \mathbf{b}, \mathbf{c}, \ldots$ (Fig. 6.15).

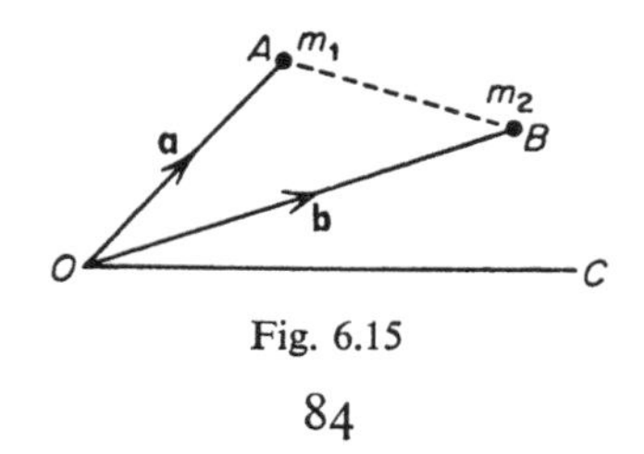

Fig. 6.15

Centre of mass of m_1 and m_2 is defined as the point P which divides AB in the ratio $m_2:m_1$.

$\therefore$ position vector of c. of m. of m_1 and $m_2 = \dfrac{m_1\mathbf{a}+m_2\mathbf{b}}{m_1+m_2}$.

C. of m. of (m_1+m_2) and m_3 divides PC in ratio $m_3:(m_1+m_2)$.

$\therefore$ position vector of c. of m. of m_1, m_2 and m_3

$$= \frac{(m_1+m_2)\dfrac{m_1\mathbf{a}+m_2\mathbf{b}}{m_1+m_2}+m_3\mathbf{c}}{m_1+m_2+m_3}$$

$$= \frac{m_1\mathbf{a}+m_2\mathbf{b}+m_3\mathbf{c}}{m_1+m_2+m_3}.$$

Hence,

$$\text{position vector of c. of m. of all particles} = \frac{m_1\mathbf{a}+m_2\mathbf{b}+m_3\mathbf{c}+\dots}{m_1+m_2+m_3+\dots}$$

$$= \frac{\Sigma m\mathbf{a}}{\Sigma m}.$$

Thus the c. of m. of a system of particles of masses $m_1, m_2, m_3, \dots$ at points $A, B, C, \dots$ is the same as the centroid of the points $A, B, C, \dots$ with associated numbers $m_1, m_2, m_3, \dots$.

Centre of gravity

The centre of gravity of a system of particles is defined as the point through which the line of action of the resultant of the system of parallel forces acting on the particles passes, the forces being proportional to the masses of the particles. With this definition the centre of gravity can be shown to be identical with the centre of mass.

Example

Masses of 5, 3, 2 *kg are situated at the points* (0, 2, 0), (6, −4, −8), (1, 6, −3). *Show that the centre of mass is the point* (2, 1, −3).

Let O be the origin and G the centre of mass.

$$\mathbf{OG} = \frac{\Sigma m\mathbf{a}}{\Sigma m}$$

$$= \frac{5(0\mathbf{i}+2\mathbf{j}+0\mathbf{k})+3(6\mathbf{i}-4\mathbf{j}-8\mathbf{k})+2(1\mathbf{i}+6\mathbf{j}-3\mathbf{k})}{10}$$

$$= \frac{20\mathbf{i}+10\mathbf{j}-30\mathbf{k}}{10}$$

$$= 2\mathbf{i}+\mathbf{j}-3\mathbf{k}.$$

Therefore centre of mass is the point (2, 1, −3).

Exercise 6

(1) If the components parallel to the x and y axes of the displacement **a** are 1 and -2 units, of the displacement **b** are -1 and 3 units and of the displacement **c** are 4 and 2 units, find the magnitude and direction of the displacements $\mathbf{a}+\mathbf{b}+\mathbf{c}$ and $\mathbf{a}-2\mathbf{b}+3\mathbf{c}$.

(2) If **a** is 4 units north, **b** is 7 units east and **c** is 4 units vertically upwards, find $\mathbf{a}+\mathbf{b}+\mathbf{c}$ and $\mathbf{a}+\mathbf{b}-\mathbf{c}$.

(3) A person travelling due east at 4 km/h finds that the wind appears to blow directly from the north. When he doubles his speed the wind appears to come from the north-east. Find the velocity of the wind.

(4) The velocity of a particle A relative to a particle B is $5\mathbf{i}+2\mathbf{j}$, and the velocity of B relative to a third particle C is $2\mathbf{i}-5\mathbf{j}$. Find the magnitude and direction of the velocity of A relative to C assuming that **i** and **j** represent velocities of 1 m/s horizontally and vertically respectively.

(5) At a certain instant two particles P and Q occupy positions A and B respectively 20 m apart. P moves towards Q with uniform velocity of 3 m/s, while Q moves in a direction perpendicular to AB with uniform velocity of 4 m/s. Determine: (i) velocity of Q relative to P; (ii) the shortest distance apart of the particles; (iii) the time taken to reach this shortest distance.

(6) Two particles move with speeds v and $2v$ respectively in opposite directions on the circumference of a circle. Show that their relative velocity has a maximum value of $3v$ when they cross one another, and a minimum value of v when they are at opposite ends of a diameter.

(7) Two particles P and Q are moving with the same speed v. P moves in a circle of centre O. Q moves along a fixed diameter AB of the circle and in the direction AB. If angle POB is θ show that the velocity of Q relative to P is $v\sqrt{(2+2\sin\theta)}$ at an angle $\frac{1}{2}\theta-\frac{1}{4}\pi$ with AB.

(8) A rigid body is rotating with an angular velocity of 6 radians per second about an axis OR where R is the point $(2, -2, 1)$. Find the angular velocity vector $\boldsymbol{\omega}$.

(9) A, B, C, D are the points $(2, -1, -1)$, $(5, 2, -1)$, $(2, -4, 2)$, $(3, -2, 3)$ respectively. Forces of 3, 1, 2 newtons act along AB, AC, AD respectively. Find their resultant.

(10) Forces of 1, 2 and 3 newtons act at the corner O of a cube one

along each of the diagonals of the faces which meet at O. Find the magnitude of the resultant and its inclination to each of the edges which meet at O.

(11) $ABCD$ is a quadrilateral. Show that the resultant of the forces **AB**, **AD**, **CB**, **CD** is **4PQ** where P, Q are the mid-points of AC, BD respectively.

(12) If G is the centroid of a triangle ABC, prove that the resultant of forces completely represented by **GA**, 2**GB**, 3**GC** is 3**GD** where D is the point in BC such that $BD = 2DC$.

(13) B and E are the mid-points of the sides AC and DF of a quadrilateral $ACDF$. Show that the system of forces completely represented by **AD**, **BF**, **CE**, **DB**, **EA**, **FC** is in equilibrium.

(14) Forces represented by **BA**, **CA** 2**BC** act along the sides of a triangle ABC. Show that their resultant is represented by 6**DE**, where D bisects BC and E is the point of trisection of CA nearer C.

(15) Show that the resultant of forces represented by **PA**, 2**PB**, 3**PC** is 6**PG** where G is the mid-point of the line joining C to a point of trisection of AB.

(16) $ABCD$ is a cyclic quadrilateral. Forces act on a particle at A in directions AB, AD their magnitudes being kDC and kBC respectively. Show that their resultant is of magnitude kBD in the direction AC.

(17) Masses of m, $2m$, ..., $6m$ are situated at the angular points O, A, B, C, D, E respectively of a regular hexagon $OABCDE$ of side a. Show that the distances of the centre of mass from OA and OD are $\frac{9\sqrt{3}a}{14}$ and $\frac{5a}{14}$ respectively.

(18) Masses of 2, 4, 6, 8 kg are situated at the points $(2, -3, 4)$, $(0, 4, 5)$, $(1, -1, 2)$, $(3, 0, -2)$ respectively. Find the co-ordinates of the centre of mass.

7

DIFFERENTIATION AND INTEGRATION

Definition of the derivative of a vector

Suppose $\mathbf{r}$ is a continuous and single-valued function of a scalar variable t. Let $\mathbf{r}$ become $\mathbf{r}+\delta\mathbf{r}$ when t becomes $t+\delta t$, δt being small.

The differential coefficient, or the derivative, of $\mathbf{r}$ with respect to t is defined as

$$\operatorname*{Lt}_{\delta t \to 0} \frac{\delta \mathbf{r}}{\delta t}, \text{ providing this limit exists.}$$

It is written as $d\mathbf{r}/dt$. Thus

$$\frac{d\mathbf{r}}{dt} = \operatorname*{Lt}_{\delta t \to 0} \frac{\delta \mathbf{r}}{\delta t}.$$

Let $\mathbf{OP} = \mathbf{r}$ and $\mathbf{OP}' = \mathbf{r}+\delta\mathbf{r}$ (Fig. 7.1).

$$\therefore \quad \mathbf{OP}+\mathbf{PP}' = \mathbf{OP}',$$

$$\therefore \quad \mathbf{r}+\mathbf{PP}' = \mathbf{r}+\delta\mathbf{r},$$

$$\therefore \quad \mathbf{PP}' = \delta\mathbf{r},$$

$$\therefore \quad \frac{\delta\mathbf{r}}{\delta t} = \frac{\mathbf{PP}'}{\delta t}.$$

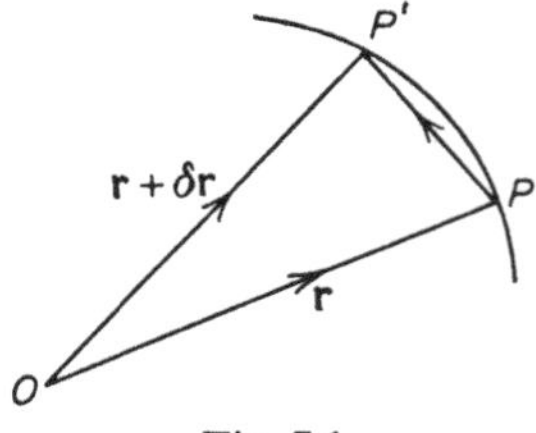

Fig. 7.1

Hence $\delta\mathbf{r}/\delta t$ is a vector whose direction is that of $\mathbf{PP}'$.

In the limit as $\delta t \to 0$ this direction is that of the tangent at P to the locus of P as t varies.

The second differential coefficient of $\mathbf{r}$ with respect to t, written as $d^2\mathbf{r}/dt^2$, is defined as the differential coefficient of $d\mathbf{r}/dt$ with respect to t.

If the scalar variable t is the time then as we shall see $d\mathbf{r}/dt$ and $d^2\mathbf{r}/dt^2$ represent the velocity and acceleration respectively, of a point at any time t. It is often convenient to write the velocity and acceleration of a point $\mathbf{r}$ as $\dot{\mathbf{r}}$ and $\ddot{\mathbf{r}}$ respectively.

Rules for the differentiation of vectors

(*a*) *Derivative of a constant vector*

$$\frac{d\mathbf{c}}{dt} = \mathbf{0} \quad (\mathbf{c} \text{ is a constant}).$$

Proof. If $\mathbf{c}$ is a constant, an increment δt in t will produce no change in $\mathbf{c}$.

$$\therefore \quad \frac{d\mathbf{c}}{dt} = \mathbf{0}.$$

(*b*) *Derivative of the product of a constant scalar and a vector*

$$\frac{d}{dt}(a\mathbf{r}) = a\frac{d\mathbf{r}}{dt}.$$

Proof. Let $\mathbf{b} = a\mathbf{r}$.

Suppose $\delta\mathbf{r}$ and $\delta\mathbf{b}$ are the increments in $\mathbf{r}$ and $\mathbf{b}$ respectively due to an increment δt in t.

Then
$$\mathbf{b}+\delta\mathbf{b} = a(\mathbf{r}+\delta\mathbf{r})$$
$$= a\mathbf{r}+a\delta\mathbf{r},$$
$$\therefore \quad \delta\mathbf{b} = a\delta\mathbf{r},$$
$$\therefore \quad \frac{\delta\mathbf{b}}{\delta t} = a\frac{\delta\mathbf{r}}{\delta t}.$$
$$\therefore \quad \underset{\delta t\to 0}{\mathrm{Lt}}\frac{\delta\mathbf{b}}{\delta t} = \underset{\delta t\to 0}{\mathrm{Lt}}\, a\frac{\delta\mathbf{r}}{\delta t},$$
$$\therefore \quad \frac{d\mathbf{b}}{dt} = a\frac{d\mathbf{r}}{dt},$$
$$\therefore \quad \frac{d}{dt}(a\mathbf{r}) = a\frac{d\mathbf{r}}{dt}.$$

(*c*) *Derivative of the sum of two vectors*

$$\frac{d}{dt}(\mathbf{r}+\mathbf{s}) = \frac{d\mathbf{r}}{dt}+\frac{d\mathbf{s}}{dt}.$$

Proof. Let $\mathbf{p} = \mathbf{r}+\mathbf{s}$

Suppose $\delta\mathbf{p}$, $\delta\mathbf{r}$ and $\delta\mathbf{s}$ are the increments in $\mathbf{p}$, $\mathbf{r}$ and $\mathbf{s}$ respectively due to an increment δt in t.

Then
$$(\mathbf{p}+\delta\mathbf{p}) = (\mathbf{r}+\delta\mathbf{r})+(\mathbf{s}+\delta\mathbf{s}),$$
$$\therefore \quad \delta\mathbf{p} = \delta\mathbf{r}+\delta\mathbf{s},$$
$$\therefore \quad \frac{\delta\mathbf{p}}{\delta t} = \frac{\delta\mathbf{r}}{\delta t}+\frac{\delta\mathbf{s}}{\delta t},$$
$$\therefore \quad \underset{\delta t\to 0}{\mathrm{Lt}} \frac{\delta\mathbf{p}}{\delta t} = \underset{\delta t\to 0}{\mathrm{Lt}} \left(\frac{\delta\mathbf{r}}{\delta t}+\frac{\delta\mathbf{s}}{\delta t}\right),$$
$$\therefore \quad \frac{d\mathbf{p}}{dt} = \frac{d\mathbf{r}}{dt}+\frac{d\mathbf{s}}{dt},$$
$$\therefore \quad \frac{d}{dt}(\mathbf{r}+\mathbf{s}) = \frac{d\mathbf{r}}{dt}+\frac{d\mathbf{s}}{dt}.$$

(*d*) *Derivative of a vector function of a scalar function*

$$\frac{d\mathbf{r}}{dt} = \frac{d\mathbf{r}}{ds}\cdot\frac{ds}{dt} \quad \text{where} \quad s = f(t).$$

Proof. Let $\mathbf{r}$ be a function of the scalar variable s and s a function of t.

Suppose $\delta\mathbf{r}$ and δs are the increments in $\mathbf{r}$ and s due to an increment δt in t.

Then
$$\frac{\delta\mathbf{r}}{\delta t} = \frac{\delta\mathbf{r}}{\delta s}\cdot\frac{\delta s}{\delta t},$$
$$\therefore \quad \underset{\delta t\to 0}{\mathrm{Lt}} \frac{\delta\mathbf{r}}{\delta t} = \underset{\delta t\to 0}{\mathrm{Lt}} \frac{\delta\mathbf{r}}{\delta s}\cdot\frac{\delta s}{\delta t},$$
$$\therefore \quad \frac{d\mathbf{r}}{dt} = \frac{d\mathbf{r}}{ds}\cdot\frac{ds}{dt}.$$

(*e*) *Derivative of the product of a variable scalar and vector*

$$\frac{d}{dt}(u\mathbf{v}) = u\frac{d\mathbf{v}}{dt}+\mathbf{v}\frac{du}{dt} \quad (u, \mathbf{v} \text{ are variables}).$$

Proof. Let u be a scalar function of t.

Let $\mathbf{a} = u\mathbf{v}$.

Suppose δu, $\delta \mathbf{v}$ and $\delta \mathbf{a}$ are the increments in u, $\mathbf{v}$ and $\mathbf{a}$ respectively due to an increment δt in t.

Then

$$\mathbf{a}+\delta\mathbf{a} = (u+\delta u)(\mathbf{v}+\delta\mathbf{v})$$

$$= u\mathbf{v}+u\delta\mathbf{v}+\mathbf{v}\delta u+\delta u\,\delta\mathbf{v},$$

$$\therefore \quad \delta\mathbf{a} = u\delta\mathbf{v}+\mathbf{v}\delta u+\delta u\,\delta\mathbf{v},$$

$$\therefore \quad \frac{\delta\mathbf{a}}{\delta t} = u\frac{\delta\mathbf{v}}{\delta t}+\mathbf{v}\frac{\delta u}{\delta t}+\frac{\delta u}{\delta t}\delta\mathbf{v},$$

$$\therefore \quad \mathop{\mathrm{Lt}}_{\delta t\to 0}\frac{\delta\mathbf{a}}{\delta t} = \mathop{\mathrm{Lt}}_{\delta t\to 0}\left(u\frac{\delta\mathbf{v}}{\delta t}+\mathbf{v}\frac{\delta u}{\delta t}+\frac{\delta u}{\delta t}\delta\mathbf{v}\right),$$

$$\therefore \quad \frac{d\mathbf{a}}{dt} = u\frac{d\mathbf{v}}{dt}+\mathbf{v}\frac{du}{dt}.$$

$$\therefore \quad \frac{d}{dt}(u\mathbf{v}) = u\frac{d\mathbf{v}}{dt}+\mathbf{v}\frac{du}{dt}.$$

Derivative of a vector in terms of its components

Let $\mathbf{r} = x\mathbf{i}+y\mathbf{j}+z\mathbf{k}$.

Since the unit vectors $\mathbf{i}$, $\mathbf{j}$ and $\mathbf{k}$ are constant in magnitude and direction

$$\frac{d\mathbf{r}}{dt} = \frac{dx}{dt}\mathbf{i}+\frac{dy}{dt}\mathbf{j}+\frac{dz}{dt}\mathbf{k}.$$

Hence the components of the derivative of a vector are the derivatives of its components for fixed directions.

Higher derivatives may be obtained in the same way. Thus

$$\frac{d^2\mathbf{r}}{dt^2} = \frac{d^2x}{dt^2}\mathbf{i}+\frac{d^2y}{dt^2}\mathbf{j}+\frac{d^2z}{dt^2}\mathbf{k}.$$

Velocity

We now consider the case of a particle moving with variable velocity. Let the position of the particle relative to an origin O at time t be $\mathbf{OP} = \mathbf{r}$ and at time $t+\delta t$ be $\mathbf{OP}' = \mathbf{r}+\delta\mathbf{r}$ (Fig. 7.2).

Then $\mathbf{PP}' = \mathbf{OP}'-\mathbf{OP} = \delta\mathbf{r}$.

The average velocity of the particle relative to O during the time interval δt is defined as $\mathbf{PP}'/\delta t$, i.e. $\delta\mathbf{r}/\delta t$.

The instantaneous velocity, or briefly the velocity, of the particle relative to O at the instant t is defined as the rate of change of its position relative to O, i.e. $d\mathbf{r}/dt$ which can be written as $\dot{\mathbf{r}}$.

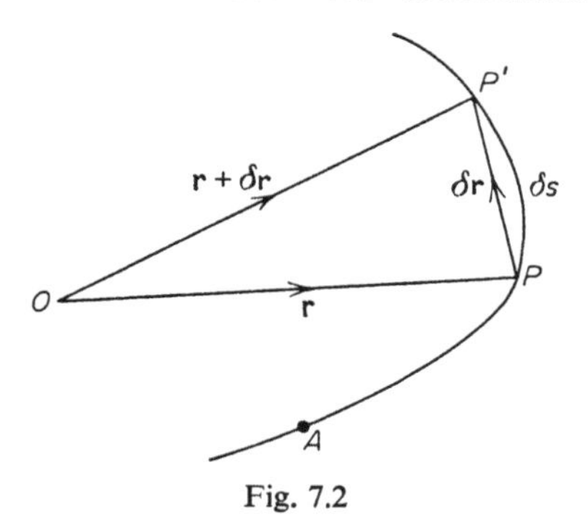

Fig. 7.2

Suppose A is any fixed point on the path of the particle and let arc $AP = s$ where s is positive if the particle after passing through A passes through P. Then we can write arc $PP' = \delta s$. We define the average speed of the particle during the interval of time δt as $\delta s/\delta t$. Now

$$\frac{d\mathbf{r}}{dt} = \underset{\delta t \to 0}{\mathrm{Lt}} \frac{\delta \mathbf{r}}{\delta t} = \underset{\delta t \to 0}{\mathrm{Lt}} \frac{\delta \mathbf{r}}{\delta s} \cdot \frac{\delta s}{\delta t}.$$

As $\delta t \to 0$, i.e. as P' approaches P, $(\delta s/\delta t) \to (ds/dt)$ which being the rate of increase of the distance of the particle from a fixed point in its path is its speed at time t. Also as $\delta t \to 0$, $|\delta \mathbf{r}|/\delta s \to 1$ and the direction of $\delta \mathbf{r}/\delta t \to$ the direction of the tangent at P, the sense of direction being the same as that of s increasing. If $\hat{\mathbf{t}}$ is the unit vector parallel to and having the same sense as this tangent we can write

$$\frac{d\mathbf{r}}{dt} = \frac{ds}{dt}\hat{\mathbf{t}}.$$

Thus the magnitude of the instantaneous velocity $\dot{\mathbf{r}}$ at the instant t is the positive value of the speed and its direction is along the tangent to the path at P.

We now show that velocity as defined by $d\mathbf{r}/dt$ is a vector. Suppose the motion of the particle P is compared by two observers moving relatively to one another. Each observer will have his own frame of reference which is fixed relative to himself. Suppose $OXYZ$, $O'X'Y'Z'$ are the two frames of reference which have no rotation relative to one another, i.e. the frames have different origins O, O' but the same initial directions (Fig. 7.3). This being so it will be sufficient when describing the relative motion to say 'relative to the origin' instead of 'relative to the frame'.

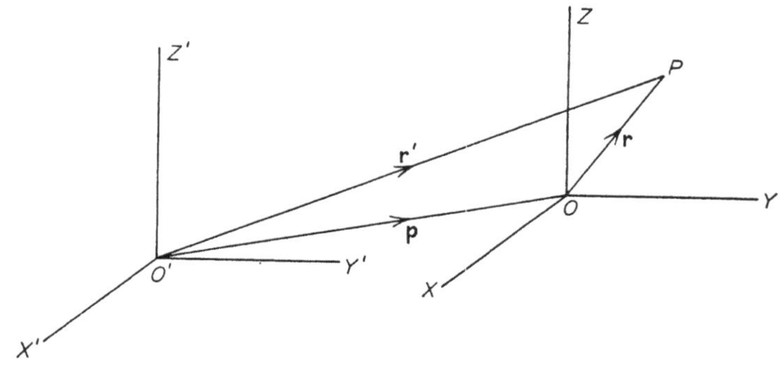

Fig. 7.3

Let $\mathbf{r}$, $\mathbf{r}'$ be the position vectors of P relative to O and O' respectively at the instant t. Also let $\mathbf{p}$ be the position vector of O relative to O' at the same time. Then we have

$$\mathbf{r}' = \mathbf{r}+\mathbf{p}.$$

This tells us that the position of P relative to O' is the vector sum of the position of P relative to O and the position of O relative to O'.

Differentiating with respect to the time t we have

$$\dot{\mathbf{r}}' = \dot{\mathbf{r}}+\dot{\mathbf{p}},$$

i.e. velocity of P relative to O' = velocity of P relative to O
+ velocity of O relative to O'.

This shows that in general velocities are added vectorially, a result we have shown previously for uniform velocities. Hence we conclude that velocity is a vector.

We note again that when $\dot{\mathbf{p}} = \mathbf{0}$ the velocity of a particle is the same relative to all fixed points.

Acceleration

Let $\mathbf{v}$ be the velocity relative to an origin O of a particle P at time t (Fig. 7.4). Take $\mathbf{v}$ to be the position vector of a point Q relative to a frame of reference with Ω as its origin. Now as $\mathbf{v}$ varies with time the point Q will also vary in position and trace out a locus which is known as the hodograph of the motion of the particle P. Suppose during the interval of time δt the particle moves from P to P' and its velocity changes from $\mathbf{v}$ to $\mathbf{v}+\delta\mathbf{v}$. Then the point Q will move along the hodograph to Q' where $\mathbf{\Omega Q'} = \mathbf{v}+\delta\mathbf{v}$ (Fig. 7.4).

The average acceleration during the interval of time δt is defined by $\mathbf{QQ'}/\delta t$, i.e. $\delta\mathbf{v}/\delta t$.

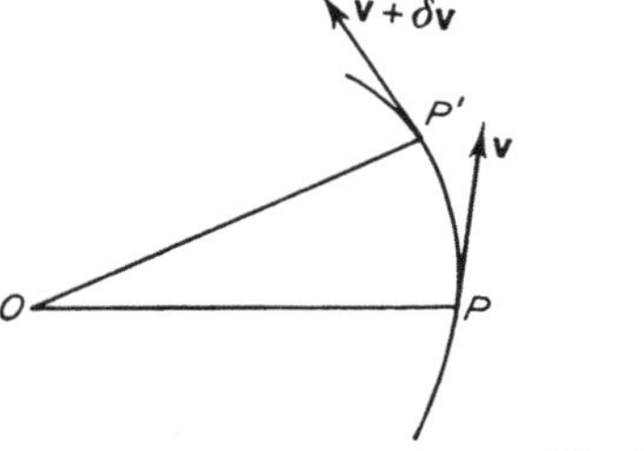

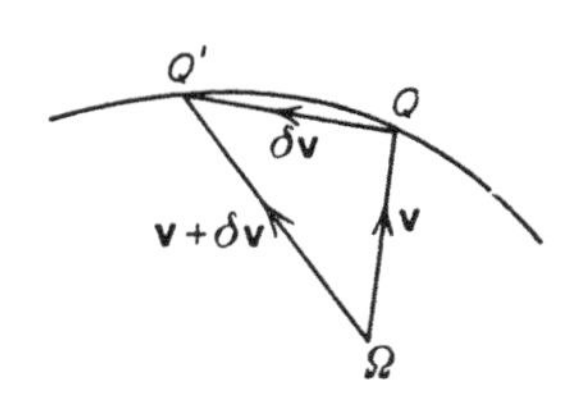

Fig. 7.4

The instantaneous acceleration, or briefly the acceleration, of the particle at the instant t is defined as the rate of change of its velocity, i.e. $d\mathbf{v}/dt$ which can be written as $\dot{\mathbf{v}}$ or $\ddot{\mathbf{r}}$.

Now

$$\frac{d\mathbf{v}}{dt} = \underset{\delta t \to 0}{\text{Lt}} \frac{\delta\mathbf{v}}{\delta t} = \underset{\delta t \to 0}{\text{Lt}} \frac{\mathbf{QQ'}}{\delta t}.$$

We see that as $\delta t \to 0$, $Q' \to Q$ and hence the direction of the instantaneous acceleration is that of the tangent at the point Q on the hodograph.

As before we compare the motion of P relative to two frames of reference. We have shown in the previous section that

$$\dot{\mathbf{r}}' = \dot{\mathbf{r}}+\dot{\mathbf{p}}.$$

Differentiating again with respect to the time t we have,

$$\ddot{\mathbf{r}}' = \ddot{\mathbf{r}}+\ddot{\mathbf{p}},$$

i.e. acceleration of P relative to O' = acceleration of P relative to O
+ acceleration of O relative to O'.

Since this shows that accelerations are added vectorially we conclude that acceleration is a vector.

We note that when $\ddot{\mathbf{p}} = \mathbf{0}$ the acceleration of a particle is the same relative to all points moving with constant velocity.

Derivative of a unit vector

Let $\mathbf{OP} = \hat{\mathbf{u}}$ be the position of a unit vector at a time t and $\mathbf{OP}' = \hat{\mathbf{u}}+\delta\hat{\mathbf{u}}$ be its position at a time $t+\delta t$ (Fig. 7.5). We then have

$$\mathbf{OP}' = \mathbf{OP}+\mathbf{PP}',$$

$$\therefore \quad \hat{\mathbf{u}}+\delta\hat{\mathbf{u}} = \hat{\mathbf{u}}+\mathbf{PP}',$$

$$\therefore \quad \delta\hat{\mathbf{u}} = \mathbf{PP}',$$

$$\therefore \quad \frac{\delta\hat{\mathbf{u}}}{\delta t} = \frac{\mathbf{PP}'}{\delta t},$$

$$\therefore \quad \frac{d\hat{\mathbf{u}}}{dt} = \operatorname*{Lt}_{\delta t\to 0} \frac{\mathbf{PP}'}{\delta t}.$$

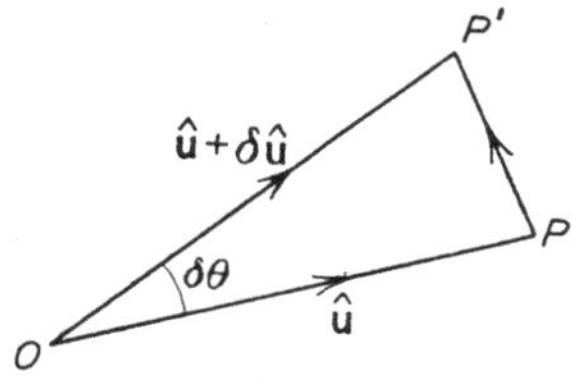

Fig. 7.5

Since $\quad |\mathbf{OP}| = |\mathbf{OP}'| = 1,$

$$|\mathbf{PP}'| = 2\sin\tfrac{1}{2}\delta\theta \text{ where angle } POP' = \delta\theta,$$

$$\therefore \quad \frac{|\mathbf{PP}'|}{\delta t} = \frac{2\sin\frac{1}{2}\delta\theta}{\delta t}$$

$$= \frac{\sin\frac{1}{2}\delta\theta}{\frac{1}{2}\delta\theta}\frac{\delta\theta}{\delta t}.$$

As $\delta t \to 0$, $\delta\theta \to 0$,

$$\frac{\sin\frac{1}{2}\delta\theta}{\frac{1}{2}\delta\theta} \to 1 \quad \text{and} \quad \frac{\delta\theta}{\delta t} \to \frac{d\theta}{dt}.$$

Hence in the limit $\mathbf{PP}'/\delta t$ has the magnitude $d\theta/dt$ and its direction is perpendicular to $\mathbf{OP}$. If $\hat{\mathbf{p}}$ is the unit vector perpendicular to the unit vector $\hat{\mathbf{u}}$ in the plane of motion and in the direction of θ increasing we have

$$\frac{d\hat{\mathbf{u}}}{dt} = \frac{d\theta}{dt}\hat{\mathbf{p}} \quad \text{or} \quad \dot{\hat{\mathbf{u}}} = \dot{\theta}\hat{\mathbf{p}}.$$

It is important to note that the direction of $\hat{\mathbf{p}}$ is the direction of $\hat{\mathbf{u}}$ when $\hat{\mathbf{u}}$ is rotated anticlockwise through a right angle, since the convention for positive angles is anticlockwise rotation.

Thus if $\hat{\mathbf{u}}$ is a unit vector which rotates in a plane with angular velocity $\dot{\theta}$ then $d\hat{\mathbf{u}}/dt$ is a vector of magnitude $\dot{\theta}$ and in a direction at right angles to $\hat{\mathbf{u}}$ in the plane.

If $\hat{\mathbf{p}}$ is rotated again in the positive direction the direction is that of $-\hat{\mathbf{u}}$. Thus

$$\dot{\hat{\mathbf{p}}} = -\dot{\theta}\hat{\mathbf{u}}.$$

Motion of a particle in two dimensions

An important application of differentiation of vectors is in the consideration of the motion of a particle in two dimensions. This motion can be referred in terms of Cartesian, polar or intrinsic co-ordinates.

Velocity and acceleration in Cartesian co-ordinates

Consider the motion of a particle in a plane curve (Fig. 7.6). Let O be a fixed point in the plane and OX, OY rectangular axes in the plane. Denoting the position vector of P by $\mathbf{r}$ and the unit vectors in the OX, OY directions by $\mathbf{i}$, $\mathbf{j}$ we have

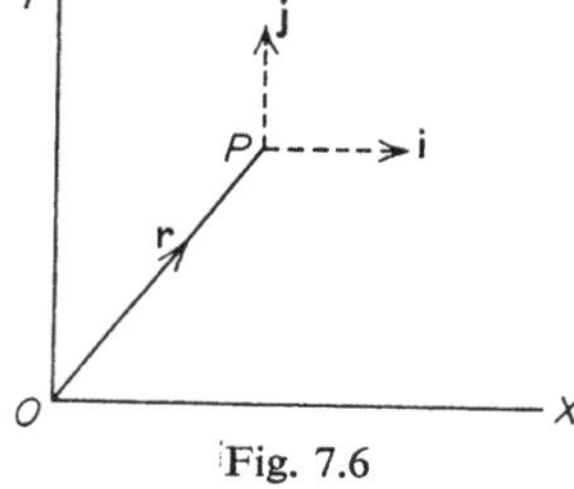

Fig. 7.6

$$\mathbf{r} = x\mathbf{i}+y\mathbf{j},$$

$$\therefore \quad \dot{\mathbf{r}} = \dot{x}\mathbf{i}+\dot{y}\mathbf{j},$$

and
$$\ddot{\mathbf{r}} = \ddot{x}\mathbf{i}+\ddot{y}\mathbf{j}.$$

Thus the velocity components parallel to the OX and OY axes are $\dot{x}$ and $\dot{y}$ respectively, and the acceleration components in these directions are $\ddot{x}$ and $\ddot{y}$.

$$\therefore \quad \text{velocity} = \sqrt{(\dot{x}^2+\dot{y}^2)},$$

$$\text{and acceleration} = \sqrt{(\ddot{x}^2+\ddot{y}^2)}.$$

Examples

(1) *The position vector of a particle at time t is given by*

$$\mathbf{r} = (\tfrac{1}{2}at^2)\,\mathbf{i}+(\tfrac{1}{2}bt^2+ut)\,\mathbf{j},$$

where a, b, u are constants and $\mathbf{i}$, $\mathbf{j}$ *are unit vectors in the direction of the x-, y-axes respectively. Obtain its velocity at time* $t = 0$ *and show that its acceleration at any time has the constant value* $\sqrt{(a^2+b^2)}$. *Also show that its path is a parabola.*

$$\mathbf{r} = \frac{at^2}{2}\mathbf{i}+\left(\frac{bt^2}{2}+ut\right)\mathbf{j}.$$

Differentiating with respect to (t),

$$\dot{\mathbf{r}} = at\mathbf{i}+(bt+u)\,\mathbf{j}.$$

When $t = 0$,
$$\dot{\mathbf{r}} = u\mathbf{j}.$$

Therefore velocity at $t = 0$ is u and in the direction of the y axis. Differentiating again

$$\ddot{\mathbf{r}} = a\mathbf{i}+b\mathbf{j}.$$

Therefore acceleration is independent of the time and is $\sqrt{(a^2+b^2)}$. We have

$$x = \frac{at^2}{2} \quad \text{and} \quad y = \frac{bt^2}{2}+ut.$$

Eliminating t,
$$y = \frac{b}{2}\frac{2x}{a}+u\sqrt{\frac{2x}{a}},$$

$$\therefore \quad y-\frac{bx}{a} = u\sqrt{\frac{2x}{a}},$$

$$\therefore \quad \left(y-\frac{bx}{a}\right)^2 = \frac{2u^2x}{a}.$$

This equation represents a parabola. Therefore the path of the particle is a parabola.

(2) *A particle moves in a plane and its co-ordinates at any time t are $(a\cos nt,\, b\sin nt)$. Show that its acceleration is always directed towards the origin of the co-ordinate system.*

Let $\mathbf{r}$ be the position vector of the particle relative to the origin and x, y the components of $\mathbf{r}$ in the direction of the unit vectors $\mathbf{i}$, $\mathbf{j}$ respectively. Then

$$\begin{aligned}
\mathbf{r} &= x\mathbf{i}+y\mathbf{j} \\
&= a\cos nt\mathbf{i}+b\sin nt\mathbf{j}, \\
\therefore \quad \dot{\mathbf{r}} &= -na\sin nt\mathbf{i}+nb\cos nt\mathbf{j} \\
\therefore \quad \ddot{\mathbf{r}} &= -n^2a\cos nt\mathbf{i}-n^2b\sin nt\mathbf{j} \\
&= -n^2(a\cos nt\mathbf{i}+b\sin nt\mathbf{j}) \\
&= -n^2\mathbf{r}.
\end{aligned}$$

Therefore its acceleration is always directed towards the origin.

Velocity and acceleration in polar co-ordinates

Let the path of a moving particle P be a plane curve (Fig. 7.7). Let O be a fixed origin and θ the inclination of OP to a fixed direction in the plane. Then the polar co-ordinates of P are (r, θ).

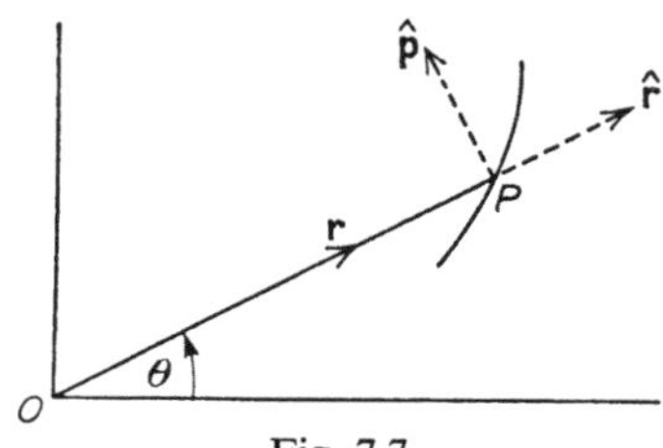

Fig. 7.7

Denote the position vector of P relative to the origin O by $\mathbf{r}$, the unit vector in the direction of $\mathbf{r}$ by $\hat{\mathbf{r}}$ and the unit vector in the plane perpendicular to $\hat{\mathbf{r}}$ by $\hat{\mathbf{p}}$, the right angle between $\hat{\mathbf{r}}$ and $\hat{\mathbf{p}}$ being measured in the conventional anticlockwise notation from $\hat{\mathbf{r}}$. It is important to note that $\hat{\mathbf{p}}$ is not in the direction of the tangent at P unless the path is a circle with O its centre.

We have

$$\mathbf{r} = r\hat{\mathbf{r}},$$

$$\therefore \quad \dot{\mathbf{r}} = \dot{r}\hat{\mathbf{r}} + r\frac{d\hat{\mathbf{r}}}{dt}$$

$$= \dot{r}\hat{\mathbf{r}} + r\dot{\theta}\hat{\mathbf{p}} \quad \text{(using differentiation of a unit vector).}$$

Therefore radial component of velocity $= \dot{r}$,

and transverse component of velocity $= r\dot{\theta}$.

For a particle moving in a circle centre O $\dot{r} = 0$, therefore the velocity $= r\dot{\theta}$ and is along the tangent.

Differentiating again,

$$\ddot{\mathbf{r}} = \ddot{r}\hat{\mathbf{r}} + \dot{r}\frac{d\hat{\mathbf{r}}}{dt} + \dot{r}\dot{\theta}\hat{\mathbf{p}} + r\ddot{\theta}\hat{\mathbf{p}} + r\dot{\theta}\frac{d\hat{\mathbf{p}}}{dt}$$

$$= \ddot{r}\hat{\mathbf{r}} + \dot{r}\dot{\theta}\hat{\mathbf{p}} + \dot{r}\dot{\theta}\hat{\mathbf{p}} + r\ddot{\theta}\hat{\mathbf{p}} - r\dot{\theta}^2\hat{\mathbf{r}}$$

$$= (\ddot{r} - r\dot{\theta}^2)\,\hat{\mathbf{r}} + (2\dot{r}\dot{\theta} + r\ddot{\theta})\,\hat{\mathbf{p}}.$$

Therefore radial component of acceleration $= \ddot{r} - r\dot{\theta}^2$,

and transverse component of acceleration $= 2\dot{r}\dot{\theta} + r\ddot{\theta}$

$$= \frac{1}{r}(2r\dot{r}\dot{\theta} + r^2\ddot{\theta})$$

$$= \frac{1}{r}\frac{d}{dt}(r^2\dot{\theta}).$$

For a particle moving in a circle centre O with uniform speed, $\dot{r} = 0$ and $\ddot{\theta} = 0$. Therefore the acceleration consists of the component $-r\dot{\theta}^2$, that is, it has an acceleration of $r\dot{\theta}^2$ directed towards the centre of the circle.

Example

A point moves along the equiangular spiral $r = e^{\theta}$ with uniform angular velocity about the origin. Prove that the acceleration is everywhere at right angles to the radius vector and proportional to its length.

Let the position vector of the point P relative to the origin be $\mathbf{r}$ and the unit vectors in the directions OP and perpendicular to OP be $\hat{\mathbf{r}}$ and $\hat{\mathbf{p}}$ respectively.

Since the angular velocity is uniform, $\dot{\theta} = \omega$ (say) and $\ddot{\theta} = 0$.

$$r = e^{\theta},$$

$$\therefore \quad \dot{r} = e^{\theta}.\dot{\theta} = r\omega,$$

and

$$\ddot{r} = \dot{r}\omega = r\omega^2.$$

Now

$$\ddot{\mathbf{r}} = (\ddot{r}-r\dot{\theta}^2)\,\hat{\mathbf{r}}+(2\dot{r}\dot{\theta}+r\ddot{\theta})\,\hat{\mathbf{p}}$$

$$= (r\omega^2-r\omega^2)\,\hat{\mathbf{r}}+2r\omega^2\hat{\mathbf{p}}$$

$$= 2r\omega^2\hat{\mathbf{p}}.$$

Therefore acceleration is at right angles to the radius vector $\mathbf{r}$ and is proportional to its length r.

Velocity and acceleration in intrinsic co-ordinates

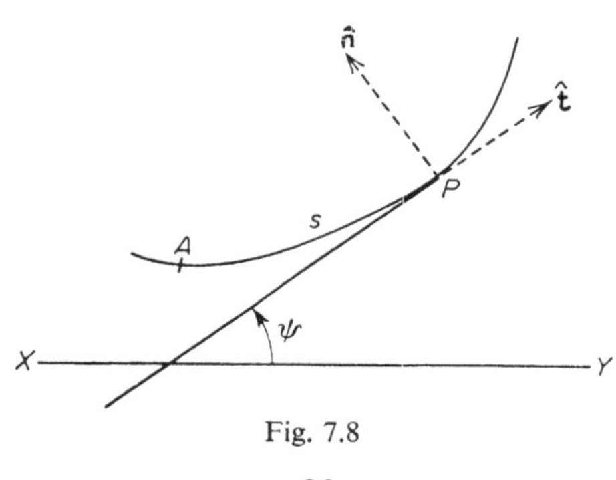

Fig. 7.8

Let A be a fixed point on the path of a moving particle and P the position of the particle at any time t (Fig. 7.8). Let arc $AP = s$ and let the tangent at P be inclined at an angle ψ to a fixed straight line XY. Then the intrinsic co-ordinates of P are (s, ψ). Let $P(s, \psi)$ be the position of the particle at a time t and $P'(s+\delta s, \psi+\delta\psi)$ its position at a time $t+\delta t$ (Fig. 7.9).

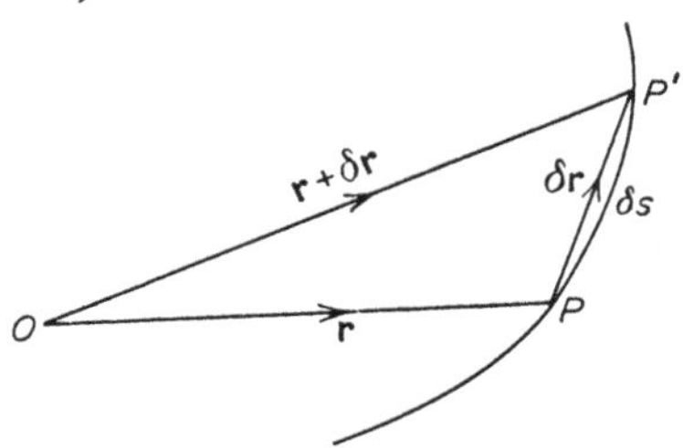

Fig. 7.9

Then $$\text{arc } PP' = \delta s.$$

Let the position vectors of P and P' relative to the origin O be $\mathbf{r}$ and $\mathbf{r}+\delta\mathbf{r}$ respectively.

The velocity $\mathbf{v}$ of P is defined as

$$\mathbf{v} = \underset{\delta t \to 0}{\text{Lt}} \frac{\mathbf{PP'}}{\delta t}$$

$$= \underset{\delta t \to 0}{\text{Lt}} \frac{\delta \mathbf{r}}{\delta t}$$

$$= \underset{\delta t \to 0}{\text{Lt}} \frac{\delta \mathbf{r}}{\delta s} \cdot \frac{\delta s}{\delta t}.$$

As $\delta t \to 0$, the modulus of $\delta\mathbf{r} \to \delta s$, i.e. $\delta r/\delta s \to 1$ and the direction of $\delta\mathbf{r} \to$ the direction of the tangent at P.

Let $\hat{\mathbf{t}}$ be the unit vector in the direction of the tangent at P. Therefore in the limit,

$$\mathbf{v} = \frac{ds}{dt}\hat{\mathbf{t}},$$

$$\therefore \quad \mathbf{v} = \dot{s}\hat{\mathbf{t}}.$$

Thus the velocity is in the direction of the tangent and there is no normal component.

Let $\hat{\mathbf{n}}$ be the unit vector in the direction of the positive normal at P.

$$\therefore \quad \frac{d\hat{\mathbf{t}}}{dt} = \frac{d\psi}{dt}\hat{\mathbf{n}} = \dot{\psi}\hat{\mathbf{n}}.$$

Differentiating $\mathbf{v} = \dot{s}\hat{\mathbf{t}}$ we get

$$\dot{\mathbf{v}} = \ddot{s}\hat{\mathbf{t}} + \dot{s}\frac{d\hat{\mathbf{t}}}{dt}$$

$$= \ddot{s}\hat{\mathbf{t}} + \dot{s}\dot{\psi}\hat{\mathbf{n}}.$$

Therefore tangential component of acceleration $= \ddot{s}$,

and normal component of acceleration $= \dot{s}\dot{\psi}$.

The normal component can be written in terms of the radius of curvature of the path at P since

$$\dot{s}\dot{\psi} = \dot{s}\frac{d\psi}{ds}\cdot\frac{ds}{dt}$$

$$= \dot{s}^2 \Big/ \frac{ds}{d\psi}$$

$$= \frac{\dot{s}^2}{\rho},$$

where ρ = radius of curvature $= ds/d\psi$ by definition.

For a particle moving with constant speed u along the circumference of a circle of radius a, $\ddot{s} = \dot{u} = 0$ and $\rho = a$. Thus the acceleration is u^2/a and is directed towards the centre of the circle.

Example

A particle moves in the curve whose equation is $s = f(\psi)$ in such a way that the tangent to the curve rotates uniformly. Prove that the normal acceleration is proportional to the radius of curvature.

$$s = f(\psi),$$

$$\therefore \quad \dot{s} = f'(\psi).\dot{\psi}, \quad \text{where} \quad f'(\psi) = \frac{ds}{d\psi}.$$

Normal acceleration $= \dot{s}\dot{\psi}$

$$= f'(\psi).\dot{\psi}^2.$$

Radius of curvature $= \dfrac{ds}{d\psi}$ (by definition)

$$= f'(\psi),$$

therefore normal acceleration $= \rho\dot{\psi}^2$.

Since $\dot{\psi}$ = constant,

normal acceleration is proportional to ρ.

Summary

Velocity components

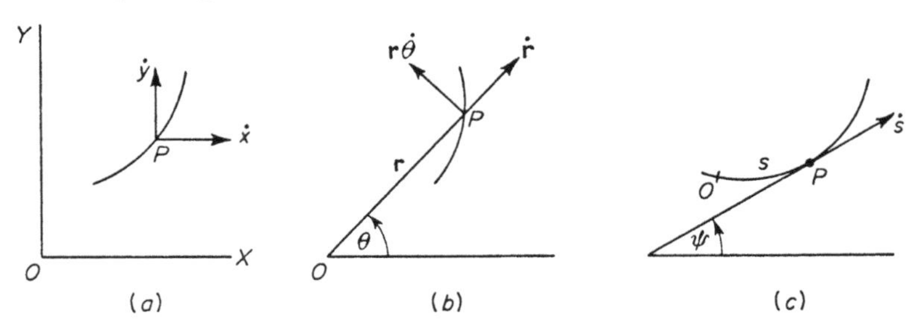

Fig. 7.10. (*a*) Cartesian, (*b*) Polar, (*c*) Intrinsic

Acceleration components

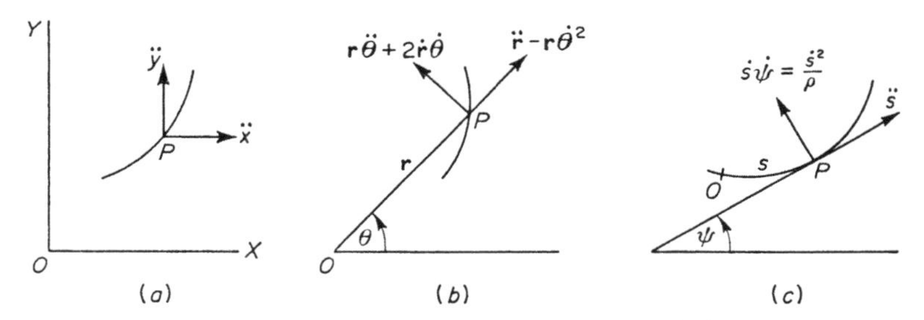

Fig. 7.11. (*a*) Cartesian, (*b*) Polar, (*c*) Intrinsic

Motion of a particle on a rotating plane

Consider the motion of a particle which is itself moving on a rotating plane. Let the plane rotate about a fixed point O with uniform angular velocity ω.

Let perpendicular axes Ox, Oy be drawn in the plane.

Let the position vector of the particle be

$$\mathbf{r} = x\mathbf{i}+y\mathbf{j},$$

where $\mathbf{i}$ and $\mathbf{j}$ are unit vectors in the Ox and Oy directions respectively which themselves rotate with the plane.

$$\therefore \quad \dot{\mathbf{r}} = \dot{x}\mathbf{i}+x\frac{d\mathbf{i}}{dt}+\dot{y}\mathbf{j}+y\frac{d\mathbf{j}}{dt}.$$

Now $$\frac{d\mathbf{i}}{dt} = \omega\mathbf{j} \quad \text{and} \quad \frac{d\mathbf{j}}{dt} = -\omega\mathbf{i},$$

from the differentiation of a unit vector.

$$\therefore \quad \dot{\mathbf{r}} = \dot{x}\mathbf{i} + \omega x\mathbf{j} + \dot{y}\mathbf{j} - \omega y\mathbf{i}$$

$$= (\dot{x} - \omega y)\mathbf{i} + (\dot{y} + \omega x)\mathbf{j}.$$

Thus the velocity of the particle consists of the two perpendicular components

$$\dot{x} - \omega y \quad \text{and} \quad \dot{y} + \omega x.$$

Differentiating again to obtain the acceleration,

$$\ddot{\mathbf{r}} = (\ddot{x} - \omega\dot{y})\mathbf{i} + (\dot{x} - \omega y)\frac{d\mathbf{i}}{dt} + (\ddot{y} + \omega\dot{x})\mathbf{j} + (\dot{y} + \omega x)\frac{d\mathbf{j}}{dt}$$

$$= (\ddot{x} - \omega\dot{y})\mathbf{i} + (\dot{x} - \omega y)\omega\mathbf{j} + (\ddot{y} + \omega\dot{x})\mathbf{j} - (\dot{y} + \omega x)\omega\mathbf{i}$$

$$= (\ddot{x} - 2\omega\dot{y} - \omega^2 x)\mathbf{i} + (\ddot{y} + 2\omega\dot{x} - \omega^2 y)\mathbf{j}.$$

The acceleration can be written as

$$\ddot{\mathbf{r}} = (\ddot{x}\mathbf{i} + \ddot{y}\mathbf{j}) + 2\omega(-\dot{y}\mathbf{i} + \dot{x}\mathbf{j}) - \omega^2(x\mathbf{i} + y\mathbf{j}).$$

$\ddot{x}\mathbf{i} + \ddot{y}\mathbf{j}$ is the acceleration of the particle with respect to the rotating co-ordinate system. If $\omega = 0$, i.e. the plane is stationary, this will be the only acceleration of the particle.

$2\omega(-\dot{y}\mathbf{i} + \dot{x}\mathbf{j})$ is the acceleration due to the velocity of the particle relative to the rotating co-ordinate system. This acceleration is known as the Coriolis acceleration and it vanishes if the particle is at rest relative to the plane.

$-\omega^2(x\mathbf{i} + y\mathbf{j}) = -\omega^2\mathbf{r}$ is the acceleration of the particle toward the centre of the rotation.

Example

An insect crawls outwards along a spoke of a bicycle wheel which is rotating uniformly at ω radians per second. If the insect crawls with uniform speed u find the magnitude and direction of its velocity and acceleration when it is a distance a from the centre.

Velocity is given by

$$\dot{\mathbf{r}} = (\dot{x} - \omega y)\mathbf{i} + (\dot{y} + \omega x)\mathbf{j},$$

where $x = a$, $\dot{x} = u$, $y = 0$, $\dot{y} = 0$, taking the rotating Ox axis as the spoke on which the insect crawls.

$$\therefore \quad \dot{\mathbf{r}} = u\mathbf{i}+\omega a\mathbf{j},$$

$$\therefore \quad \text{velocity} = \sqrt{(u^2+\omega^2a^2)} \text{ at } \tan^{-1} \omega a/u \text{ to the spoke.}$$

The acceleration is given by

$$\ddot{\mathbf{r}} = (\ddot{x}-2\omega\dot{y}-\omega^2x)\mathbf{i}+(\ddot{y}+2\omega\dot{x}-\omega^2y)\mathbf{j},$$

where $x = a$, $\dot{x} = u$, $\ddot{x} = 0$, $y = 0$, $\dot{y} = 0$, $\ddot{y} = 0$.

$$\therefore \quad \ddot{\mathbf{r}} = -\omega^2a\mathbf{i}+2\omega u\mathbf{j},$$

$$\therefore \quad \text{acceleration} = \omega\sqrt{(\omega^2a^2+4u^2)} \text{ at } \tan^{-1}-2u/\omega a \text{ to the spoke.}$$

Fig. 7.12 shows the directions of the velocity and acceleration.

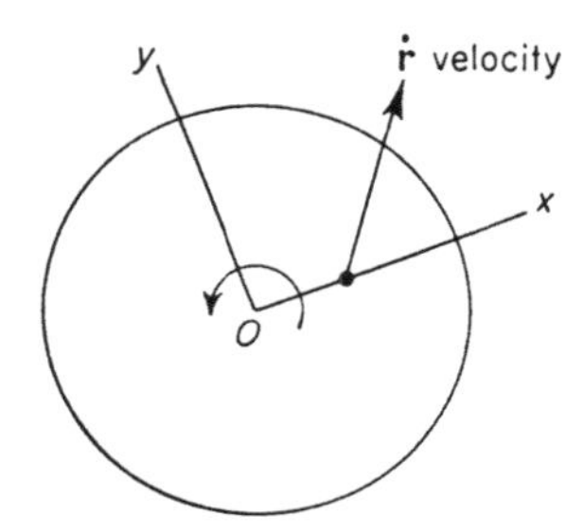

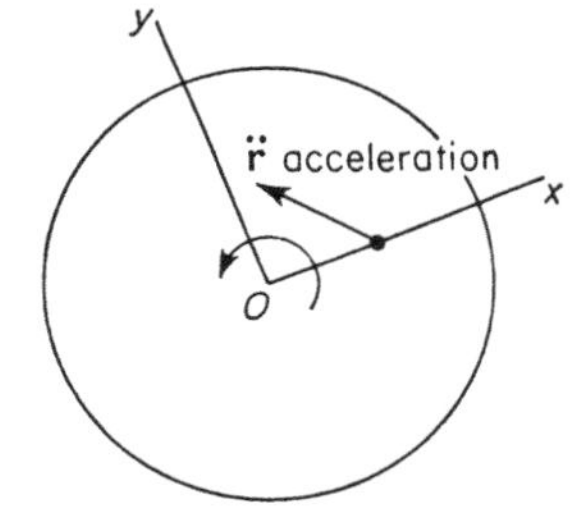

Fig. 7.12

This problem can also be solved by expressing the velocity and acceleration in terms of polar co-ordinates and this is left as an exercise.

Integration

Integration of a vector is the reverse process of the differentiation of a vector.

If
$$\frac{d\mathbf{a}}{dt} = \mathbf{b} \quad \text{then} \quad \int \mathbf{b}\,dt = \mathbf{a}+\mathbf{c},$$

where $\mathbf{c}$ is an arbitrary constant vector.

Examples

(1) *If at any time t after projection the position vector of a projectile relative to the point of projection is* $\mathbf{r}$ *and* $\mathbf{u}$ *its velocity of projection, obtain* $\mathbf{r}$ *in terms of* $\mathbf{u}$ *and* t.

If there are no resistances on the projectile and $\mathbf{g}$ is the acceleration due to gravity,

$$\ddot{\mathbf{r}} = \mathbf{g},$$

$$\therefore \quad \dot{\mathbf{r}} = \mathbf{g}t+\mathbf{a} \quad (\mathbf{a} = \text{constant of integration}).$$

When $t = 0$, $\dot{\mathbf{r}} = \mathbf{u}$,

$$\therefore \quad \mathbf{a} = \mathbf{u},$$

$$\therefore \quad \dot{\mathbf{r}} = \mathbf{g}t+\mathbf{u}.$$

Integrating again,

$$\mathbf{r} = \tfrac{1}{2}\mathbf{g}t^2+\mathbf{u}t+\mathbf{b} \quad (\mathbf{b} = \text{constant of integration}).$$

When $t = 0$, $\mathbf{r} = \mathbf{0}$ $\quad \therefore \quad \mathbf{b} = \mathbf{0}$

$$\therefore \quad \mathbf{r} = \tfrac{1}{2}\mathbf{g}t^2+\mathbf{u}t.$$

(2) *If* $d^2\mathbf{r}/dt^2 = -2\sin t\mathbf{i}-3\cos t\mathbf{j}$, *and if* $d\mathbf{r}/dt = 2\mathbf{i}$ *when* $t = 0$ *and* $\mathbf{r} = 3\mathbf{j}$ *when* $t = 0$, *show that the particle whose position vector is* $\mathbf{r}$ *is moving in an ellipse.*

$$\frac{d^2\mathbf{r}}{dt^2} = -2\sin t\mathbf{i}-3\cos t\mathbf{j},$$

$$\therefore \quad \frac{d\mathbf{r}}{dt} = 2\cos t\mathbf{i}-3\sin t\mathbf{j}+\mathbf{a} \quad (\mathbf{a} = \text{constant of integration}).$$

When $t = 0$, $d\mathbf{r}/dt = 2\mathbf{i}$,

$$\therefore \quad \mathbf{a} = \mathbf{0},$$

$$\therefore \quad \frac{d\mathbf{r}}{dt} = 2\cos t\mathbf{i}-3\sin t\mathbf{j}.$$

Integrating again,

$$\mathbf{r} = 2\sin t\mathbf{i}+3\cos t\mathbf{j}+\mathbf{b} \quad (\mathbf{b} = \text{constant of integration}).$$

When $t = 0$, $\mathbf{r} = 3\mathbf{j}$,

$$\therefore \quad \mathbf{b} = \mathbf{0},$$

$$\therefore \quad \mathbf{r} = 2\sin t\mathbf{i}+3\cos t\mathbf{j},$$

$$\therefore \quad x = 2\sin t \quad \text{and} \quad y = 3\cos t,$$

$$\therefore \quad \tfrac{1}{4}x^2+\tfrac{1}{9}y^2 = 1.$$

Therefore path of particle is an ellipse.

Exercise 7

(1) The position vector of a particle at time t is $\mathbf{r}$, $\mathbf{v}$ is its velocity and $\mathbf{a}$ is its acceleration which is constant. Show that if $\mathbf{r}_0$ and $\mathbf{v}_0$ are the initial values of $\mathbf{r}$ and $\mathbf{v}$ respectively

$$\mathbf{v} = \mathbf{v}_0+\mathbf{a}t \quad \text{and} \quad \mathbf{r}-\mathbf{r}_0 = \mathbf{v}_0 t+\tfrac{1}{2}\mathbf{a}t^2.$$

(2) The position vector of a point at any time t relative to a fixed origin is $\mathbf{a}\cos\omega t+\mathbf{b}\sin\omega t$ where $\mathbf{a}$ and $\mathbf{b}$ are constant vectors and ω is a constant scalar. Show that the acceleration is everywhere towards the origin and proportional to the distance from the origin.

(3) If $d^2\mathbf{r}/dt^2 = 4\mathbf{i}$ and if $\mathbf{r} = \mathbf{0}$ when $t = 0$ and $d\mathbf{r}/dt = 4\mathbf{j}$ when $t = 0$, prove that the point whose position vector is $\mathbf{r}$ is describing a parabola.

(4) The position vector of a particle at time t with respect to fixed axes is $(t^2+1)\mathbf{i}-2t\mathbf{j}+t^2\mathbf{k}$. Find the magnitude of the velocity and acceleration when $t = 2$.

(5) If $d^2\mathbf{r}/dt^2 = \mathbf{p}\cos\omega t$ and if $\mathbf{r} = \mathbf{0}$ when $t = 0$ and $d\mathbf{r}/dt = \mathbf{u}$ when $t = 0$, show that $\mathbf{r} = \mathbf{u}t+(\mathbf{p}/\omega^2)(1-\cos\omega t)$.

(6) If $\mathbf{p}$ and $\mathbf{q}$ are two unit vectors whose directions make angles of θ and $\theta+\frac{1}{2}\pi$ respectively with a fixed direction, prove

$$d^2\mathbf{p}/dt^2 = \mathbf{q}\ddot{\theta}-\mathbf{p}\dot{\theta}^2.$$

(7) A point moves from the origin with velocities after a time t of $a\cos t$ and $a\sin t$, parallel to the Ox and Oy axes respectively. Prove that its acceleration is a and that the equation of its path is

$$x^2+y^2 = 2ay.$$

(8) A particle is moving in a curve whose equation is $xy = c^2$. If at any point in its path its acceleration is at right angles to the x-axis, prove that the magnitude of its acceleration varies as y^3.

(9) A point P moves along the equiangular spiral $r = ae^{k\theta}$. If its acceleration is always in the direction OP where O is the origin, prove that the magnitude of its acceleration varies as r^{-3}.

(10) An insect crawls outward along a spoke of a bicycle wheel which is itself rolling on the ground with constant velocity v. If the insect crawls with constant velocity u relative to the wheel, find the radial and transverse components of the acceleration of the insect when it is a distance d from the centre of the wheel.

(11) A lamp is at a height d above the ground. A man of height h starts from a point immediately below the lamp and walks at a speed u in a circle of radius a. Show that the velocity of the end of his shadow after a time t is $ud/(d-h)$ at an angle of ut/a with his initial direction and the acceleration is $u^2d/[a(d-h)]$ at an angle of $\frac{1}{2}\pi+(ut/a)$.

(12) A smooth horizontal tube OA of length a can rotate about O. A particle is placed in the tube at a distance b from O and the tube is then set rotating with constant angular velocity ω.

Show that the particle leaves the tube after a time

$$\frac{1}{\omega}\log_e\frac{a+\sqrt{(a^2-b^2)}}{b}$$

with a velocity of $\omega\sqrt{(2a^2-b^2)}$ in a direction $\tan^{-1}a/\sqrt{(a^2-b^2)}$ with the tube.

(13) P and O are two opposite points on the banks of a river of width d, PO being perpendicular to the direction of flow of the river. A boat whose speed in still water is u starts from P and is rowed across the river with its bow always directed towards O. If the speed of the river is u show that the actual path of the boat is a parabola whose vertex is the point where the boat meets the opposite bank and whose semi latus rectum is PO. Also show that this point is at a distance $\frac{1}{2}d$ below O.

(14) A point A is moving along the circumference of a fixed circle of radius a with constant angular velocity ω about the centre. A point B is also moving along the circumference of the circle and the radii on which A and B lie are always at right angles to one another. Show that the acceleration of B relative to A is always directed towards A and is of magnitude $\sqrt{2}\omega^2a$.

(15) A particle moves on the curve $y = \log\sec x$ in such a way that the tangent rotates uniformly. Show that the magnitude of the acceleration varies as ρ^2 and is in the direction parallel to the y axis.

8

THE SCALAR PRODUCT

Introduction

The aim of this chapter is to create the algebra for a product of two vectors. We shall try to do this by developing a meaning for the product of two vectors which is capable of a geometrical interpretation.

In chapter 3 we have seen that the multiplication of a vector by a number is another vector to which we can give a precise meaning. No matter how we define the product of two vectors the product will not obviously have the same meaning as the product of two numbers, for a vector is not a number.

Consider the vectors

$$\mathbf{a} = x_1\mathbf{i}+y_1\mathbf{j}+z_1\mathbf{k} = (x_1, y_1, z_1)$$

and $$\mathbf{b} = x_2\mathbf{i}+y_2\mathbf{j}+z_2\mathbf{k} = (x_2, y_2, z_2).$$

It is reasonable to expect that any definition of the product of two vectors **a** and **b** will involve the product of their components. Now there are various product combinations of the components of **a** with the components of **b**. Our aim is to combine these components in some way which is capable of some interpretation. Once we have done this we can frame a definition and investigate the structure of our definition more closely particularly as to its meaning. Finally we have to show whether our product definition obeys the usual multiplicative laws of algebra.

Rotation of a vector

Consider the vector $\mathbf{OA} = 2\mathbf{i}+5\mathbf{j} = (2, 5)$ as in Fig. 8.1. If $\mathbf{OA}$ is rotated anticlockwise through a right angle we obtain the vector $\mathbf{OB} = (-5, 2)$. This can be seen by considering the rotation of a triangle AON through $90°$ anticlockwise when OM, MB will be the new positions of ON, NA.

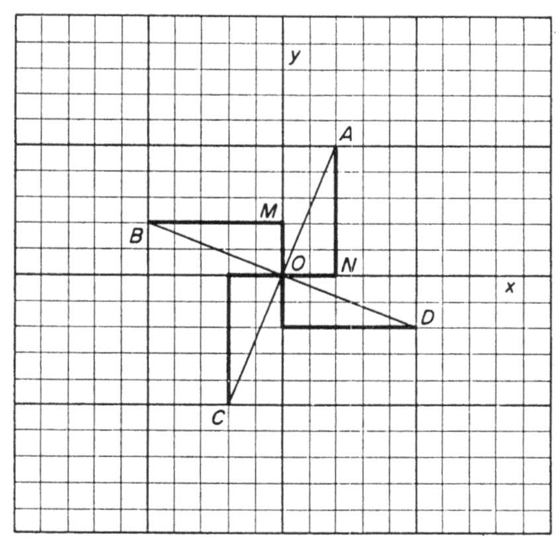

Fig. 8.1

If we rotate **OB** anticlockwise through 90° we obtain

$$\mathbf{OC} = (-2, -5).$$

A further rotation of 90° anticlockwise gives us **OD** = (5, −2).

Writing the vectors in the order of obtaining them we have

$$\mathbf{OA} = (2, 5), \quad \mathbf{OB} = (-5, 2), \quad \mathbf{OC} = (-2, -5), \quad \mathbf{OD} = (5, -2).$$

From these we can deduce that if the vector $x_1\mathbf{i}+y_1\mathbf{j} = (x_1, y_1)$ is rotated anticlockwise through 90° we obtain the vector $(-y_1, x_1)$.

Returning to the numerical examples we make the following observation about the components:

$$\mathbf{OA} = (2, 5), \qquad \mathbf{OB} = (-5, 2), \qquad (2\times -5)+(5\times 2) = 0,$$

$$\mathbf{OB} = (-5, 2), \qquad \mathbf{OC} = (-2, -5), \quad (-5\times -2)+(2\times -5) = 0,$$

$$\mathbf{OC} = (-2, -5), \quad \mathbf{OD} = (5, -2), \qquad (-2\times 5)+(-5\times -2) = 0.$$

This suggests investigating the value of $x_1x_2+y_1y_2$ for the general vectors $\mathbf{a} = (x_1, y_1)$, $\mathbf{b} = (x_2, y_2)$, where **b** is obtained by rotating **a** through 90° anticlockwise. In this case we have

$$x_2 = -y_1,$$

and

$$y_2 = x_1,$$

$$\therefore \quad x_1x_2+y_1y_2 = x_1(-y_1)+y_1x_1 = 0.$$

Thus we see that if we have two perpendicular vectors (x_1, y_1), (x_2, y_2) of equal lengths the product defined by $x_1x_2+y_1y_2$ is zero.

This is an interesting result but limited since the perpendicular vectors are equal in magnitude. It will obviously be of greater value if it is true for all perpendicular vectors irrespective of their magnitudes.

Suppose the vector (x_2, y_2) is perpendicular to the vector (x_1, y_1) and their magnitudes are not necessarily equal. In this case the length of the vector (x_2, y_2) will be k times the length of the vector formed by rotating the vector (x_1, y_1) through 90° anticlockwise, i.e. k times the length of the vector $(-y_1, x_1)$, where k is a positive or negative number. Thus vector (x_2, y_2) = vector $(-ky_1, kx_1)$ and we have $x_1x_2+y_1y_2 = x_1(-ky_1)+y_1(kx_1) = 0$.

Thus we see that if we have any two perpendicular vectors (x_1, y_1), (x_2, y_2) the value of $x_1x_2+y_1y_2$ is zero.

The scalar product: definition and notation

The study of perpendicular vectors has produced a meaning for $x_1x_2+y_1y_2$, namely that if $x_1x_2+y_1y_2 = 0$ the vectors (x_1, x_2), (y_1, y_2) are perpendicular. This leads us to the idea of $x_1x_2+y_1y_2$ as being a promising definition of a product of two vectors (x_1, y_1), (x_2, y_2). Extending this idea to vectors in space we shall consider a product of the vectors (x_1, y_1, z_1), (x_2, y_2, z_2) to be defined by

$$x_1x_2+y_1y_2+z_1z_2.$$

Now the components $x_1, x_2, y_1, y_2, z_1, z_2$ are all numbers representing the lengths of the component vectors. Therefore our product of two vectors defined by $x_1x_2+y_1y_2+z_1z_2$ is a number. We have previously discussed vector quantities. Distinct from these are quantities which possess only magnitude, such as length, mass and area and these are called scalar quantities. We can therefore think of numbers as scalar quantities or briefly scalars. For this reason our defined product $x_1x_2+y_1y_2+z_1z_2$ is called the scalar product of two vectors.

The scalar product of the vectors **a**, **b** is denoted symbolically by placing a dot between the vectors **a**, **b**, i.e. **a**.**b**. Hence it is sometimes known as the 'dot' product. The dot must never be omitted since we are not dealing with the product of two numbers.

We now state our definition of the scalar product.

Definition. *The scalar product of the vectors* $\mathbf{a} = (x_1, y_1, z_1)$, $\mathbf{b} = (x_2, y_2, z_2)$ *is defined by the number* $\mathbf{a}.\mathbf{b} = x_1x_2+y_1y_2+z_1z_2$.

Immediate consequences of the definition

(1) For the mutually perpendicular vectors

$$\mathbf{i} = (1, 0, 0), \quad \mathbf{j} = (0, 1, 0), \quad \mathbf{k} = (0, 0, 1),$$

$$\mathbf{j}.\mathbf{k} = \mathbf{k}.\mathbf{i} = \mathbf{i}.\mathbf{j} = 0.$$

(2) If $\mathbf{a} = (x, y, z)$

$$\mathbf{a}.\mathbf{a} = x^2+y^2+z^2 = a^2,$$

or writing $\mathbf{a}.\mathbf{a}$ as $\mathbf{a}^2$ we have $\mathbf{a}^2 = a^2$, i.e. the scalar product of a vector by itself is the square of its magnitude.

In particular for the unit vectors $\mathbf{i}$, $\mathbf{j}$, $\mathbf{k}$,

$$\mathbf{i}.\mathbf{i} = \mathbf{j}.\mathbf{j} = \mathbf{k}.\mathbf{k} = 1, \quad \text{i.e.} \quad \mathbf{i}^2 = \mathbf{j}^2 = \mathbf{k}^2 = 1.$$

(3) Suppose $\mathbf{a}$, $\mathbf{b}$ are parallel vectors. Let

$$\mathbf{a} = (a_1, a_2, a_3), \quad \mathbf{b} = (b_1, b_2, b_3).$$

Since $\mathbf{a}$, $\mathbf{b}$ are parallel $\quad \mathbf{b} = (ka_1, ka_2, ka_3)$,

where k is a positive or negative number.

Then

$$\begin{aligned} \mathbf{a}.\mathbf{b} &= a_1ka_1+a_2ka_2+a_3ka_3 \\ &= k(a_1^2+a_2^2+a_3^2) \\ &= ka^2. \end{aligned}$$

Now

$$b^2 = k^2(a_1^2+a_2^2+a_3^2) = k^2a^2,$$

$$\therefore \quad b = ka,$$

$$\therefore \quad \mathbf{a}.\mathbf{b} = ab,$$

that is the scalar product of parallel vectors is the product of their magnitudes.

The commutative and distributive laws

We now investigate whether the following laws hold for scalar products:

$$\mathbf{a}.\mathbf{b} = \mathbf{b}.\mathbf{a} \qquad \text{(Commutative Law),*}$$

$$\mathbf{a}.(\mathbf{b}+\mathbf{c}) = \mathbf{a}.\mathbf{b}+\mathbf{a}.\mathbf{c} \qquad \text{(Distributive Law),}$$

$$\mathbf{a}.(m\mathbf{b}) = m(\mathbf{a}.\mathbf{b}) \qquad \text{(Distributive Law).}$$

* Those familar with matrix theory will know that the Commutative Law does not hold for the multiplication of two matrices A, B, i.e. $AB \neq BA$.

Before doing so we point out that since $\mathbf{a}.(\mathbf{b}.\mathbf{c})$ has no meaning, $\mathbf{b}.\mathbf{c}$ being a number, there is no question of an associative law for scalar products.

Let

$$\mathbf{a} = (a_1, a_2, a_3),\ \mathbf{b} = (b_1, b_2, b_3),\ \mathbf{c} = (c_1, c_2, c_3).$$

We use our definition of a scalar product namely,

$$\mathbf{a}.\mathbf{b} = a_1 b_1 + a_2 b_2 + a_3 b_3.$$

The Commutative Law

$$\mathbf{a}.\mathbf{b} = a_1 b_1 + a_2 b_2 + a_3 b_3,$$

$$\mathbf{b}.\mathbf{a} = b_1 a_1 + b_2 a_2 + b_3 a_3,$$

$$\therefore \quad \mathbf{a}.\mathbf{b} = \mathbf{b}.\mathbf{a}.$$

The Distributive Laws

$$\begin{aligned}
\mathbf{b}+\mathbf{c} &= (b_1+c_1,\ b_2+c_2,\ b_3+c_3),\\
\therefore \quad \mathbf{a}.(\mathbf{b}+\mathbf{c}) &= a_1(b_1+c_1)+a_2(b_2+c_2)+a_3(b_3+c_3)\\
&= a_1 b_1 + a_1 c_1 + a_2 b_2 + a_2 c_2 + a_3 b_3 + a_3 c_3.\\
\mathbf{a}.\mathbf{b}+\mathbf{a}.\mathbf{c} &= (a_1 b_1 + a_2 b_2 + a_3 b_3)+(a_1 c_1 + a_2 c_2 + a_3 c_3),\\
\therefore \quad \mathbf{a}.(\mathbf{b}+\mathbf{c}) &= \mathbf{a}.\mathbf{b}+\mathbf{a}.\mathbf{c}.\\
m\mathbf{b} &= (mb_1,\ mb_2,\ mb_3),\\
\therefore \quad \mathbf{a}.(m\mathbf{b}) &= a_1 m b_1 + a_2 m b_2 + a_3 m b_3\\
&= m(a_1 b_1 + a_2 b_2 + a_3 b_3)\\
&= m(\mathbf{a}.\mathbf{b}).
\end{aligned}$$

In the same way by repeated application of the above results we can show that the commutative and distributive laws hold for scalar products involving the sum of several vectors. For example,

$$\begin{aligned}
(\mathbf{a}+\mathbf{b}+\mathbf{c}+\ldots).(\mathbf{p}+\mathbf{q}+\mathbf{r}+\ldots) = \mathbf{a}.\mathbf{p}&+\mathbf{a}.\mathbf{q}+\mathbf{a}.\mathbf{r}+\mathbf{b}.\mathbf{p}+\mathbf{b}.\mathbf{q}+\mathbf{b}.\mathbf{r}\\
&+\mathbf{c}.\mathbf{p}+\mathbf{c}.\mathbf{q}+\mathbf{c}.\mathbf{r}+\ldots.
\end{aligned}$$

Applications of the commutative and distributive laws

Knowing that the commutative and distributive laws apply to scalar products we can perform operations involving them just as in number algebra we can simplify the following

$$\begin{aligned}(p+q)(r+s) &= p(r+s)+q(r+s)\\ &= pr+ps+qr+qs,\\ (x+y)^2 &= (x+y)(x+y)\\ &= x(x+y)+y(x+y)\\ &= x^2+xy+yx+y^2\\ &= x^2+2xy+y^2.\end{aligned}$$

because numbers obey the commutative and distributive laws. The following two examples illustrate the simplification of scalar products:

$$\begin{aligned}(1)\ (\mathbf{a}+\mathbf{b}).(\mathbf{a}-2\mathbf{b}) &= \mathbf{a}.(\mathbf{a}-2\mathbf{b})+\mathbf{b}.(\mathbf{a}-2\mathbf{b})\\ &= \mathbf{a}.\mathbf{a}+\mathbf{a}.(-2\mathbf{b})+\mathbf{b}.\mathbf{a}+\mathbf{b}.(-2\mathbf{b})\\ &= \mathbf{a}^2-2\mathbf{a}.\mathbf{b}+\mathbf{a}.\mathbf{b}-2\mathbf{b}^2\\ &= \mathbf{a}^2-\mathbf{a}.\mathbf{b}-2\mathbf{b}^2.\end{aligned}$$

(2) *If* $\mathbf{a} = 3\mathbf{i}+5\mathbf{j}+2\mathbf{k}$ *and* $\mathbf{b} = 2\mathbf{i}+2\mathbf{j}-8\mathbf{k}$ *where* $\mathbf{i}$, $\mathbf{j}$, $\mathbf{k}$ *are three mutually perpendicular vectors, show that* $\mathbf{a}$, $\mathbf{b}$ *are perpendicular.*

$$\begin{aligned}\mathbf{a}.\mathbf{b} &= (3\mathbf{i}+5\mathbf{j}+2\mathbf{k}).(2\mathbf{i}+2\mathbf{j}-8\mathbf{k})\\ &= 6\mathbf{i}^2+6\mathbf{i}.\mathbf{j}-24\mathbf{k}.\mathbf{i}+10\mathbf{i}.\mathbf{j}+10\mathbf{j}^2-40\mathbf{j}.\mathbf{k}+4\mathbf{k}.\mathbf{i}+4\mathbf{j}.\mathbf{k}-16\mathbf{k}^2.\end{aligned}$$

Since $\mathbf{j}.\mathbf{k} = \mathbf{k}.\mathbf{i} = \mathbf{i}.\mathbf{j} = 0$ and $\mathbf{i}^2 = \mathbf{j}^2 = \mathbf{k}^2 = 1$,

$$\mathbf{a}.\mathbf{b} = 6+10-16 = 0,$$

therefore $\mathbf{a}$, $\mathbf{b}$ are perpendicular.

Otherwise, using the definition of scalar product directly

$$\begin{aligned}\mathbf{a}.\mathbf{b} &= 3\times 2+5\times 2+2\times -8\\ &= 0,\end{aligned}$$

therefore $\mathbf{a}$, $\mathbf{b}$ are perpendicular.

Meaning of the scalar product of any two vectors

We have seen that if two vectors are parallel their scalar product is the product of their lengths. It is reasonable to assume that the scalar product of two vectors in space depends on the lengths of the vectors and the angle between them. Suppose we have vectors **OP**, **OQ**. The relation between their lengths and the angle between them is given by the cosine rule. So by considering the problem of two vectors geometrically we hope to obtain further meaning of our definition of the scalar product.

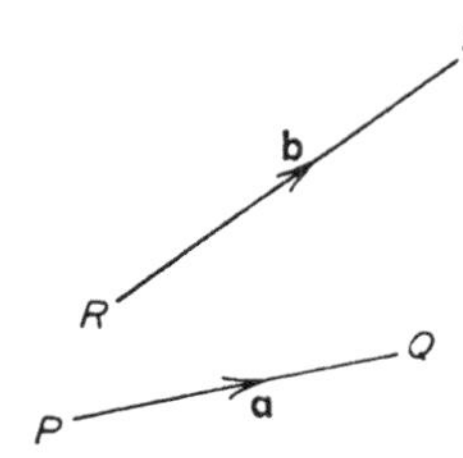

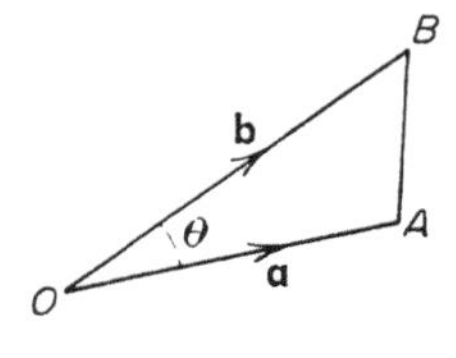

Fig. 8.2

Referring to Fig. 8.2 suppose we have two vectors

$$\mathbf{PQ} = \mathbf{a} = (x_1, y_1, z_1), \quad \mathbf{RS} = \mathbf{b} = (x_2, y_2, z_2),$$

which are inclined at an angle θ with one another. Let the two associated vectors located at the origin O be **OA**, **OB**, i.e. $\mathbf{OA} = \mathbf{a}$, $\mathbf{OB} = \mathbf{b}$, angle $AOB = \theta$.

We have

$$\mathbf{AB} = \mathbf{b}-\mathbf{a} = (x_2-x_1, y_2-y_1, z_2-z_1)$$

and applying the cosine rule to the triangle AOB,

$$AB^2 = OA^2+OB^2-2OA.OB\cos\theta,$$

$$\therefore \quad (x_2-x_1)^2+(y_2-y_1)^2+(z_2-z_1)^2$$
$$= x_1^2+y_1^2+z_1^2+x_2^2+y_2^2+z_2^2-2OA.OB\cos\theta,$$

$$\therefore \quad -2x_1x_2-2y_1y_2-2z_1z_2 = -2OA.OB\cos\theta,$$

$$\therefore \quad x_1x_2+y_1y_2+z_1z_2 = |\mathbf{a}|\,|\mathbf{b}|\cos\theta.$$

But
$$\mathbf{a}.\mathbf{b} = x_1x_2+y_1y_2+z_1z_2, \text{ by definition,}$$

$$\therefore \quad \mathbf{a}.\mathbf{b} = |\mathbf{a}|\,|\mathbf{b}|\cos\theta.$$

Another way of obtaining this result is to make use of the commutative and distributive laws which have been proved to hold for scalar products.

By the cosine rule

$$AB^2 = OA^2+OB^2-2OA.OB\cos\theta = a^2+b^2-2ab\cos\theta. \quad (1)$$

Remembering that $\mathbf{AB}.\mathbf{AB} = \mathbf{AB}^2 = AB^2$ we express AB^2 in vector form.

$$\mathbf{AB} = \mathbf{b}-\mathbf{a},$$

$$\therefore \quad \mathbf{AB}.\mathbf{AB} = (\mathbf{b}-\mathbf{a}).(\mathbf{b}-\mathbf{a}),$$

$$\therefore \quad \mathbf{AB}^2 = \mathbf{a}^2+\mathbf{b}^2-2\mathbf{a}.\mathbf{b},$$

$$\therefore \quad AB^2 = a^2+b^2-2\mathbf{a}.\mathbf{b}. \quad (2)$$

Comparing (1) and (2) $\quad \mathbf{a}.\mathbf{b} = ab\cos\theta.$

The result $\mathbf{a}.\mathbf{b} = |\mathbf{a}|\,|\mathbf{b}|\cos\theta$ expressed in words means that the scalar product of two vectors is the product of their lengths and the cosine of the included angle. Thus the scalar product has a trigonometrical meaning.

The interpretation of the scalar product given by $\mathbf{a}.\mathbf{b} = |\mathbf{a}|\,|\mathbf{b}|\cos\theta$ is of great value since we can find the angle between the vectors $\mathbf{a}$, $\mathbf{b}$ no matter where they are located by using

$$\cos\theta = \frac{\mathbf{a}.\mathbf{b}}{|\mathbf{a}|\,|\mathbf{b}|}.$$

We have previously found the angle between two vectors by using direction cosines, and it will be seen later on that the angle is more easily and neatly obtained by using the scalar product.

Care however must be exercised in dealing with the included angle θ.

Referring back to Fig. 8.2 we see that the included angle θ is the angle between the vectors **OA, OB** when **OA, OB** are both directed outward from the point O. The sense of measurement of the angle is unimportant since $\cos\theta = \cos(-\theta)$.

In Fig. 8.3 the angle between the vectors **a, b** is denoted by θ. We see that the angle is either acute or obtuse. If θ is obtuse, $\cos\theta$ is negative and the scalar product is therefore negative.

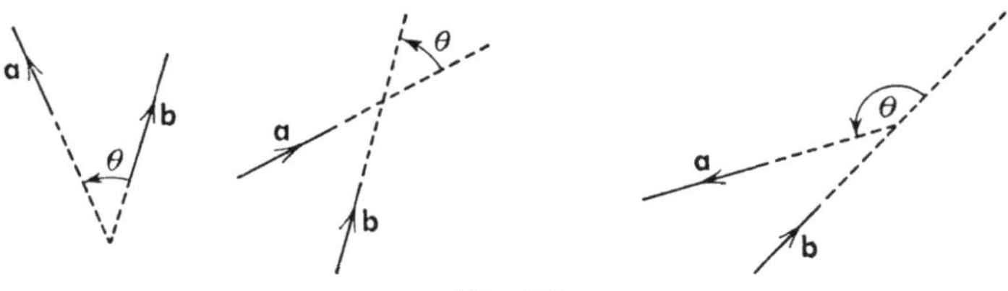

Fig. 8.3

Geometrical meaning of a.b = |a| |b| cos θ

In chapter 5 we defined the projection of **b** on **a** as $|\mathbf{b}|\cos\theta$ where θ is the angle between the vectors. Thus $|\mathbf{a}|\,|\mathbf{b}|\cos\theta$ can be regarded as the projection of **b** on **a** or of **a** on **b**. This leads to the following geometrical interpretation of the scalar product:

The scalar product of two vectors is the product of the length of one vector and the projection of the other upon it.

As we have seen the projection can be positive (θ acute) or negative (θ obtuse) and hence the scalar product is positive or negative, agreeing with the previous section.

Consequences of a.b = |a| |b| cos θ

(1) When **a**, **b** are parallel, $\cos\theta = \cos 0 = 1$,

$$\therefore \quad \mathbf{a}.\mathbf{b} = |\mathbf{a}|\,|\mathbf{b}| = ab.$$

In particular if $\mathbf{a} = \mathbf{b}$, $\quad \mathbf{a}.\mathbf{a} = a^2$.

We have had these results before by using the definition

$$\mathbf{a}.\mathbf{b} = x_1x_2+y_1y_2+z_1z_2.$$

(2) When **a**, **b** are perpendicular, $\cos\theta = \cos 90° = 0$,

$$\therefore \quad \mathbf{a}.\mathbf{b} = 0.$$

We have had this result before but only for two vectors **a**, **b** in the x-y plane. We now see that our definition $\mathbf{a}.\mathbf{b} = x_1x_2+y_1y_2+z_1z_2$ leads to the meaning of perpendicularity for any two vectors $\mathbf{a} = (x_1, y_1, z_1)$, $\mathbf{b} = (x_2, y_2, z_2)$ in space if $x_1x_2+y_1y_2+z_1z_2 = 0$.

The scalar product of a vector and the zero vector

So far we have considered $\mathbf{a} \neq \mathbf{0}$, $\mathbf{b} \neq \mathbf{0}$.

If $\mathbf{a} = \mathbf{0} = (0, 0, 0)$ then from our definition

$$\mathbf{0}.\mathbf{b} = 0+0+0 = 0.$$

Similarly, $$\mathbf{a}.\mathbf{0} = 0.$$

These results can also be obtained from $\mathbf{a}.\mathbf{b} = |\mathbf{a}|\,|\mathbf{b}|\cos\theta$ for if $\mathbf{a} = \mathbf{0}$, $|\mathbf{a}| = 0$ or if $\mathbf{b} = \mathbf{0}$, $|\mathbf{b}| = 0$ and in either case

$$|\mathbf{a}|\,|\mathbf{b}|\cos\theta = 0.$$

Thus in general $\mathbf{0}.\mathbf{p} = \mathbf{p}.\mathbf{0} = 0$.

Thus the scalar product of a vector and the zero vector is the number zero.

Alternative treatment of the scalar product

Some authors prefer a different treatment of the scalar product from the one developed here. We shall give an outline of this alternative procedure.

(1) The scalar product $\mathbf{a}.\mathbf{b}$ is defined as the number given by

$$\mathbf{a}.\mathbf{b} = |\mathbf{a}|\,|\mathbf{b}|\cos\theta,$$

where θ is the angle between the vectors $\mathbf{a}$, $\mathbf{b}$.

The special cases for perpendicular, parallel and equal vectors are then obtained by putting $\theta = 90°$, $\theta = 0$ and $\mathbf{a} = \mathbf{b}$, $\theta = 0$ respectively into the definition.

(2) The commutative and distributive laws are proved for scalar products by a method involving projections. (See Ex. 8, Question 2.)

(3) A formula for calculating the scalar product in terms of its components is deduced in the following way:

Let $$\mathbf{a} = x_1\mathbf{i}+y_1\mathbf{j}+z_1\mathbf{k}, \quad \mathbf{b} = x_2\mathbf{i}+y_2\mathbf{j}+z_2\mathbf{k}.$$

Then

$$(\mathbf{a}.\mathbf{b}) = (x_1\mathbf{i}+y_1\mathbf{j}+z_1\mathbf{k}).(x_2\mathbf{i}+y_2\mathbf{j}+z_2\mathbf{k})$$

$$= x_1x_2\mathbf{i}^2+y_1y_2\mathbf{j}^2+z_1z_2\mathbf{k}^2+\text{terms involving the scalar product of the perpendicular vectors } \mathbf{i}.\mathbf{j}, \text{ etc.}$$

Since $$\mathbf{i}^2 = \mathbf{j}^2 = \mathbf{k}^2 = 1,$$

$$\mathbf{a}.\mathbf{b} = x_1x_2+y_1y_2+z_1z_2.$$

Thus no matter how the scalar product is evolved the same relations are obtained finally. For the purpose of reference we give a summary of the important relations.

Summary

(1) The scalar product of two vectors **a, b** is a number.

(2) If $\mathbf{a} = (x_1, y_1, z_1)$, $\mathbf{b} = (x_2, y_2, z_2)$,

$$\mathbf{a}.\mathbf{b} = x_1x_2+y_1y_2+z_1z_2.$$

(3) If θ is the angle between **a, b**

$$\mathbf{a}.\mathbf{b} = |\mathbf{a}|\,|\mathbf{b}|\cos\theta.$$

(4) If **a, b** are perpendicular, $\mathbf{a}.\mathbf{b} = 0$. In particular for the mutually perpendicular unit vectors

$$\mathbf{j}.\mathbf{k} = \mathbf{k}.\mathbf{i} = \mathbf{i}.\mathbf{j} = 0.$$

(5) If **a, b** are parallel, $\mathbf{a}.\mathbf{b} = |\mathbf{a}|\,|\mathbf{b}|$.

In particular $\mathbf{a}.\mathbf{a} = \mathbf{a}^2 = a^2$.

For the unit vectors $\mathbf{i}^2 = \mathbf{j}^2 = \mathbf{k}^2 = 1$.

(6) The Commutative and Distributive Laws apply.

$$\mathbf{a}.\mathbf{b} = \mathbf{b}.\mathbf{a},$$

$$\mathbf{a}.(\mathbf{b}+\mathbf{c}) = \mathbf{a}.\mathbf{b}+\mathbf{a}.\mathbf{c},$$

$$\mathbf{a}.(m\mathbf{b}) = m(\mathbf{a}.\mathbf{b}).$$

(7) If one of the vectors is the zero vector the scalar product is 0, i.e.

$$\mathbf{0}.\mathbf{a} = \mathbf{a}.\mathbf{0} = 0.$$

General examples

The scalar product is of great use in questions involving distances and angles. The two relations

$$\mathbf{a}.\mathbf{b} = x_1x_2+y_1y_2+z_1z_2,$$

$$\mathbf{a}.\mathbf{b} = |\mathbf{a}|\,|\mathbf{b}|\cos\theta$$

are often used, as will be seen from some of the following examples:

(1) *A, B, C, D are the points* $(2, 3, 4)$, $(-1, 0, 3)$, $(2, -4, 1)$, $(1, -2, -1)$ respectively. *Show that the projections of* **AB** *on* **CD** and **CD** *on* **AB** *are* $-\frac{1}{3}$ *and* $-1/\sqrt{19}$ *respectively.*

Let **a**, **b**, **c**, **d** be the position vectors of A, B, C, D respectively relative to the origin O.

$$\mathbf{a} = (2, 3, 4),\ \mathbf{b} = (-1, 0, 3),$$

$$\therefore\quad \mathbf{AB} = \mathbf{b}-\mathbf{a} = (-3, -3, -1);$$

$$\mathbf{c} = (2, -4, 1),\ \mathbf{d} = (1, -2, -1),$$

$$\therefore\quad \mathbf{CD} = \mathbf{d}-\mathbf{c} = (-1, 2, -2),$$

$$\therefore\quad \mathbf{AB}.\mathbf{CD} = 3-6+2 = -1;$$

$$|\mathbf{AB}| = \sqrt{(3^2+3^2+1)} = \sqrt{19},$$

$$|\mathbf{CD}| = \sqrt{(1^2+2^2+2^2)} = 3.$$

$\mathbf{AB}.\mathbf{CD} = |\mathbf{AB}|\,|\mathbf{CD}|\cos\theta$, where θ is the angle between **AB**, **CD**.

Projection of **AB** on **CD** $= |\mathbf{AB}|\cos\theta = \dfrac{-1}{3} = -\dfrac{1}{3}$.

Projection of **CD** on **AB** $= |\mathbf{CD}|\cos\theta = \dfrac{-1}{\sqrt{19}} = -\dfrac{1}{\sqrt{19}}$.

(2) *Prove that the line drawn from the vertex of an isosceles triangle to the mid-point of its base is perpendicular to the base.*

Let O be the mid-point of the base BC of the isosceles triangle ABC (Fig. 8.4).

Let $\mathbf{OA} = \mathbf{p}$ and $\mathbf{OB} = \mathbf{q}$,

$$\therefore\quad \mathbf{OC} = -\mathbf{q}.$$

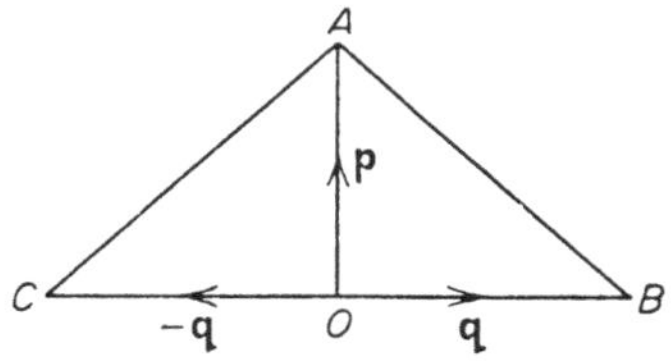

We now make use of the fact that $BA = CA$.

Fig. 8.4

$$\mathbf{BA} = \mathbf{p}-\mathbf{q},$$

$$\mathbf{CA} = \mathbf{p}+\mathbf{q}.$$

$$\mathbf{BA}.\mathbf{BA} = (\mathbf{p}-\mathbf{q}).(\mathbf{p}-\mathbf{q}) = \mathbf{p}^2-2\mathbf{p}.\mathbf{q}+\mathbf{q}^2,$$

$$\mathbf{CA}.\mathbf{CA} = (\mathbf{p}+\mathbf{q}).(\mathbf{p}+\mathbf{q}) = \mathbf{p}^2+2\mathbf{p}.\mathbf{q}+\mathbf{q}^2.$$

Now $\quad \mathbf{BA}.\mathbf{BA} = BA^2$ and $\mathbf{CA}.\mathbf{CA} = CA^2$,

since $\quad \mathbf{a}.\mathbf{a} = |\mathbf{a}|\,|\mathbf{a}|\cos 0 = |\mathbf{a}|^2 = a^2.$

Also since $BA = CA$ we have

$$p^2 - 2\mathbf{p}.\mathbf{q} + q^2 = p^2 + 2\mathbf{p}.\mathbf{q} + q^2,$$

$$\therefore \quad 4\mathbf{p}.\mathbf{q} = 0,$$

$$\therefore \quad \mathbf{p}.\mathbf{q} = 0$$

$$\therefore \quad \mathbf{p} \text{ is perpendicular to } \mathbf{q} \quad (\mathbf{p} \neq \mathbf{0}, \mathbf{q} \neq \mathbf{0}),$$

$$\therefore \quad AO \text{ is perpendicular to } BC.$$

(3) *AD is the median of a triangle. Prove* $AB^2 + AC^2 = 2AD^2 + \frac{1}{2}BC^2$ (Apollonius's Theorem).

Let $\mathbf{b}$, $\mathbf{c}$ be the position vectors of B, C relative to the origin A (Fig. 8.5).

Then $\mathbf{AD} = \frac{1}{2}(\mathbf{b}+\mathbf{c})$ and $\mathbf{BC} = \mathbf{c}-\mathbf{b}$.

We make use of the result $\mathbf{a}.\mathbf{a} = \mathbf{a}^2 = a^2$.

$$\mathbf{AD}.\mathbf{AD} = \tfrac{1}{4}(\mathbf{b}+\mathbf{c}).(\mathbf{b}+\mathbf{c}),$$

and

$$\mathbf{BC}.\mathbf{BC} = (\mathbf{c}-\mathbf{b}).(\mathbf{c}-\mathbf{b}),$$

$$\therefore \quad AD^2 = \tfrac{1}{4}(b^2 + c^2 + 2\mathbf{b}.\mathbf{c}),$$

$$\therefore \quad 4AD^2 = b^2 + c^2 + 2\mathbf{b}.\mathbf{c}.$$

Also

$$BC^2 = b^2 + c^2 - 2\mathbf{b}.\mathbf{c}.$$

Adding

$$4AD^2 + BC^2 = 2b^2 + 2c^2 = 2AB^2 + 2AC^2,$$

$$\therefore \quad AB^2 + AC^2 = 2AD^2 + \tfrac{1}{2}BC^2.$$

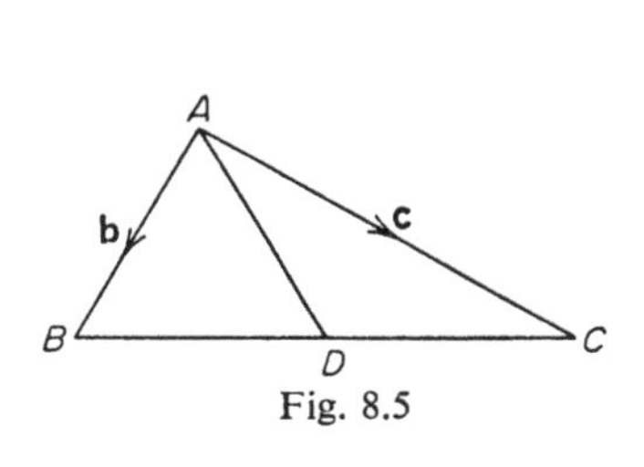

Fig. 8.5

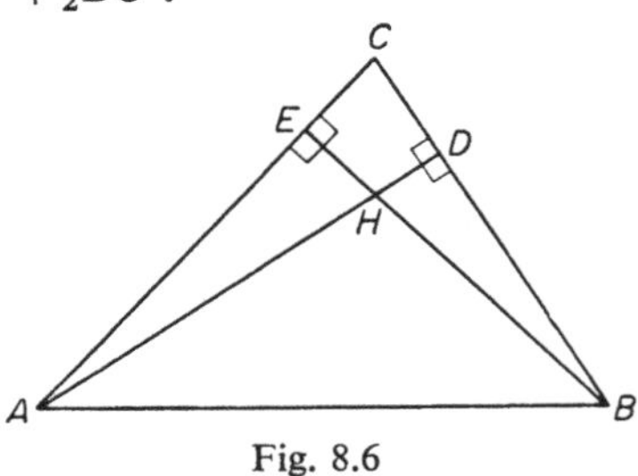

Fig. 8.6

(4) *Prove that the altitudes of a triangle are concurrent*

Let the altitudes AD, BE of a triangle meet at H (Fig. 8.6).

Let the position vectors of A, B, C with H as the origin be $\mathbf{a}$, $\mathbf{b}$, $\mathbf{c}$ respectively.

Since **HA** is perpendicular to **BC**,

$$\mathbf{HA}.\mathbf{BC} = 0,$$
$$\therefore \quad \mathbf{a}.(\mathbf{c}-\mathbf{b}) = 0,$$
$$\therefore \quad \mathbf{a}.\mathbf{c}-\mathbf{a}.\mathbf{b} = 0,$$
$$\therefore \quad \mathbf{a}.\mathbf{c} = \mathbf{a}.\mathbf{b}. \tag{1}$$

(Note here that we cannot cancel the vector **a** from each side since obviously **b** and **c** need not be equal. In general cancellation of a common vector from each side of a scalar product equation is not allowed.)

Similarly, since **HB** is perpendicular to **AC**

$$\mathbf{HB}.\mathbf{AC} = 0,$$
$$\therefore \quad \mathbf{b}.(\mathbf{c}-\mathbf{a}) = 0,$$
$$\therefore \quad \mathbf{b}.\mathbf{c} = \mathbf{a}.\mathbf{b}. \tag{2}$$

From (1) and (2) above we obtain

$$\mathbf{a}.\mathbf{c} = \mathbf{b}.\mathbf{c},$$
$$\therefore \quad \mathbf{c}.(\mathbf{a}-\mathbf{b}) = 0,$$
$$\therefore \quad \mathbf{HC}.\mathbf{BA} = 0,$$

$\therefore$ **HC** is perpendicular to **BA**,

$\therefore$ the altitude from C on to AB passes through H,

$\therefore$ the altitudes are concurrent.

(5) *In any triangle ABC,* $\mathbf{BC} = \mathbf{a}$, $\mathbf{CA} = \mathbf{b}$, $\mathbf{AB} = \mathbf{c}$. *Prove that* $|\mathbf{a}|^2 = |\mathbf{b}|^2+|\mathbf{c}|^2+2\mathbf{b}.\mathbf{c}$. *Comment on this result and discuss the result when* $\mathbf{b}.\mathbf{c} = 0$.

In Fig. 8.7,

$$\mathbf{a} = -\mathbf{b}-\mathbf{c} = -(\mathbf{b}+\mathbf{c}),$$
$$\therefore \quad \mathbf{a}.\mathbf{a} = -(\mathbf{b}+\mathbf{c}).-(\mathbf{b}+\mathbf{c})$$
$$= (\mathbf{b}+\mathbf{c}).(\mathbf{b}+\mathbf{c})$$
$$= \mathbf{b}.\mathbf{b}+\mathbf{c}.\mathbf{c}+2\mathbf{b}.\mathbf{c}.$$

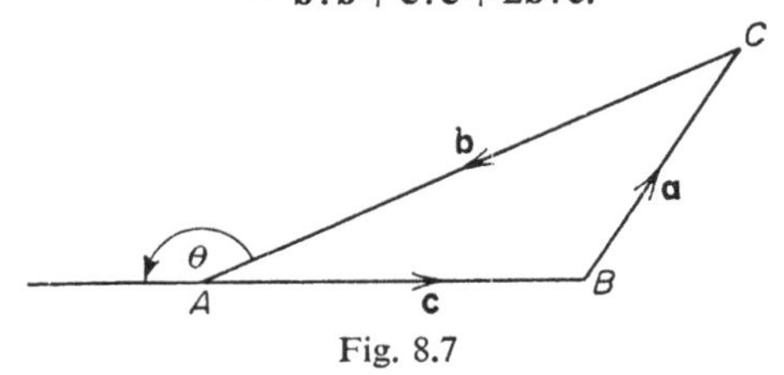

Fig. 8.7

Since $\mathbf{a}.\mathbf{a} = |\mathbf{a}|^2, \quad \mathbf{b}.\mathbf{b} = |\mathbf{b}|^2, \quad \mathbf{c}.\mathbf{c} = |\mathbf{c}|^2,$

$$|\mathbf{a}|^2 = |\mathbf{b}|^2+|\mathbf{c}|^2+2\mathbf{b}.\mathbf{c}.$$

Writing $|\mathbf{a}| = a, \quad |\mathbf{b}| = b, \quad |\mathbf{c}| = c$

we have $a^2 = b^2+c^2+2\mathbf{b}.\mathbf{c}.$

Geometrically this is the Extension of Pythagoras's Theorem in vector form.

It is also the vector form of the cosine rule which can be put into the usual form by writing

$$\begin{aligned} 2\mathbf{b}.\mathbf{c} &= 2|\mathbf{b}|\,|\mathbf{c}|\cos\theta = 2bc\cos\theta \\ &= 2bc\cos(180^\circ - A) \\ &= -2bc\cos A, \\ \therefore \quad a^2 &= b^2+c^2-2bc\cos A. \end{aligned}$$

When $\mathbf{b}.\mathbf{c} = 0$, $\mathbf{b}$ is perpendicular to $\mathbf{c}$, i.e. angle $BAC = 90^\circ$. The vector form of the cosine rule then becomes

$$a^2 = b^2+c^2,$$

which is in agreement with the well-known theorem of Pythagoras.

(6) *The position vectors of the points A, B, C are* $(8, 4, -3)$, $(6, 3, -4)$, $(7, 5, -5)$ *respectively. Find the angle between* **AB** *and* **BC**. *Hence find the area of triangle ABC.*

$$\begin{aligned} \mathbf{AB} = \mathbf{b}-\mathbf{a} \quad &= (6, 3, -4)-(8, 4, -3) \\ &= (-2, -1, -1), \\ \therefore \quad |\mathbf{AB}| &= \sqrt{(2^2+1^2+1^2)} = \sqrt{6}. \\ \mathbf{BC} = \mathbf{c}-\mathbf{b} \quad &= (7, 5, -5)-(6, 3, -4) \\ &= (1, 2, -1), \\ \therefore \quad |\mathbf{BC}| &= \sqrt{(1^2+2^2+1^2)} = \sqrt{6}. \\ \mathbf{AB}.\mathbf{BC} &= |\mathbf{AB}|\,|\mathbf{BC}|\cos\theta, \\ \therefore \quad (-2, -1, -1).(1, 2, -1) &= \sqrt{6}.\sqrt{6}\cos\theta, \\ \therefore \quad -2-2+1 &= 6\cos\theta, \\ \therefore \quad \cos\theta &= -\tfrac{3}{6} = -\tfrac{1}{2}, \\ \therefore \quad \theta &= 120^\circ. \end{aligned}$$

This is the angle between **AB** and **BC**,

$$\therefore \quad \text{angle between } \mathbf{BA} \text{ and } \mathbf{BC} = 60°.$$

$$\therefore \quad \text{angle } ABC = 60°.$$

$$\text{Area of triangle } ABC = \tfrac{1}{2}.BA\,.\,BC\sin B$$

$$= \tfrac{1}{2}.\sqrt{6}.\sqrt{6}.\frac{\sqrt{3}}{2}$$

$$= \frac{3\sqrt{3}}{2}.$$

(7) *OABC is a tetrahedron. If the edge OC is perpendicular to the edge AB and the edge OB is perpendicular to the edge AC, show that the edge OA is perpendicular to the edge BC.*

Let the position vectors of A, B, C relative to O as origin be **a, b, c** respectively (Fig. 8.8).

Since **OC** is perpendicular to **AB** we have

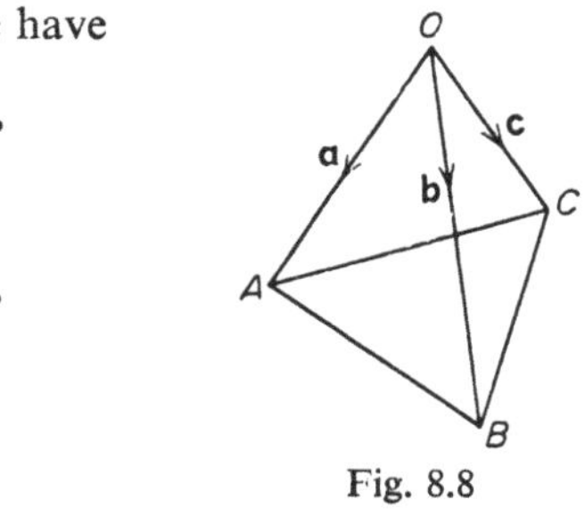

Fig. 8.8

$$\mathbf{OC}.\mathbf{AB} = \mathbf{c}.(\mathbf{b}-\mathbf{a}) = 0,$$

$$\therefore \quad \mathbf{b}.\mathbf{c} = \mathbf{a}.\mathbf{c}.$$

Similarly, $\mathbf{OB}.\mathbf{AC} = \mathbf{b}.(\mathbf{c}-\mathbf{a}) = 0,$

$$\therefore \quad \mathbf{b}.\mathbf{c} = \mathbf{a}.\mathbf{b},$$

$$\therefore \quad \mathbf{a}.\mathbf{c} = \mathbf{a}.\mathbf{b},$$

$$\therefore \quad \mathbf{a}.(\mathbf{c}-\mathbf{b}) = 0,$$

$$\therefore \quad \mathbf{OA}.\mathbf{BC} = 0,$$

$\therefore$ OA is perpendicular to the edge BC.

(8) *P and Q are two points on the earth's surface with latitudes* 60°, 30° N. *and longitudes* 80° *and* 20° E. *respectively. Find the distance of P from Q measured along the great circle through P and Q, assuming the earth is a sphere of radius* $6{\cdot}37 \times 10^3$ *km.*

Let the equator lie on the XY plane and Greenwich meridian on the XZ plane. Let R be the radius of the earth.

Referring to Fig. 8.9, P is the point (ON, NM, MP), i.e. point ($R\cos 60° \cos 80°$, $R\cos 60° \sin 80°$, $R\sin 60°$).

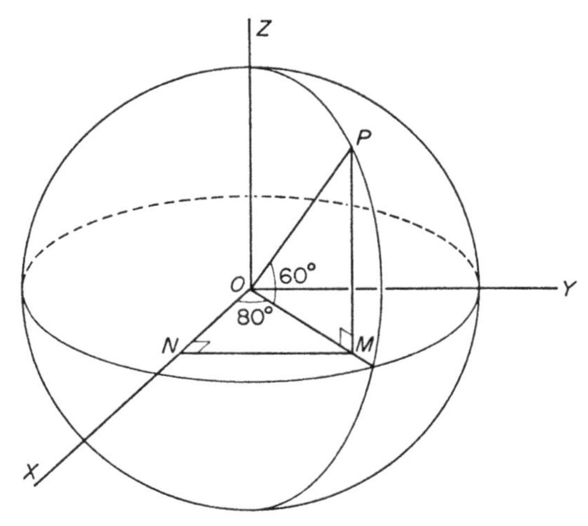

Fig. 8.9

Similarly, Q is the point

$$(R\cos 30^\circ \cos 20^\circ, \quad R\cos 30^\circ \sin 20^\circ, \quad R\sin 30^\circ).$$

If we know the angle $POQ = \theta$ in radians the length of the arc PQ along the great circle is given by $R\theta$.

The angle θ is the angle between the vectors **OP**, **OQ** and is given by the scalar product relation

$$\mathbf{OP}.\mathbf{OQ} = |\mathbf{OP}|\,|\mathbf{OQ}|\cos\theta.$$

Now

$$\begin{aligned}\mathbf{OP}.\mathbf{OQ} &= (R\cos 60^\circ \cos 80^\circ, R\cos 60^\circ \sin 80^\circ, R\sin 60^\circ).\\ &\quad (R\cos 30^\circ \cos 20^\circ, R\cos 30^\circ \sin 20^\circ, R\sin 30^\circ)\\ &= R^2 (\cos 60^\circ \cos 30^\circ \cos 80^\circ \cos 20^\circ\\ &\quad + \cos 60^\circ \cos 30^\circ \sin 80^\circ \sin 20^\circ\\ &\quad + \sin 60^\circ \sin 30^\circ)\\ &= R^2[\cos 60^\circ \cos 30^\circ \cos(80^\circ - 20^\circ) + \sin 60^\circ \sin 30^\circ]\\ &= R^2.\left(\frac{1}{2}.\frac{\sqrt{3}}{2}.\frac{1}{2} + \frac{\sqrt{3}}{2}.\frac{1}{2}\right)\\ &= \frac{3\sqrt{3}}{8} R^2.\end{aligned}$$

Also

$$|\mathbf{OP}|\,|\mathbf{OQ}|\cos\theta = R^2\cos\theta,$$

$$\therefore \quad \cos\theta = \frac{3\sqrt{3}}{8},$$

$$\therefore \quad \theta = 0{\cdot}8639 \text{ radians.}$$

$$\therefore \quad \text{distance} = R\theta$$

$$= 6{\cdot}37\times10^3\times0{\cdot}8639 \text{ km}$$

$$= 5{\cdot}5\times10^3 \text{ km.}$$

Differentiation of the scalar product

By proceeding as in chapter 7 we show that if $\mathbf{a} = f(t)$, $\mathbf{b} = f(t)$ then

$$\frac{d}{dt}(\mathbf{a}.\mathbf{b}) = \mathbf{a}.\frac{d\mathbf{b}}{dt}+\mathbf{b}.\frac{d\mathbf{a}}{dt}.$$

Let $y = \mathbf{a}.\mathbf{b}$ and δy, $\delta\mathbf{a}$, $\delta\mathbf{b}$ be the increments in y, $\mathbf{a}$, $\mathbf{b}$ respectively due to an increment of δt in t. Then

$$y+\delta y = (\mathbf{a}+\delta\mathbf{a}).(\mathbf{b}+\delta\mathbf{b})$$

$$= \mathbf{a}.\mathbf{b}+\mathbf{a}.\delta\mathbf{b}+\mathbf{b}.\delta\mathbf{a}+\delta\mathbf{a}.\delta\mathbf{b}.$$

But

$$y = \mathbf{a}.\mathbf{b},$$

$$\therefore \quad \delta y = \mathbf{a}.\delta\mathbf{b}+\mathbf{b}.\delta\mathbf{a}+\delta\mathbf{a}.\delta\mathbf{b},$$

$$\therefore \quad \underset{\delta t\to0}{\mathrm{Lt}}\frac{\delta y}{\delta t} = \underset{\delta t\to0}{\mathrm{Lt}}\left(\mathbf{a}.\frac{\delta\mathbf{b}}{\delta t}+\mathbf{b}.\frac{\delta\mathbf{a}}{\delta t}+\delta\mathbf{a}.\frac{\delta\mathbf{b}}{\delta t}\right),$$

$$\therefore \quad \frac{dy}{dt} = \mathbf{a}.\frac{d\mathbf{b}}{dt}+\mathbf{b}.\frac{d\mathbf{a}}{dt}.$$

The following is an important special case. If $\mathbf{a} \neq \mathbf{0}$,

$$\frac{d}{dt}(\mathbf{a}^2) = \frac{d}{dt}(\mathbf{a}.\mathbf{a}) = 2\mathbf{a}.\frac{d\mathbf{a}}{dt}.$$

But

$$\mathbf{a}^2 = a^2 \quad \text{and} \quad \frac{d}{dt}(a^2) = 2a\frac{da}{dt}.$$

$$\therefore \quad \mathbf{a}.\frac{d\mathbf{a}}{dt} = a\frac{da}{dt}.$$

Further if $\mathbf{a}$ is a vector of constant length but of changing direction, a is a constant and $da/dt = 0$,

$$\therefore \quad \mathbf{a}.\frac{d\mathbf{a}}{dt} = 0,$$

$$\therefore \quad \mathbf{a} \quad \text{and} \quad \frac{d\mathbf{a}}{dt} \text{ are perpendicular.}$$

Thus the derivative of a vector of constant length is perpendicular to the vector.

A special case of this result is when $\mathbf{a}$ is a unit vector, i.e. of constant length 1 and we have shown in chapter 7 that the derivative of a unit vector is perpendicular to the vector.

Integration

Regarding integration as the reverse of differentiation we have the following immediately from the previous section:

(1) $\displaystyle\int\left(\mathbf{a}.\frac{d\mathbf{b}}{dt}+\mathbf{b}.\frac{d\mathbf{a}}{dt}\right) dt = \mathbf{a}.\mathbf{b}+c.$

(2) $\displaystyle\int 2\mathbf{a}.\frac{d\mathbf{a}}{dt}\, dt = \mathbf{a}.\mathbf{a}+c = \mathbf{a}^2+c.$

The arbitrary constant c is a number since $\mathbf{a}.\mathbf{b}$ and $\mathbf{a}^2$ are numbers.

Examples

(1) *Differentiate the following in which* $\mathbf{r}$ *is a function of the time* t *and* $\mathbf{a}$, $\mathbf{b}$ *are constant vectors*:

$$\text{(i)}\ \mathbf{r}^2+\frac{1}{\mathbf{r}^2}, \quad \text{(ii)}\ \mathbf{r}.\frac{d\mathbf{r}}{dt}, \quad \text{(iii)}\ \left(\frac{d\mathbf{r}}{dt}\right)^2, \quad \text{(iv)}\ \frac{\mathbf{r}+\mathbf{a}}{\mathbf{r}^2+\mathbf{a}^2}, \quad (v)\ (\mathbf{a}.\mathbf{r})\mathbf{b}.$$

(i) $\displaystyle\frac{d}{dt}\left(\mathbf{r}^2+\frac{1}{\mathbf{r}^2}\right) = 2\mathbf{r}.\frac{d\mathbf{r}}{dt}-\frac{2}{\mathbf{r}^4}\mathbf{r}.\frac{d\mathbf{r}}{dt} = 2\mathbf{r}.\dot{\mathbf{r}}-\frac{2\mathbf{r}.\dot{\mathbf{r}}}{\mathbf{r}^4} = 2r\dot{r}-\frac{2\dot{r}}{r^3}.$

(ii) $\displaystyle\frac{d}{dt}\left(\mathbf{r}.\frac{d\mathbf{r}}{dt}\right) = \frac{d\mathbf{r}}{dt}.\frac{d\mathbf{r}}{dt}+\mathbf{r}.\frac{d^2\mathbf{r}}{dt^2} = \dot{\mathbf{r}}^2+\mathbf{r}.\ddot{\mathbf{r}}.$

(iii) $\displaystyle\frac{d}{dt}\left(\frac{d\mathbf{r}}{dt}\right)^2 = \frac{d}{dt}\left(\frac{d\mathbf{r}}{dt}.\frac{d\mathbf{r}}{dt}\right) = 2\frac{d\mathbf{r}}{dt}.\frac{d^2\mathbf{r}}{dt^2} = 2\dot{\mathbf{r}}.\ddot{\mathbf{r}}.$

(iv) $\displaystyle\frac{d}{dt}\left(\frac{\mathbf{r}+\mathbf{a}}{\mathbf{r}^2+\mathbf{a}^2}\right) = \frac{(\mathbf{r}^2+\mathbf{a}^2)d\mathbf{r}/dt-(\mathbf{r}+\mathbf{a})2\mathbf{r}.d\mathbf{r}/dt}{(\mathbf{r}^2+\mathbf{a}^2)^2}$

$$= \frac{\dot{\mathbf{r}}}{r^2+a^2}-\frac{2r\dot{r}(\mathbf{r}+\mathbf{a})}{(r^2+a^2)^2}.$$

(v) $\displaystyle\frac{d}{dt}(\mathbf{a}.\mathbf{r})\mathbf{b} = \left(\mathbf{a}.\frac{d\mathbf{r}}{dt}\right)\mathbf{b} = (\mathbf{a}.\dot{\mathbf{r}})\mathbf{b}.$

(2) *If* $d^2\mathbf{r}/dt^2 + n^2\mathbf{r} = \mathbf{0}$, *show that* $(d\mathbf{r}/dt)^2 = c - n^2\mathbf{r}^2$ *where* n, c *are constants.*

$$\frac{d^2\mathbf{r}}{dt^2} + n^2\mathbf{r} = \mathbf{0},$$

$$\therefore \quad \frac{d^2\mathbf{r}}{dt^2} = -n^2\mathbf{r}.$$

The right-hand side cannot be integrated immediately since we do not know $\mathbf{r}$ in terms of t.

To integrate the equation both sides are multiplied by $2(d\mathbf{r}/dt)$ to form scalar products. We then have

$$2\frac{d\mathbf{r}}{dt}\cdot\frac{d^2\mathbf{r}}{dt^2} = -2n^2\mathbf{r}\,.\frac{d\mathbf{r}}{dt},$$

$$\therefore \quad \int 2\frac{d\mathbf{r}}{dt}\cdot\frac{d^2\mathbf{r}}{dt^2}\,dt = -n^2\int 2\mathbf{r}\,.\frac{d\mathbf{r}}{dt}\,dt.$$

The left-hand integral is $(d\mathbf{r}/dt)^2$ from Ex. 1 (iii) and the right-hand integral is $\mathbf{r}^2$.

$$\therefore \quad (d\mathbf{r}/dt)^2 = -n^2\mathbf{r}^2 + c,$$

where c is a constant number.

Work done by a constant force

We now consider an application of the scalar product in Mechanics.

Suppose a particle at A is displaced to B by a constant force $\mathbf{P}$, the angle between $\mathbf{P}$ and $\mathbf{AB}$ being θ (Fig. 8.10). The force $\mathbf{P}$ is then said to do work on the particle.

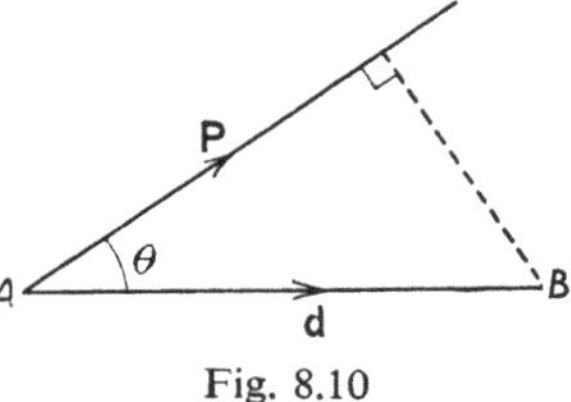

Fig. 8.10

The work done by the force on the particle is defined as the product of the magnitude of the force and the resolved part of the displacement in the direction of the force.

If the work done $= W$ and $\mathbf{AB} = \mathbf{d}$ then

$$W = Pd\cos\theta,$$

where $P = |\mathbf{P}|$ and $d = |\mathbf{d}|$.

This is a scalar quantity. Since $d\cos\theta$ is the projection of $\mathbf{d}$ on $\mathbf{P}$ we have

$$W = \mathbf{P}.\mathbf{d}.$$

From this it is evident that if $\mathbf{P}$ and $\mathbf{d}$ are perpendicular the work done is zero.

Consider several forces $\mathbf{P}_1, \mathbf{P}_2, \ldots$ acting on the particle. Let the displacement be $\mathbf{d}$. Then the total work done by the forces is

$$\mathbf{P}_1.\mathbf{d}+\mathbf{P}_2.\mathbf{d}+\ldots = (\mathbf{P}_1+\mathbf{P}_2+\ldots).\mathbf{d}.$$

Now $\mathbf{P}_1+\mathbf{P}_2+\ldots$ is the sum of the forces, that is, it is the resultant of the forces.

Thus the total work done is the same as the work done by the resultant of the system of forces.

Suppose
$$\mathbf{P} = X\mathbf{i}+Y\mathbf{j}+Z\mathbf{k} \quad \text{and} \quad \mathbf{d} = x\mathbf{i}+y\mathbf{j}+z\mathbf{k}.$$

Then work done by $\mathbf{P}$ is given by

$$\begin{aligned} W &= \mathbf{P}.\mathbf{d} \\ &= (X\mathbf{i}+Y\mathbf{j}+Z\mathbf{k}).(x\mathbf{i}+y\mathbf{j}+z\mathbf{k}) \\ &= Xx+Yy+Zz. \end{aligned}$$

Example

Find the work done by the forces 3, 6 *newtons acting in the directions of the vectors* (1, 2, −2), (−2, −1, 2) *if a particle is displaced from a point* $A(3, 5, -2)$ *to a point* $B(-1, 3, 2)$, *the unit of length being* 1 *metre.*

The unit vectors in the given directions are

$$\tfrac{1}{3}(\mathbf{i}+2\mathbf{j}-2\mathbf{k}) \quad \text{and} \quad \tfrac{1}{3}(-2\mathbf{i}-\mathbf{j}+2\mathbf{k}).$$

The forces are

$$\mathbf{P}_1 = \mathbf{i}+2\mathbf{j}-2\mathbf{k} \quad \text{and} \quad \mathbf{P}_2 = 2(-2\mathbf{i}-\mathbf{j}+2\mathbf{k}).$$

Let the resultant of $\mathbf{P}_1$, $\mathbf{P}_2$ be $\mathbf{R}$.

$$\therefore \quad \mathbf{R} = -3\mathbf{i}+2\mathbf{k}.$$

The displacement $\mathbf{d}$ is given by $\mathbf{AB} = -4\mathbf{i}-2\mathbf{j}+4\mathbf{k}.$

The work done is given by

$$\begin{aligned} W &= \mathbf{R}.\mathbf{d} \\ &= (-3\mathbf{i}+2\mathbf{k}).(-4\mathbf{i}-2\mathbf{j}+4\mathbf{k}) \\ &= (12+8)\ \text{joules} \\ &= 20\ \text{joules}. \end{aligned}$$

Work done by a variable force

Let MN be the path along which a particle travels under the action of a variable force (Fig. 8.11). Take several points on the curve and join them up as shown forming sides of a polygon.

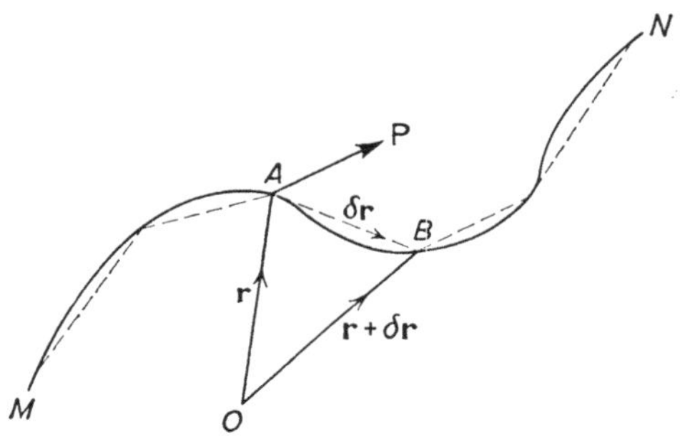

Fig. 8.11

Suppose at any time the particle is at the point A whose position vector relative to a fixed origin O is $\mathbf{r}$ and a short time later is at the point B, position vector $\mathbf{r}+\delta\mathbf{r}$. Then the particle has undergone a displacement $\mathbf{AB} = \delta\mathbf{r}$.

The work done by the force producing the displacement $\delta\mathbf{r}$ is approximately $\mathbf{P}.\delta\mathbf{r}$ where $\mathbf{P}$ is the value of the variable force at A. The work done by the variable force as the particle moves from M to N along the curved path is approximately the sum of all such terms as $\mathbf{P}.\delta\mathbf{r}$, i.e. $\Sigma(\mathbf{P}.\delta\mathbf{r})$, formed by considering the work done by the variable force producing displacements along the sides of the polygon.

If the number of the sides of the polygon is increased indefinitely then the length of each side tends to zero, and the work done by the variable force is the limiting value of the above sum. This is written in the usual way as

$$\int \mathbf{P}.d\mathbf{r}.$$

Thus if W is the work done from M to N we have

$$W = \int \mathbf{P}.d\mathbf{r}.$$

Suppose $\quad \mathbf{P} = X\mathbf{i}+Y\mathbf{j}+Z\mathbf{k} \quad$ and $\quad \mathbf{r} = x\mathbf{i}+y\mathbf{j}+z\mathbf{k}.$

Then
$$W = \int \mathbf{P}.d\mathbf{r}$$
$$= \int (X\mathbf{i}+Y\mathbf{j}+Z\mathbf{k}).d(x\mathbf{i}+y\mathbf{j}+z\mathbf{k})$$
$$= \int (X\,dx+Y\,dy+Z\,dz).$$

Example

Find the work done by the field of force:

$$\text{(i) } \mathbf{P} = x^2\mathbf{i}+y^2\mathbf{j}, \qquad \text{(ii) } \mathbf{P} = y^2\mathbf{i}+x^2\mathbf{j}$$

on a particle constrained to move from (0, 0) *to* (2, 4) *along the parabola* $y = x^2$.

(i) $$\mathbf{P} = x^2\mathbf{i}+y^2\mathbf{j} \quad \text{and} \quad \mathbf{r} = x\mathbf{i}+y\mathbf{j}.$$

Work done

$$= \int \mathbf{P}.d\mathbf{r}$$
$$= \int_{(0,\,0)}^{(2,\,4)} (x^2\mathbf{i}+y^2\mathbf{j}).d(x\mathbf{i}+y\mathbf{j})$$
$$= \int_{(0,\,0)}^{(2,\,4)} (x^2dx+y^2dy)$$
$$= \int_0^2 (x^2dx+x^4.2x\,dx), \quad \text{since} \quad y = x^2 \quad \text{and} \quad dy = 2x\,dx$$
$$= \int_0^2 (x^2+2x^5)\,dx$$
$$= \left[\frac{x^3}{3}+\frac{x^6}{3}\right]_0^2$$
$$= \tfrac{1}{3}(8+64)$$
$$= \tfrac{72}{3}$$
$$= 24.$$

(ii) $$\mathbf{P} = y^2\mathbf{i}+x^2\mathbf{j} \quad \text{and} \quad \mathbf{r} = x\mathbf{i}+y\mathbf{j}.$$

Work done

$$= \int \mathbf{P}.d\mathbf{r}$$

$$= \int_{(0,\,0)}^{(2,\,4)} (y^2\mathbf{i}+x^2\mathbf{j}).(x\mathbf{i}+y\mathbf{j})$$

$$= \int_{(0,\,0)}^{(2,\,4)} (y^2\,\mathrm{d}x+x^2\,dy)$$

$$= \int_0^2 (x^4\,dx+x^2.2x\,dx), \quad \text{since} \quad y = x^2 \quad \text{and} \quad dy = 2xdx$$

$$= \int_0^2 (x^4+2x^3)dx$$

$$= \left[\frac{x^5}{5}+\frac{x^4}{2}\right]_0^2$$

$$= \tfrac{32}{5}+\tfrac{16}{2} = \tfrac{72}{5}.$$

Exercise 8

(1) If $\mathbf{a} = 2\mathbf{i}-2\mathbf{j}+\mathbf{k}$, $\mathbf{b} = \mathbf{i}+2\mathbf{j}-3\mathbf{k}$, $\mathbf{c} = 2\mathbf{i}-\mathbf{j}+4\mathbf{k}$, obtain $(\mathbf{b}+\mathbf{c})$, $3\mathbf{b}$ and the projections of $\mathbf{b}$, $\mathbf{c}$, $(\mathbf{b}+\mathbf{c})$, $3\mathbf{b}$ on $\mathbf{a}$. Hence verify that $p(\mathbf{b}+\mathbf{c}) = p(\mathbf{b})+p(\mathbf{c})$ and $p(3\mathbf{b}) = 3p(\mathbf{b})$ where $p(\mathbf{x})$ denotes the projection of $\mathbf{x}$ on $\mathbf{a}$.

(2) If $p(\mathbf{x})$ denotes the projection of $\mathbf{x}$ on $\mathbf{a}$ deduce that

$$\mathbf{a}.(\mathbf{b}+\mathbf{c}) = \mathbf{a}.\mathbf{b}+\mathbf{a}.\mathbf{c} \quad \text{and} \quad \mathbf{a}.(k\mathbf{b}) = k(\mathbf{a}.\mathbf{b})$$

by multiplying by $\mathbf{a}$ the results

$$p(\mathbf{b}+\mathbf{c}) = p(\mathbf{b})+p(\mathbf{c}) \quad \text{and} \quad p(k\mathbf{b}) = kp(\mathbf{b}).$$

(3) If $\mathbf{p} = -3\mathbf{i}+4\mathbf{j}+5\mathbf{k}$, $\mathbf{q} = 2\mathbf{i}-\mathbf{j}+3\mathbf{k}$, $\mathbf{r} = 4\mathbf{i}+3\mathbf{j}-2\mathbf{k}$ evaluate $\mathbf{p}.\mathbf{q}$ and $\mathbf{p}.\mathbf{r}$ and verify that $\mathbf{p}.(\mathbf{q}+\mathbf{r}) = \mathbf{p}.\mathbf{q}+\mathbf{p}.\mathbf{r}$. Also simplify $(\mathbf{p}.\mathbf{q})\mathbf{r}+(\mathbf{p}.\mathbf{r})\mathbf{q}$.

(4) If angle $ABC = 90°$ and $\mathbf{a}$, $\mathbf{b}$, $\mathbf{c}$ are the position vectors of A, B, C respectively, prove that $\mathbf{b}^2+2\mathbf{a}.\mathbf{c} = \mathbf{b}.\mathbf{c}+\mathbf{a}.\mathbf{c}+\mathbf{a}.\mathbf{b}$.

(5) In triangle AOB, angle $AOB = 90°$. If P and Q are points of trisection on AB, prove that $OP^2+OQ^2 = \frac{5}{9}AB^2$.

(6) $ABCD$ is a trapezium with $AB = a$, $DC = 2a$, $DA = b$. E is a point in BC such that $BE = \frac{1}{3}BC$. Show that $\mathbf{AC}.\mathbf{DE} = \frac{2}{3}(4a^2-b^2)$. Comment on the case of $b = 2a$.

(7) A, B, C are the points $(1, 2, 3)$, $(0, 6, 11)$, $(5, -2, 10)$ respectively. Show that the angle between the vectors $\mathbf{AB}$, $\mathbf{AC}$ is $\cos^{-1}\frac{4}{9}$. Hence find the shortest distance from B to the line AC.

(8) Using scalar products prove that the angle between two lines with direction cosines (l_1, m_1, n_1) and (l_2, m_2, n_2) is

$$\cos^{-1}(l_1 l_2+m_1 m_2+n_1 n_2).$$

(*Hint.* Consider unit vectors $\hat{\mathbf{u}}_1$, $\hat{\mathbf{u}}_2$ at angle θ to each other.)

(9) Prove that the diagonals of a rhombus are perpendicular.

(10) Prove that the acute angle between the diagonals of a cube is $\cos^{-1}\frac{1}{3}$.

(11) Find the unit vectors which are perpendicular to the vectors

$$\mathbf{a} = 2\mathbf{i}-3\mathbf{j}+6\mathbf{k} \quad \text{and} \quad \mathbf{b} = -6\mathbf{i}+2\mathbf{j}+3\mathbf{k}.$$

(12) Prove that the sum of the squares of the diagonals of a parallelogram is equal to twice the sum of the squares on two adjacent sides.

(13) Prove that the angle in a semicircle is a right angle.

(14) $ABCD$ is any quadrilateral with P, Q the mid-points of the diagonals AC, BD respectively. Prove that

$$AB^2+BC^2+CD^2+DA^2 = AC^2+BD^2+4PQ^2.$$

(15) $OABC$ is a tetrahedron. The edges OA, OB are perpendicular to the edges BC, AC respectively. Prove that the edge OC is perpendicular to the edge AB and that

$$OA^2+BC^2 = OB^2+AC^2 = OC^2+AB^2.$$

(16) G is the centroid of a triangle ABC and X is any point in or out of the plane ABC. Prove that

$$XA^2+XB^2+XC^2 = GA^2+GB^2+GC^2+3XG^2.$$

(17) Find the angles, the sides and the area of the triangle whose vertices are at the points $A(1, 1, -1)$, $B(2, -1, 1)$, $C(-1, 1, 1)$.

(18) If $\mathbf{v} = \mathbf{v}_0+\mathbf{a}t$ and $\mathbf{r}-\mathbf{r}_0 = \mathbf{v}_0 t+\frac{1}{2}\mathbf{a}t^2$ prove that

$$\mathbf{v}^2 = \mathbf{v}_0^2+2\mathbf{a}.(\mathbf{r}-\mathbf{r}_0).$$

(19) $\mathbf{u}$, $\mathbf{v}$ are two unit vectors lying in the $\mathbf{i}-\mathbf{j}$ plane and making angles of θ, ϕ respectively with $\mathbf{i}$. Show that $\mathbf{u} = \mathbf{i}\cos\theta+\mathbf{j}\sin\theta$ and $\mathbf{v} = \mathbf{i}\cos\phi+\mathbf{j}\sin\phi$. Obtain $\mathbf{u}.\mathbf{v}$ and hence show

$$\cos(\theta-\phi) = \cos\theta\cos\phi+\sin\theta\sin\phi.$$

(20) O is the circumcentre of triangle ABC, and R is the radius of the circumcircle. Show that

$$\mathbf{OB}.\mathbf{OC} = R^2\cos 2A,\quad \mathbf{OA}.\mathbf{OC} = R^2\cos 2B,\quad \mathbf{OA}.\mathbf{OB} = R^2\cos 2C.$$

Hence prove $bc\cos A = R^2(1+\cos 2A-\cos 2B-\cos 2C)$.

(21) (i) If $\mathbf{r}.(d\mathbf{r}/dt) = 0$ show that $|\mathbf{r}|$ is constant. (ii) If

$$\mathbf{r} = \mathbf{i}\cos nt+\mathbf{j}\sin nt$$

show that $d(\mathbf{r}^2)/dt = 0$.

(22) If $\hat{\mathbf{u}}$ is a unit vector show that $\hat{\mathbf{u}}.(d\hat{\mathbf{u}}/dt) = 0$ and

$$\hat{\mathbf{u}}.\left(\hat{\mathbf{u}}+\frac{d^2\hat{\mathbf{u}}}{dt^2}\right)+\left(\frac{d\hat{\mathbf{u}}}{dt}\right)^2 = 1.$$

(23) Prove that

$$\int \mathbf{u}.\frac{d\mathbf{v}}{dt}\,dt = \mathbf{u}.\mathbf{v}-\int \mathbf{v}.\frac{d\mathbf{u}}{dt}\,dt.$$

(24) Forces of 6, 12, 9 newtons in the directions of the vectors $4\mathbf{i}+4\mathbf{j}-7\mathbf{k}$, $7\mathbf{i}-4\mathbf{j}+4\mathbf{k}$, $-4\mathbf{i}+7\mathbf{j}-4\mathbf{k}$ respectively act on a particle producing a displacement of $(2\mathbf{i}+3\mathbf{j}+6\mathbf{k})$ m. Calculate the work done.

(25) Calculate the work done by the field of force

$$\mathbf{P} = (x+y)^2\,\mathbf{i}+(x-y)^2\mathbf{j}$$

on a particle moving from (1, 0) to (0, 1) along the circle

$$x = \cos t,\quad y = \sin t.$$

(26) Calculate the work done by the field of force

$$\mathbf{P} = \frac{x\mathbf{i}+y\mathbf{j}}{\sqrt{(x^2+y^2)}}$$

on a particle moving around a square whose sides are $x = 0$, $x = 1$, $y = 0$, $y = 1$.

(*Hint*. Consider the field of force along each side and hence calculate the work done on each of the 4 legs round the square, working anticlockwise.)

MISCELLANEOUS EXERCISES

(1) A vector of magnitude PQ in a direction P to Q is represented by $\mathbf{PQ}$.

(*a*) $\mathbf{OB}$ makes an angle of 30° with the x axis and $\mathbf{OC}$ makes an angle of 120° with the x axis. Calculate the magnitudes of $\mathbf{OB}$ and $\mathbf{OC}$ if $2\mathbf{OB}+3\mathbf{OC} = 6\mathbf{i}+4\mathbf{j}$, where $\mathbf{i}$ and $\mathbf{j}$ are unit vectors along the x axis and y axis respectively.

(*b*) The points H and K are the middle points of the sides BC and CD respectively of a parallelogram $ABCD$. Prove that

$$3(\mathbf{AB}+\mathbf{AC}+\mathbf{AD}) = 4(\mathbf{AH}+\mathbf{AK}). \qquad \text{(C.)}$$

(2) (*a*) A vector of magnitude OP in the direction from O to P is represented by $\mathbf{OP}$. If $\mathbf{OP}-3\mathbf{OQ}+2\mathbf{OR} = 0$ show that P, Q, R are collinear.

(*b*) A unit vector parallel to the x axis is represented by $\mathbf{i}$ and a unit vector parallel to the y axis by $\mathbf{j}$.

If $\mathbf{OP} = a\mathbf{i}+s\mathbf{j}$ and $\mathbf{OQ} = -a\mathbf{i}+t\mathbf{j}$, where a is a constant and s and t are variables, show that the locus of P and Q are parallel straight lines. In this case find $\mathbf{OQ}$ when $\mathbf{OP} = 2\mathbf{i}+3\mathbf{j}$ and OQ is perpendicular to OP. (C.)

(3) (*a*) Unit vectors along the Ox and Oy axes are represented by $\mathbf{i}$ and $\mathbf{j}$ respectively, and $\mathbf{OP}$ represents a vector of magnitude OP in the direction from O to P. A triangle PQR is formed by the extremities of the vectors $\mathbf{OP}$, $\mathbf{OQ}$ and $\mathbf{OR}$, where

$$\mathbf{OP} = 2\mathbf{i}, \quad \mathbf{OQ} = -3\mathbf{i}+\mathbf{j} \quad \text{and} \quad \mathbf{OR} = a\mathbf{i}+b\mathbf{j},$$

a and b being constants. There are particles of masses m, $2m$ and $3m$ at the points P, Q and R respectively. If the centre of mass of the particles is at G where $\mathbf{OG} = \frac{1}{6}(5\mathbf{i}+8\mathbf{j})$, find a and b.

(*b*) If $\mathbf{OA} = 2\mathbf{i}+3\mathbf{j}$ and $\mathbf{OB} = \lambda(-\mathbf{i}+5\mathbf{j})$, find $\mathbf{OM}$ where M is the middle point of AB, and find the locus of M as λ varies. (C.)

(4) Particles of equal mass m are placed, one each at the n points whose position vectors are $\mathbf{r}_1, \mathbf{r}_2, \ldots, \mathbf{r}_n$. Prove that the centre of mass of the system of particles is at the point whose position vector is $(\mathbf{r}_1+\mathbf{r}_2+\ldots+\mathbf{r}_n)/n$.

In a system of four particles of unit mass the position vectors of the particles are **a, b, c, d**. In a second system three particles of unit mass have position vectors $\frac{3}{2}\mathbf{a}$, $3\mathbf{b}$, $\frac{1}{2}(\mathbf{c}+\mathbf{d})$; and in a third system two particles of unit mass have position vectors of $2\mathbf{a}$ and $5\mathbf{b}$. Show that the centres of mass of the three systems are collinear. (C.)

(5) A, B, C are three given points and x, y are scalar constants. Prove that the sum of the vectors $x\mathbf{AB}$ and $y\mathbf{AC}$ is $(x+y)\mathbf{AD}$, the point D being such that $x\mathbf{BD} = y\mathbf{DC}$.

Three forces are represented completely by the vectors $p\mathbf{AD}$, $q\mathbf{BE}$, $r\mathbf{CF}$ where D, E, F are points on the sides BC, CA, AB respectively of the triangle ABC such that

$$\frac{BD}{DC} = l, \quad \frac{CE}{EA} = m, \quad \frac{AF}{FB} = n.$$

Show that the three forces are equivalent to three forces acting along the sides of the triangle, and find these equivalent forces.

If $p = q = r$ and $l = m = n$ show that the system reduces to a couple and find its magnitude in terms of p, l and the area of the triangle. (C.)

(6) Find the radial and transverse components of acceleration referred to polar co-ordinates (r, θ) of a particle moving in a plane.

P is a point on the curve $r = a+b\cos\theta$, O is the origin and Q is the point on the initial line such that $OQ = PQ$ and the angle $POQ = \theta$. A particle describes the curve in such a manner that the radius vector OP rotates with constant angular velocity ω. Show that the radial and transverse components of acceleration of the particle when at P are $-\omega^2(a+2b\cos\theta)$ and $-2b\omega^2\sin\theta$ respectively.

Show also that the acceleration of the particle may be written as $a\omega^2\lambda+2b\omega^2\mu$ where λ, μ are unit vectors in the directions PO, PQ respectively. (C.)

(7) (i) If $\mathbf{OA} = \mathbf{a}$, $\mathbf{OB} = \mathbf{b}$ and $\mathbf{OC} = \mathbf{c}$ show that the condition for $\mathbf{OA}$ to be perpendicular to $\mathbf{BC}$ is $\mathbf{a}.(\mathbf{c}-\mathbf{b}) = 0$. Also show that if $\mathbf{OB}$ is perpendicular to $\mathbf{CA}$, then $\mathbf{OC}$ is perpendicular to $\mathbf{AB}$.

(ii) Given that $\mathbf{c} = \mathbf{a}+\mathbf{b}$ expand the right-hand side of each of the equations

$$\mathbf{c}.\mathbf{c} = (\mathbf{a}+\mathbf{b}).\mathbf{c}, \quad \mathbf{c}.\mathbf{c} = (\mathbf{a}+\mathbf{b}).(\mathbf{a}+\mathbf{b}).$$

State the equivalent formulae in trigonometry.

(8) Prove that the sum of the squares on the edges of any tetrahedron is equal to four times the sum of the squares on the joins of the mid-points of opposite edges.

(9) $OABC$ is a tetrahedron. If G is the centroid of ABC prove that $3\mathbf{OG} = \mathbf{OA}+\mathbf{OB}+\mathbf{OC}$. Furthermore, if OA, OB, OC and the angles BOC, COA, AOB are denoted by a, b, c and α, β, γ respectively, prove that

$$(3OG)^2 = a^2+b^2+c^2+2bc\cos\alpha+2ca\cos\beta+2ab\cos\gamma.$$

(10) ABC is a triangle. P is a point on BC such that $BP:PC = l:m$. Q is a point on CA such that $CQ:QA = n:l$. If O is any fixed point obtain $\mathbf{OP}$ in terms of $\mathbf{OB}$, $\mathbf{OC}$ and $\mathbf{OQ}$ in terms of $\mathbf{OA}$, $\mathbf{OC}$. By eliminating $\mathbf{OC}$, show that R divides AB in the ratio $-m:n$ where R is the point of intersection of AB and PQ. Hence verify Menelaus's Theorem, namely,

$$\frac{BP}{PC}\cdot\frac{CQ}{QA}\cdot\frac{AR}{RB} = -1.$$

9

THE STRAIGHT LINE AND THE PLANE

Introduction

In this chapter we shall consider two important loci—the straight line and the plane.

Consider a variable point which moves under certain specified conditions. The Cartesian form of the locus of the point is the equation connecting its co-ordinates, namely x, y if the locus is two-dimensional, or x, y, z if the locus is three-dimensional. The vector form of the locus of the point is the equation connecting its position vector relative to some origin with other given vectors.

If the point $P(x, y, z)$ is on the locus and $\mathbf{r}$ is its position vector relative to the origin O, then

$$\mathbf{r} = x\mathbf{i}+y\mathbf{j}+z\mathbf{k}.$$

Furthermore, if x, y, z are expressed in terms of a parameter u, that is,

$$x = f(u), \quad y = \phi(u), \quad z = \psi(u),$$

we may write

$$\mathbf{r} = f(u)\mathbf{i}+\phi(u)\mathbf{j}+\psi(u)\mathbf{k}.$$

Here we have expressed the vector equation of the locus in terms of the parameter u.

In the following treatment of loci we shall take P to be the variable point and $\mathbf{r}$ to be its position vector relative to the origin. For brevity we shall also write $P(\mathbf{r})$ to mean the point P whose position vector is $\mathbf{r}$. So as to be in keeping with the usual notation in co-ordinate geometry we shall let the co-ordinates of the variable point $P(\mathbf{r})$ be x, y, z, that is, $\mathbf{r} = (x, y, z)$, and those of a given point say $A(\mathbf{r}_1)$ on the locus be x_1, y_1, z_1, that is, $\mathbf{r}_1 = (x_1, y_1, z_1)$.

The straight line

Direction-vector of a line

The direction of a line may be specified by referring to any other line parallel to it. In particular, if we assign a direction and sense in describing a line then these may be specified by a vector.

In two dimensions the direction of a line is given by its gradient, which is defined as the tangent of the angle which the line makes with the positive x-axis, the angle being measured in the usual way from

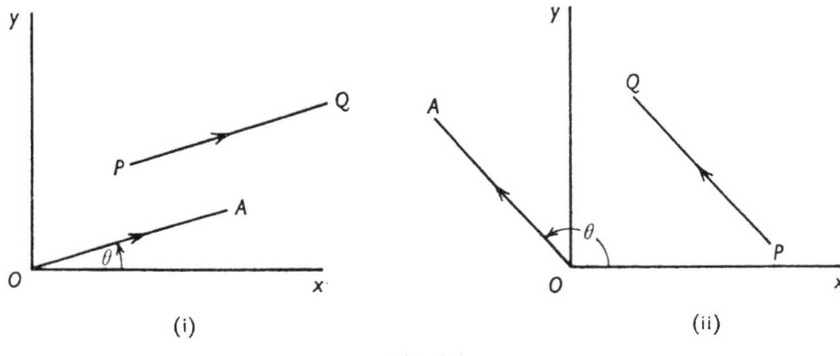

Fig. 9.1

the positive x-axis. Referring to Fig. 9.1 PQ is any straight line. From the origin O, draw **OA** parallel to PQ. Then the gradient of OA, namely $\tan\theta$, gives the gradient of PQ. In (i) the gradient is positive, whereas in (ii) the gradient is negative.

In space, the analogue of gradient involves the cosines of the angles made by the line with the positive Ox, Oy, Oz axes. These angles are known as direction angles.

In Fig. 9.2 let PQ be a straight line in space, described in the sense from P to Q. From O draw a unit vector **OA** in the same direction and sense as **PQ**. Let l, m, n be the orthogonal projections of **OA** on the axes Ox, Oy, Oz respectively, and α, β, γ the angles made by **OA** with the unit vectors **i**, **j**, **k**. We have

$$\begin{aligned} l &= \mathbf{OA}.\mathbf{i} \\ &= |\mathbf{OA}|\,|\mathbf{i}|\cos\alpha \\ &= \cos\alpha \end{aligned}$$

since $|\mathbf{OA}| = |\mathbf{i}| = 1$. Similarly

$$m = \cos\beta, \quad n = \cos\gamma.$$

The numbers l, m, n are known as the direction cosines of the vector **OA** and hence of **PQ**. Since l, m, n are the orthogonal projections of **OA** they are the components of **OA**; that is,

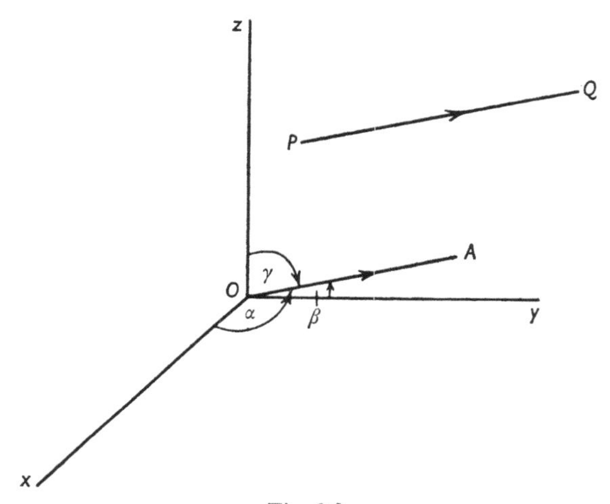

Fig. 9.2

$$\mathbf{OA} = l\mathbf{i}+m\mathbf{j}+n\mathbf{k}.$$

Hence $$|\mathbf{OA}|^2 = l^2+m^2+n^2.$$

But $$|\mathbf{OA}| = 1.$$

Therefore $$l^2+m^2+n^2 = 1.$$

Often it is convenient to specify the direction of a vector by a set of three numbers a, b, c which are proportional to the direction cosines l, m, n. Such numbers are called either the direction ratios or the direction components of the vector.

It is possible to refer to the direction cosines of a line instead of the direction ratios of a vector, but since the direction of a line may be specified by either of two unit vectors having opposite senses, we have two sets of direction angles, namely

$$\alpha, \beta, \gamma \quad \text{and} \quad 180°-\alpha, 180°-\beta, 180°-\gamma,$$

and two sets of direction cosines, namely

$$\cos\alpha,\quad \cos\beta,\quad \cos\gamma \quad\text{and}\quad -\cos\alpha,\quad -\cos\beta,\quad -\cos\gamma.$$

Thus the direction cosines of a line are not unique and we overcome this situation by considering a vector in one of the directions. We shall also find it convenient to use direction ratios instead of direction cosines since the latter often involve irrational square roots.

Suppose we have a line L which is parallel to the vector $\mathbf{l}$. We shall refer to the vector $\mathbf{l}$ as the direction-vector of the line L. The components of $\mathbf{l}$ are the direction ratios of the vector. If the direction-vector $\mathbf{l}$ is a unit vector then its components are the direction cosines of the vector; that is, the direction ratios of a unit vector are its direction cosines.

Vector equation of the straight line through a given point and in a given direction

Let $A(\mathbf{r}_1)$ be a fixed point and $\mathbf{l} \neq \mathbf{0}$ be a fixed vector. We require the equation of the straight line L which passes through A and which is parallel to $\mathbf{l}$ (Fig. 9.3).

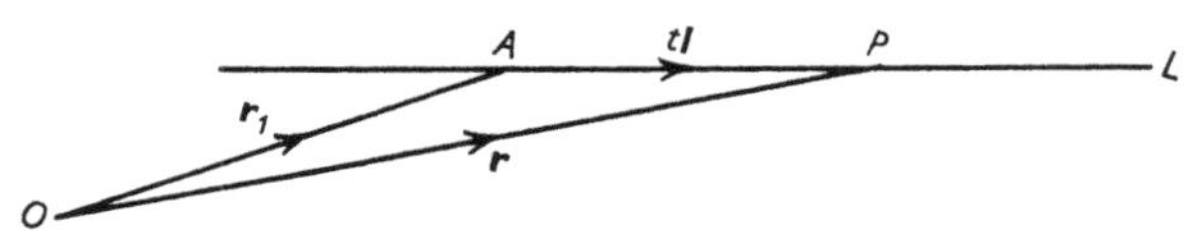

Fig. 9.3

Let $P(\mathbf{r})$ be any variable point on the line L. Since $\mathbf{AP}$ is parallel to $\mathbf{l}$ we have

$$\mathbf{AP} = t\mathbf{l}$$

where t is a number dependent on the position of P. Therefore

$$\mathbf{r}-\mathbf{r}_1 = t\mathbf{l},$$

that is,

$$\mathbf{r} = \mathbf{r}_1+t\mathbf{l}. \qquad (9{\cdot}1)$$

This is the parametric equation of the required line L since for any point P on L there is a unique value of t, and for any value of t there is a unique point P on L. For suppose we consider a particular value of t, say t'. Then the point P' given by (9·1) has position vector $\mathbf{r}'$ where

$$\mathbf{r}' = \mathbf{r}_1+t'\mathbf{l}.$$

Therefore $\mathbf{r}' - \mathbf{r}_1 = t'\mathbf{l}$,

that is, $\mathbf{AP}' = t'\mathbf{l}$.

Hence the line AP' passes through A and is parallel to $\mathbf{l}$; consequently P' lies on L. In general, any point P, such that $\mathbf{AP} = t\mathbf{l}$ and whose parameter t satisfies (9·1), lies on L. The parameter t may assume any value including zero, in which case P is the point A.

If
$$\mathbf{r} = (x, y, z), \quad \mathbf{r}_1 = (x_1, y_1, z_1), \quad \mathbf{l} = (a, b, c),$$
equation (9·1) is
$$(x, y, z) = (x_1, y_1, z_1) + t(a, b, c),$$
from which we obtain
$$\left.\begin{aligned} x &= x_1 + ta, \\ y &= y_1 + tb, \\ z &= z_1 + tc. \end{aligned}\right\} \tag{9·2}$$

This is the parametric form of the equations of the straight line through the point (x, y, z) and having direction ratios a, b, c. From (9·2) we deduce

$$\frac{x - x_1}{a} = \frac{y - y_1}{b} = \frac{z - z_1}{c} \quad (a \neq 0, b \neq 0, c \neq 0), \tag{9·3}$$

which is the standard Cartesian form of the equations of the straight line through the point (x_1, y_1, z_1) and having the direction ratios a, b, c. It should be noted that two equations are required to represent a straight line in space.

Examples

(1) *Find the equations of the line through the point* (1, 2, −3) *and parallel to* $3\mathbf{i} - 2\mathbf{j} + 4\mathbf{k}$. *Verify that the point* (−2, 4, −7) *is on the line.*

The vector equation of the straight line is

$$\mathbf{r} = (x, y, z) = (1, 2, -3) + t(3, -2, 4),$$

where t is a parameter.

Hence the standard Cartesian form of the equations of the straight line is
$$\frac{x-1}{3} = \frac{y-2}{-2} = \frac{z+3}{4}.$$

Substituting $x = -2$ in these equations we find $y = 4$, $z = -7$. Therefore the point $(-2, 4, -7)$ is on the line.

(2) *Show that the locus whose equations are*

$$\frac{3x+1}{2} = \frac{2-y}{3} = \frac{2z+4}{5}$$

is the straight line through the point $(-\frac{1}{3}, 2, -2)$ *and parallel to the vector* $4\mathbf{i}-18\mathbf{j}+15\mathbf{k}$.

The Cartesian form

$$\frac{3x+1}{2} = \frac{2-y}{3} = \frac{2z+4}{5}$$

may be written as the vector equation

$$\mathbf{r} = (x, y, z) = (-\tfrac{1}{3}, 2, -2)+t(\tfrac{2}{3}, -3, \tfrac{5}{2})$$

where t is a parameter.

Hence the locus is the straight line through the point $(-\frac{1}{3}, 2, -2)$ and having $\frac{2}{3}\mathbf{i}-3\mathbf{j}+\frac{5}{2}\mathbf{k}$ or $4\mathbf{i}-18\mathbf{j}+15\mathbf{k}$ as its direction-vector.

(3) *Find the equations of the straight line joining the points*

$$A\,(2, 1, -1) \quad \textit{and} \quad B\,(0, 6, 3).$$

The direction-vector of the line is given by

$$\mathbf{AB} = (-2, 5, 4).$$

The vector equation of the line is

$$\mathbf{r} = (x, y, z) = (2, 1, -1)+t(-2, 5, 4)$$

where t is a parameter.

Hence the standard Cartesian form of the equations of the line is

$$\frac{x-2}{-2} = \frac{y-1}{5} = \frac{z+1}{4}. \tag{9·4}$$

The vector equation of the line can also be written

$$\mathbf{r} = (x, y, z) = (0, 6, 3)+t(-2, 5, 4),$$

and the Cartesian form is therefore

$$\frac{x}{-2} = \frac{y-6}{5} = \frac{z-3}{4}. \tag{9·5}$$

Thus the equations of a given line are not unique. Suppose we are given the equations of two lines. To decide whether they are distinct or not we proceed as follows:

(i) Obtain their direction-vectors.

(ii) Find a point on one of them and check if it lies on the other line.

If the direction-vectors are parallel *and* the same point lies on both the lines then the lines are the same.

Equation (9·4) may be put into the form of equation (9·5) by subtracting 1 from each of the three expressions of (9·4).

(4) *Find the perpendicular distance of the point A* (5, 4, 1) *from the line*

$$\frac{x-6}{5} = \frac{y+15}{1} = \frac{z-14}{8}$$

and find the co-ordinates of the foot N of the perpendicular.

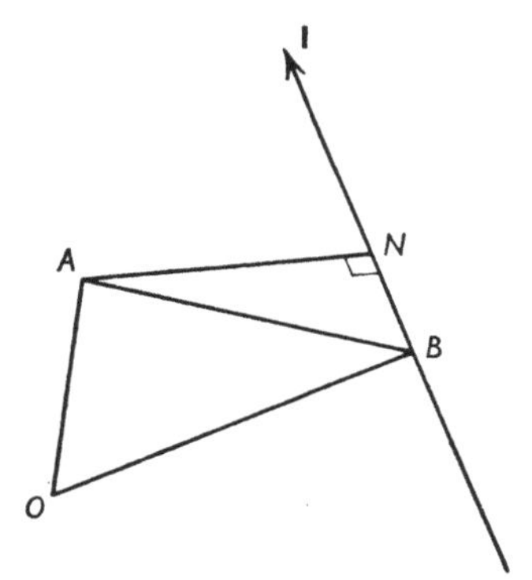

Fig. 9.4

From the equation we see that the point (6, −15, 14) is on the line. Referring to Fig. 9.4 let B be this point on the line. The direction-vector $\mathbf{l}$ of the line is given by

$$\mathbf{l} = (5, 1, 8).$$

The projection of **AB** on $\mathbf{l}$ is BN and this is given by

$$BN = \mathbf{AB}.\frac{\mathbf{l}}{|\mathbf{l}|}$$

$$= (1, -19, 13).\frac{(5, 1, 8)}{\sqrt{90}}$$

$$= \sqrt{90}.$$

From Pythagoras' theorem

$$AN^2 = AB^2 - BN^2$$

$$= (1+361+169)-90 = 441.$$

Therefore $$AN = 21,$$

i.e. the length of the perpendicular is 21 units.

The equation of a straight line is given by

$$\mathbf{r} = \mathbf{r}_1 + t\mathbf{l}.$$

Taking $$\mathbf{ON} = \mathbf{r}, \quad \mathbf{OB} = \mathbf{r}_1$$

we have $$\mathbf{ON} = (6,\ -15,\ 14)+t(5,\ 1,\ 8),$$
$$= (6+5t,\ -15+t,\ 14+8t),$$

where t is a parameter to be determined.

Now
$$\mathbf{AN} = \mathbf{ON}-\mathbf{OA},$$
$$= (1+5t,\ -19+t,\ 13+8t).$$

Since **AN** is perpendicular to **l** we have
$$\mathbf{AN}.\mathbf{l} = 0.$$

Therefore $$5(1+5t)+(-19+t)+8(13+8t) = 0,$$

giving $$t = -1.$$

Hence $$\mathbf{ON} = (1,\ -16,\ 6),$$

i.e. N is the point $(1,\ -16,\ 6)$.

Vector equation of the straight line through two given points

Suppose that $A(\mathbf{r}_1)$, $B(\mathbf{r}_2)$ are the two given points. The direction-vector of the line joining A, B is given by **AB** where
$$\mathbf{AB} = \mathbf{r}_2-\mathbf{r}_1.$$

The equation of the line passing through $A(\mathbf{r}_1)$ and having direction-vector **l** is
$$\mathbf{r} = \mathbf{r}_1+t\mathbf{l}.$$

Therefore the equation of the line passing through $A(\mathbf{r}_1)$ and having direction-vector **AB** is
$$\mathbf{r} = \mathbf{r}_1+t(\mathbf{r}_2-\mathbf{r}_1),$$

i.e. $$\mathbf{r} = (1-t)\mathbf{r}_1+t\mathbf{r}_2. \qquad (9{\cdot}6)$$

If $\quad \mathbf{r} = (x, y, z),\quad \mathbf{r}_1 = (x_1, y_1, z_1),\quad \mathbf{r}_2 = (x_2, y_2, z_2),$

we have the equivalent parametric form
$$x = x_1+(x_2-x_1)t,$$
$$y = y_1+(y_2-y_1)t,$$
$$z = z_1+(z_2-z_1)t,$$

or the equivalent Cartesian form
$$\frac{x-x_1}{x_2-x_1} = \frac{y-y_1}{y_2-y_1} = \frac{z-z_1}{z_2-z_1}.$$

Writing $t = p/(p+q)$, equation (9·6) becomes

$$\mathbf{r} = \frac{q\mathbf{r}_1 + p\mathbf{r}_2}{p+q},$$

expressing the fact that the point $P(\mathbf{r})$ divides the join of $A(\mathbf{r}_1)$ and $B(\mathbf{r}_2)$ in the ratio $p:q$.

If we look at equation (9·6) more closely we see that the algebraic sum of the coefficients of the vectors is zero. Thus a necessary condition for three points to be collinear is that the algebraic sum of the coefficients of their position vectors vanishes. This is also a sufficient condition. For, supposing the condition is satisfied in a linear relation involving $\mathbf{r}$, $\mathbf{r}_1$, $\mathbf{r}_2$ in which the coefficient of $\mathbf{r}$ has been made unity, we can write

$$\mathbf{r} = p\mathbf{r}_1 + (1-p)\mathbf{r}_2,$$

i.e.

$$\mathbf{r} = \mathbf{r}_2 + p(\mathbf{r}_1 - \mathbf{r}_2).$$

Thus $P(\mathbf{r})$ lies on the straight line through $A(\mathbf{r}_1)$, $B(\mathbf{r}_2)$.

The angle between two straight lines

Definition. *The angle between two straight lines is defined as the angle between their direction-vectors.*

This definition covers the case of two skew lines in space, that is, two non-parallel lines which do not meet; for the angle between two skew lines is defined as the angle between two coplanar lines which are respectively parallel to the skew lines.

Let $\mathbf{l}_1$, $\mathbf{l}_2$ be the direction-vectors of two lines. Then the angle θ between them is given by

$$\cos\theta = \frac{\mathbf{l}_1 . \mathbf{l}_2}{|\mathbf{l}_1|\,|\mathbf{l}_2|}.$$

If the lines are parallel we have

$$\mathbf{l}_1 = \lambda \mathbf{l}_2$$

where λ is a constant, and if the lines are perpendicular we have

$$\mathbf{l}_1 . \mathbf{l}_2 = 0.$$

Exercise 9(*a*)

(1) Find (*a*) the vector equation, (*b*) the Cartesian equations, of the straight line passing through the point $A\,(2, -1, 3)$ and parallel to the line through the points $B\,(3, 2, -1)$, $C\,(-1, 1, 2)$. Verify that the point $(-2, -2, 6)$ is on the required line.

(2) Obtain the co-ordinates of the point of intersection of the lines

$$\frac{x-2}{5} = \frac{y+5}{-9} = \frac{z}{7},$$

$$x+1 = y-6 = \frac{z+3}{2}.$$

(3) Find the equations of the straight line joining the points $A\,(2, -3, 4)$, $B\,(1, 6, -1)$. Obtain the point of intersection of this line with the xOy plane.

(4) Determine whether the lines

$$\frac{x-2}{-3} = \frac{y+1}{1} = \frac{z+3}{2},$$

$$\frac{2x-7}{-6} = \frac{2y+3}{2} = \frac{z+4}{2}$$

are the same or not.

(5) Find the equations of the line through the point $A\,(6, 2, -4)$ and having direction ratios $(2, -1, 3)$. If this line meets the yOz plane in the point B find (i) the length of AB, (ii) the angle AB makes with the yOz plane.

(6) Find the co-ordinates of the two points on the line

$$\frac{x+3}{2} = \frac{y-1}{3} = \frac{z-2}{-1}$$

which are $\sqrt{33}$ units of length from the point $(2, 3, 4)$.

(7) Find the vertices and the angles of the triangle whose sides are the lines

$$\frac{x-2}{-3} = \frac{y-1}{3} = \frac{z-3}{2},$$

$$\frac{x+1}{-4} = \frac{y-4}{2} = \frac{z-5}{7},$$

$$\frac{x-2}{1} = \frac{y-1}{1} = \frac{z-3}{-5}.$$

(8) Two forces $\mathbf{F}_1 = \mathbf{i}+\mathbf{j}+\mathbf{k}$ and $\mathbf{F}_2 = \mathbf{i}+2\mathbf{j}-\mathbf{k}$ act through points whose position vectors are $\mathbf{S}_1 = \mathbf{i}+\mathbf{j}+2\mathbf{k}$ and $\mathbf{S}_2 = p\mathbf{j}+5\mathbf{k}$ respectively, relative to a fixed point and three mutually perpendicular vectors $\mathbf{i}$, $\mathbf{j}$ and $\mathbf{k}$. If the lines of action of $\mathbf{F}_1$ and $\mathbf{F}_2$ intersect, find p, and find the vector equation of the line of action of the resultant of $\mathbf{F}_1$ and $\mathbf{F}_2$. (L.)

(9) Verify that the point $(6, -5, -1)$ lies on the line

$$\mathbf{r} = 2(1+t)\mathbf{i}-(1+2t)\mathbf{j}-(3-t)\mathbf{k}.$$

If Q is the foot of the perpendicular from the point $P(4, 7, -9)$ to the line, find

(i) the co-ordinates of Q,
(ii) the length of PQ,
(iii) the equations of the perpendicular.

The shortest distance between two skew lines

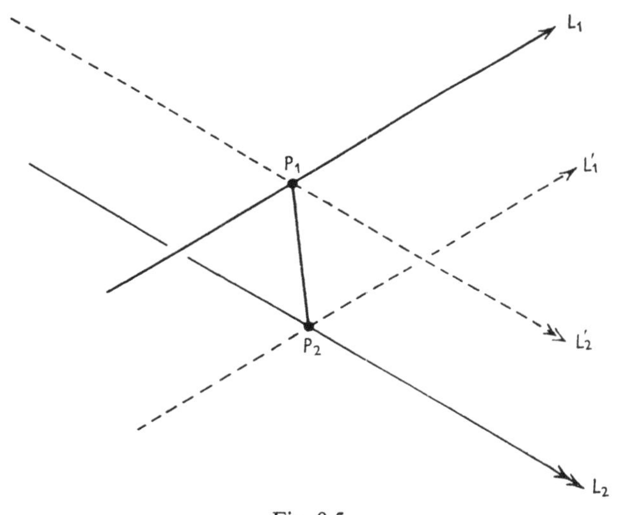

Fig. 9.5

Non-intersecting, non-parallel straight lines in space are known as skew lines. A property of skew lines is that they have a common perpendicular, the length of which is the least distance between the lines. For, consider the skew lines L_1 and L_2. Through the point P_1 on L_1 draw the line L_2' parallel to L_2 such that the plane L_2L_2' is perpendicular to the plane L_1L_2' (Fig. 9.5). Also through the point P_2 on L_2 draw the line L_1' parallel to L_1 such that the plane L_1L_1' is perpendicular to the plane L_2L_1'. It follows the planes L_1L_2' (π_1) and L_2L_1' (π_2) are parallel, and therefore the planes L_1L_1' and L_2L_2' are perpendicular to both π_1 and π_2. Hence P_1P_2 is perpendicular to π_1 and π_2, and therefore to L_1 and L_2. Now the distance from any point in π_1 to any point in π_2 is least when the points are the feet of the common perpendicular. But the lines L_1, L_2 passing through P_1, P_2 are contained in π_1, π_2 and therefore the common perpendicular P_1P_2 is the shortest distance between the lines.

Example

Find (i) *the shortest distance,* (ii) *the co-ordinates of the feet of the common perpendicular,* (iii) *the equations of the common perpendicular, for the skew lines*

$$L_1 \equiv \frac{x+7}{-8} = \frac{y-5}{3} = \frac{z-4}{1}$$

and

$$L_2 \equiv \frac{x+4}{4} = \frac{y}{3} = \frac{z-19}{-2}.$$

If $\mathbf{l}_1$, $\mathbf{l}_2$ are the direction-vectors of L_1, L_2 respectively, we have

$$\mathbf{l}_1 = (-8, 3, 1), \quad \mathbf{l}_2 = (4, 3, -2).$$

The points $(-7, 5, 4)$, $(-4, 0, 19)$ are on L_1, L_2 and the position vectors of any points $P_1(\mathbf{r}_1)$, $P_2(\mathbf{r}_2)$ on L_1, L_2 respectively are

$$\mathbf{r}_1 = (-7, 5, 4)+s(-8, 3, 1) = (-7-8s, 5+3s, 4+s),$$

$$\mathbf{r}_2 = (-4, 0, 19)+t(4, 3, -2) = (-4+4t, 3t, 19-2t),$$

where s, t are parameters.

Therefore
$$\mathbf{P}_1\mathbf{P}_2 = (3+8s+4t, \ -5-3s+3t, \ 15-s-2t).$$

We require P_1P_2 to be perpendicular to both L_1, L_2. This is so when

$$\mathbf{P}_1\mathbf{P}_2 . \mathbf{l}_1 = 0 \quad \text{and} \quad \mathbf{P}_1\mathbf{P}_2 . \mathbf{l}_2 = 0.$$

Evaluating the scalar products we obtain

$$-24-74s-25t = 0 \quad \text{and} \quad -33+25s+29t = 0.$$

The solution of these equations is

$$s = -1, \quad t = 2.$$

By substituting these values we obtain the co-ordinates of the feet of the perpendicular, and they are

$$P_1\,(1, 2, 3), \quad P_2\,(4, 6, 15).$$

The shortest distance is given by

$$|\mathbf{P}_1\mathbf{P}_2| = \sqrt{(3^2+4^2+12^2)} = 13.$$

The common perpendicular passes through the point (1, 2, 3) and has direction-vector (3, 4, 12). Hence the equations of the common perpendicular are

$$\frac{x-1}{3} = \frac{y-2}{4} = \frac{z-3}{12}.$$

Exercise 9(*b*)

(1) Find the length and the equations of the shortest distance between the lines

$$\frac{x-6}{2} = \frac{y+4}{5} = \frac{z-2}{1},$$

$$\frac{x+1}{-4} = \frac{y-9}{5} = \frac{z-5}{7}.$$

(2) The points A (6, 0, 0), B (0, 4, 0), C (0, 0, 0) and D (3, 2, 1) are the corners of a tetrahedron. Find the equations of the line that intersects and is perpendicular to AC and BD. Also determine the shortest distance between AC and BD.

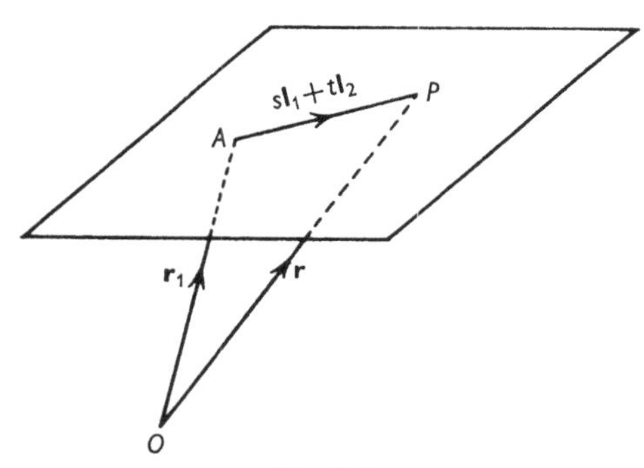

Fig. 9.6

The plane

Vector equation of the plane through a given point and parallel to two given lines

Let $A(\mathbf{r}_1)$ be the given point, $P(\mathbf{r})$ any point in the plane and $\mathbf{l}_1$, $\mathbf{l}_2$ the direction-vectors of the two given lines (Fig. 9.6). Since $\mathbf{AP}$ is coplanar with $\mathbf{l}_1$ and $\mathbf{l}_2$ we may write

$$\mathbf{AP} = s\mathbf{l}_1 + t\mathbf{l}_2$$

where s, t are the numbers depending on the position of P as it moves over the plane. Therefore

$$\mathbf{r}-\mathbf{r}_1 = s\mathbf{l}_1+t\mathbf{l}_2,$$

i.e.

$$\mathbf{r} = \mathbf{r}_1+s\mathbf{l}_1+t\mathbf{l}_2.$$

This is the vector equation of the required plane in terms of the parameters s and t. For any point P on the plane there are unique values of s and t, and, by an argument similar to that on page 140, for any particular values of s and t there is a unique point in the given plane.

Example

Find the equation of the plane containing the line

$$\frac{x-1}{3} = \frac{y+4}{4} = \frac{z-1}{1}$$

and which is parallel to the line

$$\frac{x}{4} = \frac{y}{3} = \frac{z}{12}.$$

The point $(1, -4, 1)$ is in the plane. The plane is parallel to the direction-vectors of the lines, that is, to the vectors $(3, 4, 1)$ and $(4, 3, 12)$. The vector equation of the plane is therefore

$$\mathbf{r} = (x, y, z) = (1, -4, 1)+s(3, 4, 1)+t(4, 3, 12).$$

Hence

$$x = 1+3s+4t, \quad y = -4+4s+3t, \quad z = 1+s+12t.$$

On eliminating s and t we obtain

$$45x-32y-7z = 166$$

as the required equation of the plane.

Vector equation of the plane containing three given non-collinear points

Let $A(\mathbf{r}_1)$, $B(\mathbf{r}_2)$, $C(\mathbf{r}_3)$ be the given points on the required plane and $P(\mathbf{r})$ be any point on the plane (Fig. 9.7). Since the vectors **AP**, **AB**, **AC** are coplanar we may write

$$\mathbf{AP} = s\mathbf{AB}+t\mathbf{AC}$$

where s, t are numbers depending on the position of P, as P moves across the plane. Therefore

$$\mathbf{r}-\mathbf{r}_1 = s(\mathbf{r}_2-\mathbf{r}_1)+t(\mathbf{r}_3-\mathbf{r}_1).$$

The required equation of the plane is

$$\mathbf{r} = \mathbf{r}_1+s(\mathbf{r}_2-\mathbf{r}_1)+t(\mathbf{r}_3-\mathbf{r}_1)$$

or

$$\mathbf{r} = (1-s-t)\mathbf{r}_1+s\mathbf{r}_2+t\mathbf{r}_3.$$

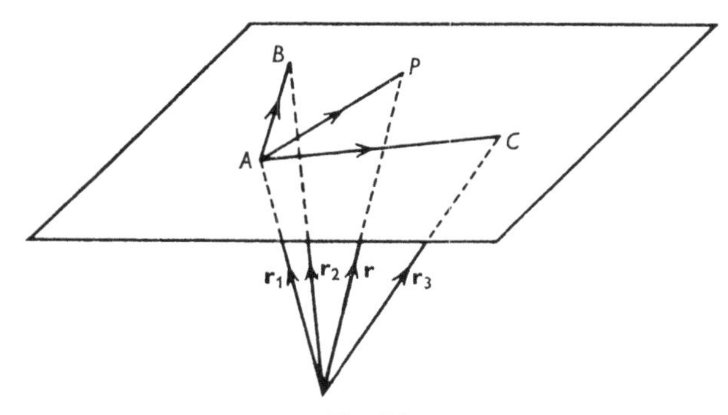

Fig. 9.7

This equation connects the position vectors of four coplanar points and we observe that in this equation the algebraic sum of the coefficients of the vectors is zero. Thus a necessary condition for four points to be coplanar is that the algebraic sum of the coefficients of their position vectors vanishes. This is also a sufficient condition; for suppose we have the condition satisfied in a linear relation in which the coefficient of $\mathbf{r}$ has been made unity, then we have

$$\mathbf{r} = (1-p-q)\mathbf{r}_1+p\mathbf{r}_2+q\mathbf{r}_3,$$
$$= \mathbf{r}_1+p(\mathbf{r}_2-\mathbf{r}_1)+q(\mathbf{r}_3-\mathbf{r}_1),$$

showing that the point $P(\mathbf{r})$ is in the plane containing the points $A(\mathbf{r}_1)$, $B(\mathbf{r}_2)$, $C(\mathbf{r}_3)$.

Example

A plane passes through the points $A\,(1, 2, 3)$, $B\,(-1, 2, 0)$ *and* $C\,(2, -1, -1)$. *Find its equation.*

The vector equation of the plane is

$$\mathbf{r} = (x, y, z) = (1-s-t)(1, 2, 3)+s(-1, 2, 0)+t(2, -1, -1),$$

where s and t are parameters.

Hence $$x = 1-s-t-s+2t = 1-2s+t,$$
$$y = 2-2s-2t+2s-t = 2-3t,$$
$$z = 3-3s-3t-t = 3-3s-4t.$$

On eliminating s and t we obtain

$$9x+11y-6z = 13$$

as the required equation of the plane.

Normal-vector of a plane

Consider a plane π and let A be a point in the plane. Let $\mathbf{n}$ be a vector such that $\mathbf{n}$ is perpendicular to the vector $\mathbf{AP}$ for all points P in the plane. We say that the vector $\mathbf{n}$ is perpendicular to the plane π, and we shall refer to the vector $\mathbf{n}$ as the *normal-vector* of the plane. It follows that $k\mathbf{n}$ (where k is a number different from zero) is also a normal-vector of the plane. As before we may associate the terms direction ratios and direction cosines with the components of the normal-vector.

Vector equation of the plane containing a given point and normal to a given line

Let $\mathbf{n}$ be the normal-vector of the plane, A a fixed point and P any point on the plane. Taking O as the origin, let $\mathbf{OA} = \mathbf{r}_1$, $\mathbf{OP} = \mathbf{r}$ (Fig. 9.8). Since $\mathbf{AP}$ is perpendicular to the normal-vector of the plane we have

$$\mathbf{AP}.\mathbf{n} = 0.$$

But $$\mathbf{AP} = \mathbf{r}-\mathbf{r}_1.$$

Therefore $$(\mathbf{r}-\mathbf{r}_1).\mathbf{n} = 0. \tag{9·7}$$

This is the vector equation of the plane containing the point whose position vector is $\mathbf{r}$ and having $\mathbf{n}$ as the normal-vector. In terms of Cartesian co-ordinates, if

$$\mathbf{r} = (x, y, z), \quad \mathbf{r}_1 = (x_1, y_1, z_1), \quad \mathbf{n} = (a, b, c),$$

we have $$a(x-x_1)+b(y-y_1)+c(z-z_1) = 0 \tag{9·8}$$

which is the equation of the plane containing the point (x_1, y_1, z_1) and with direction ratios a, b, c.

If the normal-vector is the unit vector $\hat{\mathbf{n}}$ then the projection of $\mathbf{r}_1$ on $\hat{\mathbf{n}}$, namely $\mathbf{r}_1 . \hat{\mathbf{n}}$, is numerically equal to the perpendicular distance of the plane from the origin. Since this distance is constant we may write (9·7) as

$$\mathbf{r}.\hat{\mathbf{n}} = p \tag{9·9}$$

where p is a constant, numerically equal to the perpendicular distance of the plane from the origin. More generally we may write

$$\mathbf{r}.\mathbf{n} = k, \tag{9·10}$$

where k is a constant.

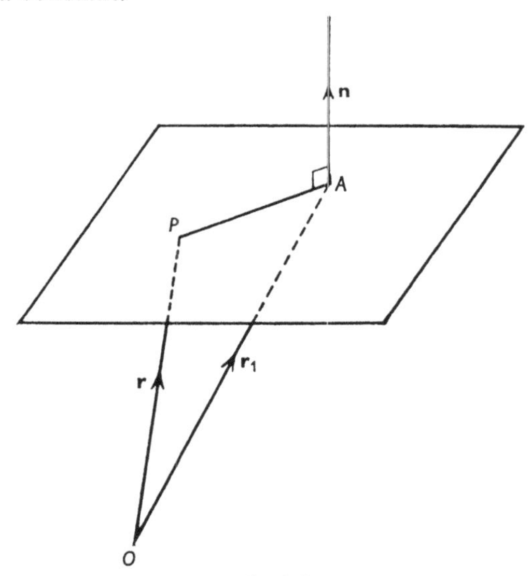

Fig. 9.8

The Cartesian equivalent of (9·9) is

$$lx + my + nz = p, \tag{9·11}$$

where l, m, n are the direction cosines of $\hat{\mathbf{n}}$, that is, of the plane. Equation (9·11) is usually known as the perpendicular form of the equation of the plane.

The Cartesian equivalent of (9·10) is

$$ax + by + cz = k, \tag{9·12}$$

where a, b, c are the direction ratios of the plane.

Equations (9·8), (9·11), (9·12) are linear in the co-ordinates x, y, z. We shall now show that any linear equation in x, y, z represents a plane. The equation

$$ax+by+cz = k$$

may be written in the vector form

$$\mathbf{n}.\mathbf{r} = k \quad (\mathbf{n} \neq \mathbf{0}). \tag{9·13}$$

Consider the point A whose position vector $\mathbf{r}_1$ is given by

$$\mathbf{r}_1 = \frac{k\mathbf{n}}{|\mathbf{n}|^2}.$$

This point obviously satisfies (9·13) since

$$\mathbf{n}.\frac{k\mathbf{n}}{|\mathbf{n}|^2} = k.$$

We may therefore write

$$\mathbf{n}.\mathbf{r}_1 = k. \tag{9·14}$$

Hence from (9·13) and (9·14) we have

$$\mathbf{n}.\mathbf{r} = \mathbf{n}.\mathbf{r}_1,$$

i.e.

$$\mathbf{n}.(\mathbf{r}-\mathbf{r}_1) = 0. \tag{9·15}$$

Equation (9·15) represents the locus of a variable point $P(\mathbf{r})$ such that **AP** is always perpendicular to $\mathbf{n}$, that is, it is the equation of the plane through $A(\mathbf{r}_1)$ with normal-vector $\mathbf{n}$.

Referring to (9·10), (9·12) it is important to notice that given the equation of a plane we may write the normal-vector; for in each case the components of the normal-vector are the coefficients of $\mathbf{r}$ and x, y, z on the left-hand side of the equation. In (9·10) the normal-vector is $\mathbf{n}$ and in (9·12) the normal-vector is (a, b, c).

Of the various forms of the equations of the plane the form

$$\mathbf{n}.\mathbf{r} = k,$$

i.e.

$$ax+by+cz = k$$

is the most useful. The determination of the equation of the plane generally falls into two distinct parts, namely

(*a*) finding the normal-vector,

(*b*) finding the constant k.

Examples

(1) *Find the equation of the plane containing the point* (1, −1, 2) *and perpendicular to the vector* $3\mathbf{i}-2\mathbf{j}+\mathbf{k}$.

Method I. The point $\mathbf{r}_1 = (1, -1, 2)$ lies on the plane. The normal-vector of the plane is $\mathbf{n} = (3, -2, 1)$.

The equation of the plane is given by

$$\mathbf{n}.(\mathbf{r}-\mathbf{r}_1) = 0,$$

i.e.

$$\mathbf{n}.\mathbf{r} = \mathbf{n}.\mathbf{r}_1.$$

Therefore $(3, -2, 1).(x, y, z) = (3, -2, 1).(1, -1, 2)$.

Hence the required equation is

$$3x-2y+z = 7.$$

Method II. The normal-vector of the plane is $\mathbf{n} = (3, -2, 1)$.

The equation of the plane is given by

$$\mathbf{n}.\mathbf{r} = k$$

where k is a constant to be determined.

Since the point (1, −1, 2) satisfies this equation, we have

$$\mathbf{n}.(1, -1, 2) = k.$$

Therefore $(3, -2, 1).(x, y, z) = (3, -2, 1).(1, -1, 2)$.

Hence the required equation is

$$3x-2y+z = 7.$$

Method III. The direction ratios of the normal to the plane are $a = 3, b = -2, c = 1$. The plane contains the point $x_1 = 1, y_1 = -1, z_1 = 2$.

The equation of the plane is given by

$$a(x-x_1)+b(y-y_1)+c(z-z_1) = 0.$$

On substitution, the required equation of the plane is found to be

$$3x-2y+z = 7.$$

Method IV. The direction ratios of the normal to the plane are $a = 3, b = -2, c = 1$.

The equation of the plane is given by

$$ax+by+cz = k,$$

i.e. $$3x-2y+z = k,$$

where k is a constant to be determined.

Since the point $(1, -1, 2)$ satisfies this equation we have

$$k = 7,$$

and the required equation of the plane is therefore

$$3x-2y+z = 7.$$

(2) *Find the co-ordinates of the foot of the perpendicular from the point* $(3, -1, 2)$ *to the plane* $5x-6y-30z = 23$.

The direction-vector of the perpendicular is given by the normal-vector of the plane, that is, by $(5, -6, -30)$. Since the point $(3, -1, 2)$ is on this perpendicular, the vector equation of the perpendicular is

$$\mathbf{r} = (3, -1, 2)+t(5, -6, -30),$$

i.e. $$\mathbf{r} = (3+5t, -1-6t, 2-30t),$$

where t is a parameter.

The equation of the plane may be written

$$(5, -6, -30).\mathbf{r} = 23.$$

Hence the required value of the parameter t for the foot of the perpendicular is given by

$$5(3+5t)-6(-1-6t)-30(2-30t) = 23.$$

On solving we obtain $$t = \tfrac{2}{31}.$$

Hence the point

$$(3+\tfrac{10}{31}, -1-\tfrac{12}{31}, 2-\tfrac{60}{31}) \quad \text{that is} \quad (\tfrac{103}{31}, -\tfrac{43}{31}, \tfrac{2}{31})$$

is the foot of the perpendicular.

(3) *Find the equations of the projection of the line*

$$\frac{x-1}{2} = \frac{y}{-1} = \frac{z+2}{1}$$

on the plane $$2x+y-3z = 9.$$

The point $(1, 0, -2)$ is on the line, The direction-vector of the line is $(2, -1, 1)$. Hence the vector equation of the line is

$$\mathbf{r} = (1, 0, -2)+t(2, -1, 1) = (1+2t, -t, -2+t).$$

Let $P(1+2t, -t, -2+t)$ be any point on the line, and let Q be the foot of the perpendicular from P to the plane. The normal-vector of the plane is $\mathbf{n} = (2, 1, -3)$. Since $\mathbf{PQ}$ and $\mathbf{n}$ are parallel, we may write

$$\mathbf{PQ} = k(2, 1, -3)$$

where k is some constant to be determined.

Now

$$\mathbf{OQ} = \mathbf{OP}+\mathbf{PQ} = (1+2t+2k, \ -t+k, \ -2+t-3k).$$

Since Q lies on the plane

$$2x+y-3z = 9,$$

we have $\quad 2(1+2t+2k)+(-t+k)-3(-2+t-3k) = 9.$

The solution of this equation is

$$k = \tfrac{1}{14}.$$

Hence

$$\mathbf{OQ} = (2t+\tfrac{8}{7}, \ -t+\tfrac{1}{14}, \ t-\tfrac{31}{14}) = (\tfrac{8}{7}, \tfrac{1}{14}, -\tfrac{31}{14})+t(2, -1, 1).$$

Since the locus of Q defines the projection of the line on the plane we obtain as the equations of the projection

$$\frac{x-\frac{8}{7}}{2} = \frac{y-\frac{1}{14}}{-1} = \frac{z+\frac{31}{14}}{1}.$$

Exercise 9(*c*)

(1) Find the equation of the plane which contains the lines

$$\frac{x-1}{-1} = \frac{y-2}{2} = \frac{z-3}{4},$$

$$\frac{x+1}{2} = \frac{y-4}{-2} = \frac{z-2}{1}.$$

(2) Find the equation of the plane containing the points $(1, 1, -2)$, $(2, 5, 3)$, $(-2, -1, 3)$.

(3) Find the equation of the plane containing the point $(-2, -5, 2)$ and the line

$$\frac{x-7}{1} = \frac{y}{3} = \frac{z-1}{-5}.$$

(4) Find the equation of the plane which passes through the point $(2, 1, -3)$ and is normal to the vector $\mathbf{i}-2\mathbf{j}-4\mathbf{k}$.

(5) Show that $\mathbf{r} = \lambda\mathbf{i}+\mu\mathbf{j}+(\lambda a+\mu b+c)\mathbf{k}$

represents the equation of a plane, where a, b, c are constants and λ, μ are parameters.

(6) Find the equation of the plane containing the point (1, 2, 3) and which is normal to the line

$$\frac{x+2}{2} = \frac{y+7}{3} = \frac{z+5}{-2}.$$

Find the point of intersection of the line and the plane.

(7) Find the length of the perpendicular from the point (2, 3, 4) to the plane $3x+4y-12z = 9$.

(8) Show that the length of the perpendicular from the point $\mathbf{r}_1$ to the plane $\mathbf{n}.\mathbf{r} = k$ is given by

$$\frac{|\mathbf{n}.\mathbf{r}_1-k|}{|\mathbf{n}|}.$$

(9) Find the condition for the line $\mathbf{r} = \mathbf{a}+\mathbf{b}t$ to be parallel to the plane $\mathbf{n}.\mathbf{r} = k$. Show that this condition is satisfied by the line

$$\frac{x-4}{2} = \frac{y-3}{5} = \frac{z+1}{-2}$$

and the plane $\qquad x-2y-4z = 5.$

Find the distance of the line from the plane.

(10) Prove that the line $\mathbf{r} = \mathbf{a}+\mathbf{b}t$ intersects the plane $\mathbf{n}.\mathbf{r} = k$ at the point whose position vector is

$$\mathbf{a}+\frac{k-\mathbf{a}.\mathbf{n}}{\mathbf{b}.\mathbf{n}}\mathbf{b},$$

provided $\mathbf{b}.\mathbf{n} \neq 0$.

Deduce that the position vector of the foot of the perpendicular from the point $\mathbf{a}$ to the plane $\mathbf{n}.\mathbf{r} = k$ is given by

$$\mathbf{a}+\frac{k-\mathbf{a}.\mathbf{n}}{\mathbf{n}^2}\mathbf{n}.$$

(11) Find the equations of the projection of the straight line

$$\frac{x+1}{3} = \frac{y-2}{2} = \frac{z-3}{-1}$$

on the plane $\qquad x+y+2z = 4.$

Find the projection of the point (1, 2, 3) on this plane.

(12) The internal bisector of the angle A of a triangle ABC meets the side BC in P. Prove that $BP:PC = AB:AC$.

The angle between two planes

Definition. *The angle between two planes is defined as the angle between two lines, one drawn in each plane, at right angles to the line of intersection of the two planes.*

Let π_1 and π_2 be two planes and AB their line of intersection, Fig. 9.9(i). P_1O and P_2O are lines in π_1 and π_2 respectively, such that each is perpendicular to AB. Then angle P_1OP_2 is the angle between

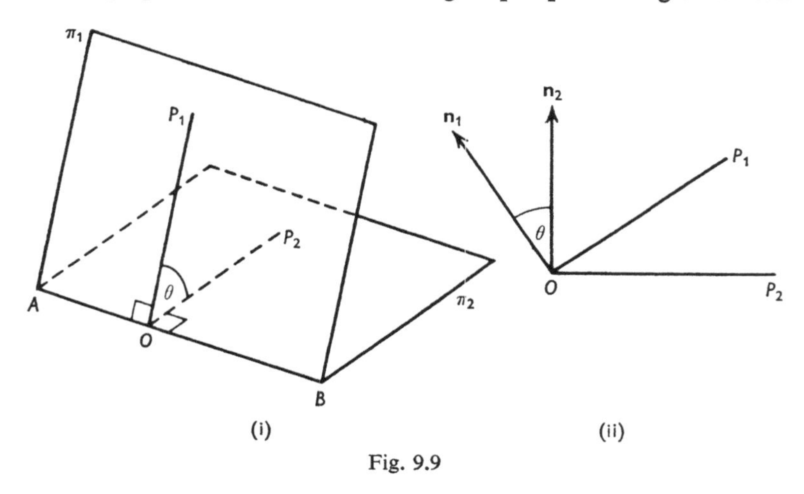

Fig. 9.9

the planes. In Fig. 9.9(ii) P_1O and P_2O are the intersections of the planes with the plane of the paper, the line of intersection AOB being perpendicular to the plane of the paper. The normal-vectors $\mathbf{n}_1$, $\mathbf{n}_2$ of the planes then lie in the plane of the paper. The angle between $\mathbf{n}_1$ and $\mathbf{n}_2$ is clearly seen to be equal to the angle between P_1O and P_2O. Hence we have the alternative definition:

The angle between two planes is defined as the angle between their normal-vectors.

The angle θ between the planes $\mathbf{n}_1 . \mathbf{r} = k_1$ and $\mathbf{n}_2 . \mathbf{r} = k_2$ is given by

$$\cos\theta = \frac{\mathbf{n}_1 . \mathbf{n}_2}{|\mathbf{n}_1||\mathbf{n}_2|}.$$

If the planes are parallel we have

$$\mathbf{n}_1 = \lambda\mathbf{n}_2,$$

where λ is a constant and if the planes are perpendicular we have

$$\mathbf{n}_1 . \mathbf{n}_2 = 0.$$

The angle between a line and a plane

Definition. *The angle between a line and a plane is defined as the angle between the line and the projection of the line on the plane.*

In Fig. 9.10, AB is the line, AN is its projection on the plane, θ is

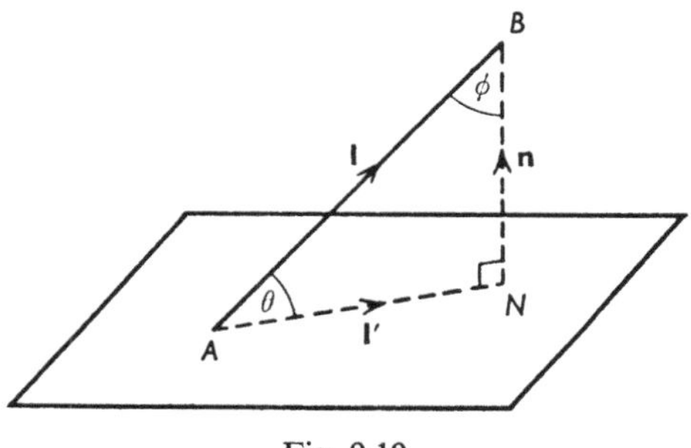

Fig. 9.10

the required angle and ϕ is the angle between the line and the perpendicular BN to the plane, Hence if $\mathbf{l}$, $\mathbf{n}$ are the direction-vector and normal-vector of the line and plane respectively, we have

$$\cos\phi = \frac{\mathbf{l}.\mathbf{n}}{|\mathbf{l}|\,|\mathbf{n}|}.$$

But
$$\phi = \tfrac{1}{2}\pi - \theta.$$

Hence
$$\sin\theta = \frac{\mathbf{l}.\mathbf{n}}{|\mathbf{l}|\,|\mathbf{n}|}.$$

When $\theta = 0$, the line is parallel to the plane and we have

$$\mathbf{l}.\mathbf{n} = 0 \quad \text{(line parallel to plane)}.$$

When $\theta = \frac{1}{2}\pi$, the line is perpendicular to the plane and this occurs when

$$\mathbf{l} = \lambda\mathbf{n} \quad \text{(line perpendicular to plane)},$$

where λ is a parameter.

Exercise 9(d)

(1) Find the acute angle between the planes

(i) $2x-y+z+8 = 0, \quad 3x+2y-z-1 = 0,$

(ii) $5x-y-2z = 5, \quad y = 0.$

(2) Find the acute angle between the line

$$\frac{y-1}{3} = \frac{y+7}{4} = \frac{z-5}{12}$$

and the plane $\qquad 2x-6y+3z = 8.$

(3) A plane makes an angle of 60° with the line $x = y = z$ and 45° with the line $x = 0$, $y = z$. Find the angle which it makes with the plane $x = 0$.

10

OTHER LOCI

Introduction

In this chapter we shall consider the vector equations of some two-dimensional loci (circle, parabola, ellipse, rectangular hyperbola) and of some three-dimensional loci (circular helix, sphere).

After the development of the vector product, further consideration of the line and plane is made in chapters 11 and 12, and for this reason parts of this chapter may be postponed if desired.

The circle

Definition. *The locus of a point which moves in a plane, so that its distance from a fixed point in the plane is constant, is a circle.*

The fixed point is known as the centre and the constant distance is known as the radius,

Consider the circle in the x–y plane, with centre at the origin O and radius a. Let P be a point on the circle, and let the position vector of P relative to O be $\mathbf{r}$. If OP makes an angle θ with the x-axis (Fig. 10.1), we have

$$\mathbf{OP} = |\mathbf{r}|\cos\theta\mathbf{i}+|\mathbf{r}|\sin\theta\mathbf{j},$$

that is,

$$\mathbf{r} = a\cos\theta\mathbf{i}+a\sin\theta\mathbf{j}. \qquad (10{\cdot}1)$$

This is, in terms of the parameter θ, the vector equation of the circle in the x–y plane, with centre at the origin and radius a.

This equation holds for all values of θ. For

$$|\mathbf{r}|^2 = a^2\cos^2\theta+a^2\sin^2\theta,$$

that is,

$$|\mathbf{r}| = a,$$

which is independent of θ, showing that equation (10·1) represents a circle for all values of θ.

Now consider the circle in space, radius a, centre C and in the plane containing the perpendicular unit vectors $\hat{\mathbf{u}}$ and $\hat{\mathbf{v}}$ (Fig. 10.2).

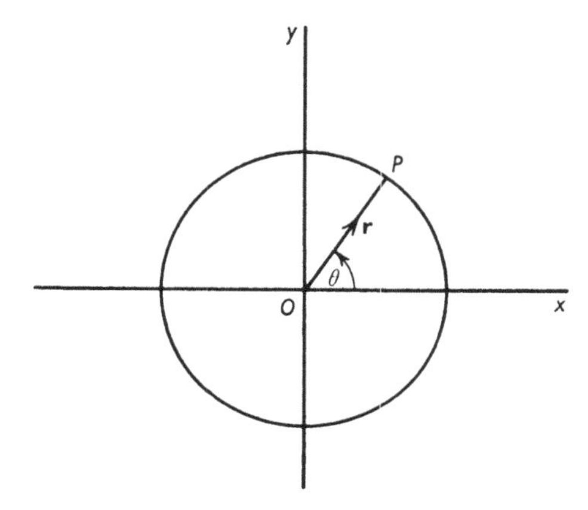

Fig. 10.1

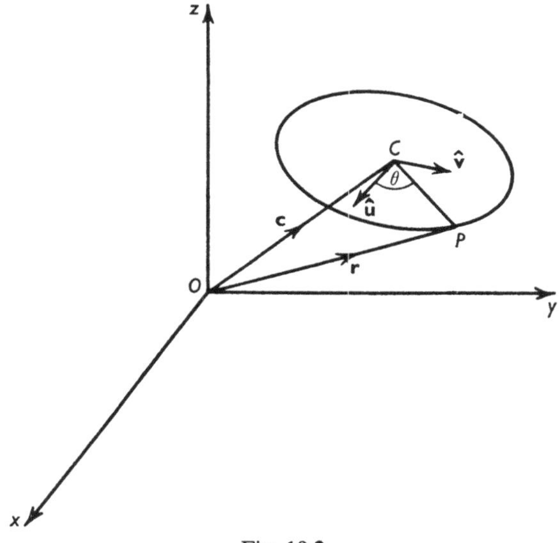

Fig. 10.2

Let P be any point on the circle and $\mathbf{c}$, $\mathbf{r}$ the position vectors of C, P relative to the origin O. Let θ be the angle between $\mathbf{CP}$ and $\hat{\mathbf{u}}$, Then

$$\mathbf{CP} = a\cos\theta\hat{\mathbf{u}} + a\sin\theta\hat{\mathbf{v}}.$$

But $$\mathbf{OP} = \mathbf{OC} + \mathbf{CP}.$$

Therefore $$\mathbf{r} = \mathbf{c} + a\cos\theta\hat{\mathbf{u}} + a\sin\theta\hat{\mathbf{v}}. \tag{10·2}$$

This is the vector equation, in terms of the parameter θ, of the circle in space, centre $\mathbf{c}$, radius a and in the plane containing the perpendicular unit vectors $\hat{\mathbf{u}}$ and $\hat{\mathbf{v}}$.

The vector equation of a circle, in a non-parametric form, will now be considered.

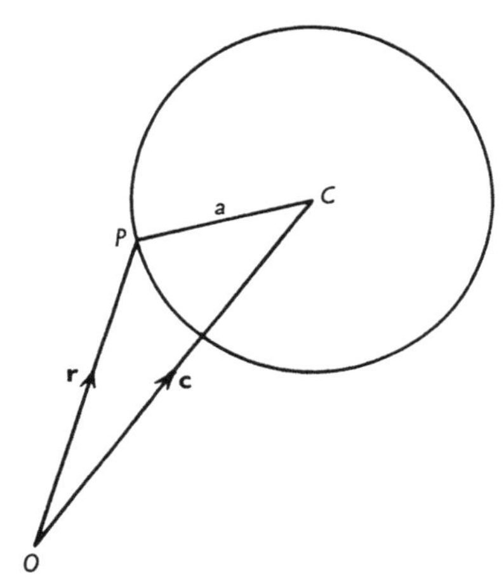

Fig. 10.3

Consider a circle centre C and radius a (Fig. 10.3). Let P be any point on the circumference and $\mathbf{c}$, $\mathbf{r}$ the position vectors of C, P relative to an origin O contained in the plane of the circle. From the definition

$$CP^2 = a^2.$$

Now $$\mathbf{CP} = \mathbf{r} - \mathbf{c}.$$

Therefore $$(\mathbf{r} - \mathbf{c})^2 = a^2 \quad (\text{since } \mathbf{CP}^2 = CP^2),$$

that is, $$\mathbf{r}^2 - 2\mathbf{r}.\mathbf{c} + \mathbf{c}^2 = a^2. \tag{10·3}$$

This is the equation of the circle relative to the origin O. Note that in this equation the position vectors $\mathbf{r}$, $\mathbf{c}$ are in the plane of the circle.

If we now choose a two-dimensional co-ordinate system with O as origin and the Ox and Oy axes in the plane of the circle, we may write

$$\mathbf{r} = (x, y), \quad \mathbf{c} = (h, k).$$

Equation (10·3) now becomes

$$x^2+y^2-2(xh+yk)+h^2+k^2 = a^2,$$

i.e.
$$(x-h)^2+(y-k)^2 = a^2, \tag{10·4}$$

which is the well-known form of the equation of the circle with centre at the point (h, k) and with radius a.

When the centre C of the circle is at the origin O we have $\mathbf{c} = \mathbf{0}$ and equation (10·3) reduces to

$$\mathbf{r}^2 = a^2. \tag{10·5}$$

The Cartesian counterpart of (10·5) is

$$x^2+y^2 = a^2.$$

Example

$P(\mathbf{r})$ is a variable point. If

$$\mathbf{r} = (1+4\cos\theta-6\sin\theta/\surd 5)\mathbf{i}+(-2-4\cos\theta+15\sin\theta/\surd 5)\mathbf{j}$$
$$+(3+7\cos\theta+12\sin\theta/\surd 5)\mathbf{k},$$

show that the locus of P is a circle. Find the centre, radius and the equation of the plane of the circle.

The equation can be written

$$\mathbf{r} = (\mathbf{i}-2\mathbf{j}+3\mathbf{k})+\cos\theta(4\mathbf{i}-4\mathbf{j}+7\mathbf{k})+3\sin\theta\left(-\frac{2}{\surd 5}\mathbf{i}+\frac{5}{\surd 5}\mathbf{j}+\frac{4}{\surd 5}\mathbf{k}\right)$$
$$= (\mathbf{i}-2\mathbf{j}+3\mathbf{k})+9\cos\theta(\tfrac{4}{9}\mathbf{i}-\tfrac{4}{9}\mathbf{j}+\tfrac{7}{9}\mathbf{k})$$
$$+9\sin\theta\left(-\frac{2}{3\surd 5}\mathbf{i}+\frac{5}{3\surd 5}\mathbf{j}+\frac{4}{3\surd 5}\mathbf{k}\right).$$

Let $C(\mathbf{c})$ be the point $(1, -2, 3)$ and $\hat{\mathbf{u}}$, $\hat{\mathbf{v}}$ be the unit vectors

$$(\tfrac{4}{9}\mathbf{i}-\tfrac{4}{9}\mathbf{j}+\tfrac{7}{9}\mathbf{k}), \quad \left(-\frac{2}{3\surd 5}\mathbf{i}+\frac{5}{3\surd 5}\mathbf{j}+\frac{4}{3\surd 5}\mathbf{k}\right)$$

respectively. Then the equation can be written

$$\mathbf{r}-\mathbf{c} = 9\cos\theta\hat{\mathbf{u}}+9\sin\theta\hat{\mathbf{v}},$$

that is
$$\mathbf{CP} = 9\cos\theta\hat{\mathbf{u}}+9\sin\theta\hat{\mathbf{v}}.$$

Hence **CP** is in the plane containing C and parallel to the plane containing the unit vectors $\hat{\mathbf{u}}$, $\hat{\mathbf{v}}$. Furthermore, since C is a fixed point and $|\mathbf{CP}| = 9$, the locus of P is a circle, centre C and radius 9 units.

Let $a\mathbf{i}+b\mathbf{j}+c\mathbf{k}$ be a normal-vector of the plane of the circle. Then since the normal-vector is perpendicular to each of the unit vectors, we have

$$(4\mathbf{i}-4\mathbf{j}+7\mathbf{k}).(a\mathbf{i}+b\mathbf{j}+c\mathbf{k}) = 0$$

and
$$(-2\mathbf{i}+5\mathbf{j}+4\mathbf{k}).(a\mathbf{i}+b\mathbf{j}+c\mathbf{k}) = 0.$$

Therefore $\quad 4a-4b+7c = 0 \quad$ and $\quad -2a+5b+4c = 0.$

From these we obtain $\quad a = \frac{17}{10}b, \quad c = -\frac{2}{5}b.$

Hence $17\mathbf{i}+10\mathbf{j}-4\mathbf{k}$ is a normal-vector of the plane.

The equation of the plane is

$$17x+10y-4z = k.$$

But the point $(1, -2, 3)$ is in the plane. Hence the equation of the plane is

$$17x+10y-4z = 17(1)+10(-2)-4(3),$$

that is
$$17x+10y-4z = -15.$$

The parabola

Definition. *The locus of a point which moves in a plane so that its distance from a fixed point (the focus) is equal to its distance from a fixed straight line (the directrix) is a parabola.*

Consider the parabola in the x–y plane, with vertex at the origin O and focus at the point $S(a, 0)$, Fig. 10.4. The Cartesian equation of the parabola is $y^2 = 4ax$. The co-ordinates of any point $P(\mathbf{r})$ on the parabola may be written $(at^2, 2at)$, where t is a parameter. Then

$$\mathbf{r} = at^2\mathbf{i}+2at\mathbf{j}$$

is the vector equation, in terms of the parameter t, of the parabola in the x–y plane, with vertex at the origin and focus at $(a, 0)$.

Again, if $P(\mathbf{r})$ is any point on a parabola in space with vertex at the point $C(\mathbf{c})$ and $\hat{\mathbf{u}}$, $\hat{\mathbf{v}}$ are perpendicular unit vectors in the plane of the parabola, $\hat{\mathbf{u}}$ being parallel to the axis of the parabola (Fig. 10.5), we have

$$\mathbf{r} = \mathbf{c}+at^2\hat{\mathbf{u}}+2at\hat{\mathbf{v}}. \tag{10·6}$$

This is the vector equation, in terms of the parameter t, of the parabola in space, vertex $\mathbf{c}$, latus rectum $4a$ and in the plane con-

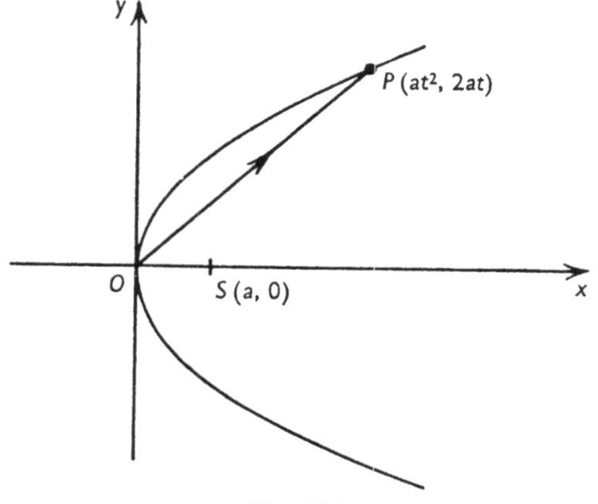

Fig. 10.4

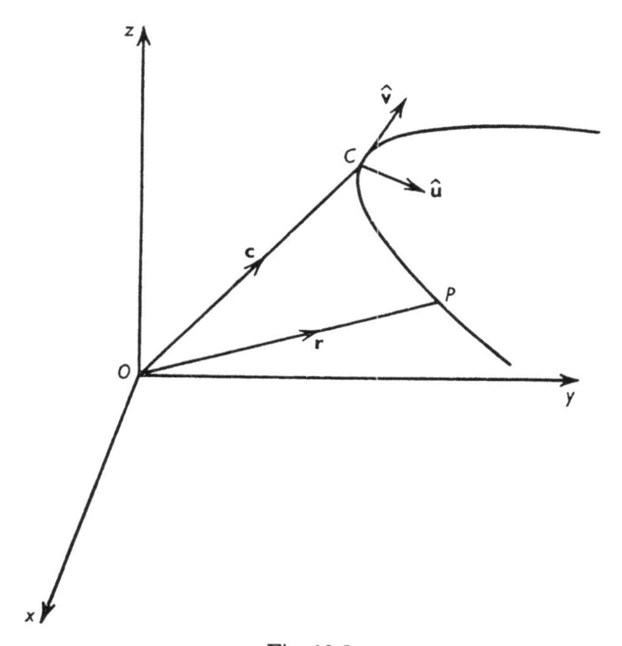

Fig. 10.5

taining the perpendicular unit vectors $\hat{\mathbf{u}}$, $\hat{\mathbf{v}}$, the direction of $\hat{\mathbf{u}}$ being that of the axis of the parabola.

Alternatively, starting from the definition of a parabola, we can obtain its equation in a non-parametric form.

In Fig. 10.6, let A be the focus, DD' the directrix and AK the perpendicular from A to DD'. Take as origin O the midpoint of AK. Obviously O is a point on the locus. Let $P(\mathbf{r})$ be any point on the

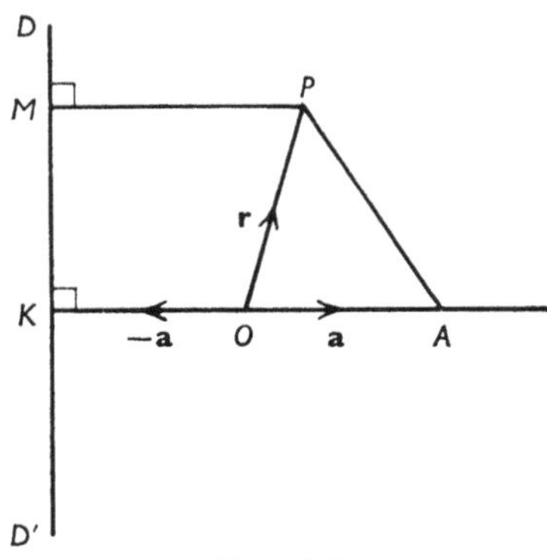

Fig. 10.6

locus and let PM be the perpendicular to DD'. We have from the definition

$$MP = AP,$$

that is, $$\mathbf{MP}^2 = \mathbf{AP}^2.$$

Now $$MP = KO + \text{projection of } OP \text{ on } OA.$$

Therefore $$\mathbf{MP} = \mathbf{a} + \frac{(\mathbf{r}\,.\,\mathbf{a})}{\mathbf{a}^2}\,\mathbf{a},$$

where $\mathbf{OA} = \mathbf{a}$.

Also $$\mathbf{AP} = \mathbf{r} - \mathbf{a}.$$

Since $\mathbf{MP}^2 = \mathbf{AP}^2$ we have

$$\left\{\mathbf{a} + \frac{(\mathbf{r}\,.\,\mathbf{a})}{\mathbf{a}^2}\,\mathbf{a}\right\}^2 = (\mathbf{r} - \mathbf{a})^2.$$

Expanding $$\mathbf{a}^2 + 2\mathbf{r}\,.\,\mathbf{a} + \frac{(\mathbf{r}\,.\,\mathbf{a})^2}{\mathbf{a}^2} = \mathbf{r}^2 - 2\mathbf{r}\,.\,\mathbf{a} + \mathbf{a}^2,$$

giving $$\mathbf{r}^2 = 4\mathbf{r}\,.\,\mathbf{a} + \frac{(\mathbf{r}\,.\,\mathbf{a})^2}{\mathbf{a}^2}. \tag{10·7}$$

This is the equation of the parabola whose vertex is at the origin. In Cartesian form taking

$$\mathbf{r} = (x, y) \quad \text{and} \quad \mathbf{a} = (a, 0)$$

we have
$$x^2+y^2 = 4ax+\frac{a^2x^2}{a^2},$$

i.e.
$$y^2 = 4ax, \tag{10·8}$$

which is the standard form of the equation of the parabola with its vertex at the origin and focus at the point $(a, 0)$.

The parametric equation equivalent to (10·7) is

$$\mathbf{r} = at^2\mathbf{i}+2at\mathbf{j}$$

and the parametric equations equivalent to (10·8) are

$$x = at^2, \quad y = 2at,$$

where t is a scalar parameter.

Example

Show that the curve $\mathbf{r} = bt\mathbf{i}+(c+dt)t\mathbf{j}$

is a parabola, where b, c, d are constants and t is a parameter.

The equation of the curve is

$$\begin{aligned}\mathbf{r} &= bt\mathbf{i}+(dt^2+ct)\mathbf{j}\\ &= bt\mathbf{i}+d\left(t^2+\frac{c}{d}t\right)\mathbf{j}\\ &= bt\mathbf{i}+d\left(t+\frac{c}{2d}\right)^2\mathbf{j}-\frac{c^2}{4d}\mathbf{j}.\end{aligned}$$

Let $s = t+(c/2d)$. Therefore

$$\begin{aligned}\mathbf{r} &= \left(s-\frac{c}{2d}\right)b\mathbf{i}+ds^2\mathbf{j}-\frac{c^2}{4d}\mathbf{j}\\ &= \left(-\frac{bc}{2d}\mathbf{i}-\frac{c^2}{4d}\mathbf{j}\right)+bs\mathbf{i}+ds^2\mathbf{j}\\ &= \left(-\frac{bc}{2d}\mathbf{i}-\frac{c^2}{4d}\mathbf{j}\right)+2\left(\frac{b^2}{4d}\right)\left(\frac{2sd}{b}\right)\mathbf{i}+\frac{b^2}{4d}\left(\frac{2sd}{b}\right)^2\mathbf{j}\\ &= \mathbf{v}+2ap\mathbf{i}+ap^2\mathbf{j},\end{aligned}$$

where $\mathbf{v}$ is the constant vector

$$-\frac{bc}{2d}\mathbf{i}-\frac{c^2}{4d}\mathbf{j},$$

a the constant scalar $b^2/4d$ and p the parameter

$$\frac{2sd}{b} = \frac{2d}{b}\left(t+\frac{c}{2d}\right).$$

Thus the curve is a parabola, in the x–y plane, vertex at the point $(-bc/2d, -c^2/4d)$, latus rectum b^2/d, axis parallel to the y-axis.

Alternatively, we have

$$x = bt \quad\text{and}\quad y = ct+dt^2.$$

Eliminating t we obtain

$$\left(x+\frac{bc}{2d}\right)^2 = \frac{b^2}{d}\left(y+\frac{c^2}{4d}\right)$$

which is the Cartesian equation of the parabola.

The path of a projectile

It is well known that the locus of a projectile moving freely under uniform gravity only is a parabola. We shall obtain the vector equation of this locus and show it is a parabola.

First of all consider the motion of a particle under constant acceleration $\mathbf{f}$. Its equation of motion is

$$\ddot{\mathbf{r}} = \mathbf{f},$$

where $\mathbf{r}$ is its position vector at any time t. Integrating with respect to t gives

$$\dot{\mathbf{r}} = t\mathbf{f}+\mathbf{u},$$

where $\mathbf{u}$ is the arbitrary constant representing the initial velocity, i.e. when $t = 0$, $\dot{\mathbf{r}} = \mathbf{u}$. Integrating again gives

$$\mathbf{r} = \tfrac{1}{2}t^2\mathbf{f}+t\mathbf{u} \tag{10·9}$$

the arbitrary constant being the zero vector since $\mathbf{r} = \mathbf{0}$ when $t = 0$.

Equation (10·9) is the locus of a particle moving with constant acceleration. In general as we move from point to point gravity varies both in direction and magnitude. We shall here consider a projectile moving under uniform gravity, that is, the acceleration of gravity **g** is constant in direction and magnitude throughout the motion of the projectile. This being so we have

$$\mathbf{f} = \mathbf{g} = -|\mathbf{g}|\mathbf{j}$$

taking **j** to be the unit vector in the direction of the upward vertical. The initial velocity **u** in terms of components is given by

$$\mathbf{u} = |\mathbf{u}|\cos\alpha\mathbf{i} + |\mathbf{u}|\sin\alpha\mathbf{j},$$

where α is the angle of projection measured from the horizontal (Fig. 10.7).

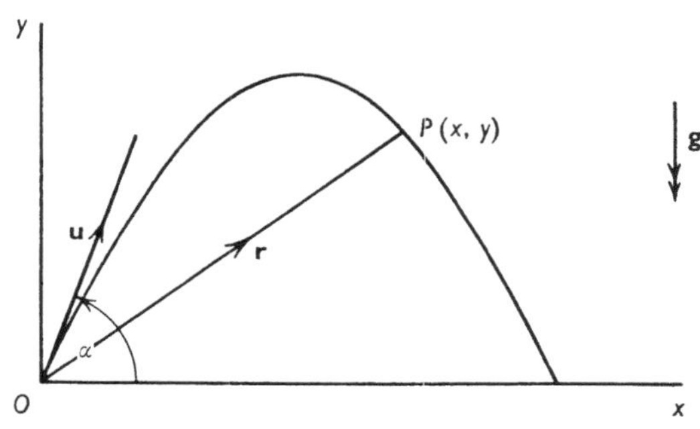

Fig. 10.7

Hence
$$\mathbf{r} = -\tfrac{1}{2}|\mathbf{g}|t^2\mathbf{j} + |\mathbf{u}|t\cos\alpha\mathbf{i} + |\mathbf{u}|t\sin\alpha\mathbf{j}$$
$$= |\mathbf{u}|t\cos\alpha\mathbf{i} + (|\mathbf{u}|\sin\alpha - \tfrac{1}{2}|\mathbf{g}|t)t\mathbf{j}.$$

This is the equation of the path of the projectile in terms of the time-parameter t.

Proceeding as in the example on page 169 we obtain

$$\mathbf{r} = \left(\frac{u^2\sin\alpha\cos\alpha}{g}\mathbf{i} + \frac{u^2\sin^2\alpha}{2g}\mathbf{j}\right) + 2\left(-\frac{u^2\cos^2\alpha}{2g}\right)\left(-\frac{gs}{u\cos\alpha}\right)\mathbf{i}$$
$$+\left(-\frac{u^2\cos^2\alpha}{2g}\right)\left(-\frac{gs}{u\cos\alpha}\right)^2\mathbf{j},$$

where $u = |\mathbf{u}|$.

Thus the trajectory is a parabola, vertex at the point

$$\left(\frac{u^2\sin\alpha\cos\alpha}{g}, \frac{u^2\sin^2\alpha}{2g}\right), \quad \text{semi-latus rectum} \quad \frac{u^2\cos^2\alpha}{g}$$

and axis in the direction of the negative y-axis.

By eliminating t from the parametric equations

$$x = ut\cos\alpha, \quad y = (u\sin\alpha - \tfrac{1}{2}gt)t$$

we obtain the Cartesian form of the equation of the parabola, namely

$$y = x\tan\alpha - \frac{gx^2}{2u^2\cos^2\alpha}.$$

This may be written as

$$y - \frac{u^2\sin^2\alpha}{2g} = -\frac{g}{2u^2\cos^2\alpha}\left(x - \frac{u^2\sin\alpha\cos\alpha}{g}\right)^2.$$

The ellipse

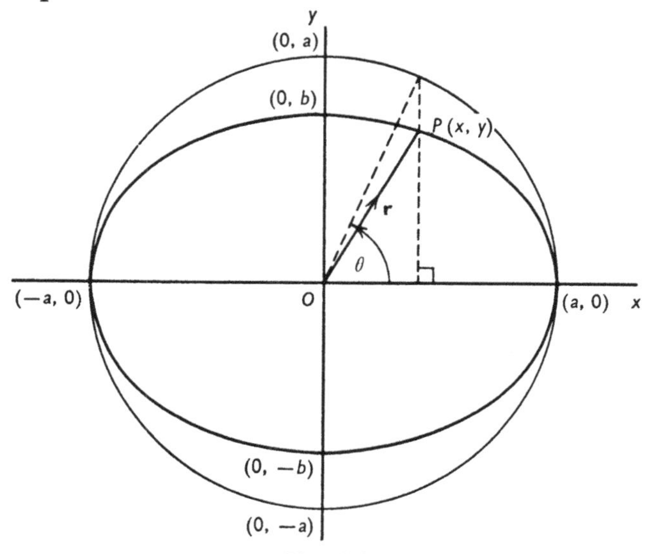

Fig. 10.8

The Cartesian form of the equation of an ellipse with centre at the origin and major and minor axes $2a$ and $2b$ respectively is

$$\frac{x^2}{a^2} + \frac{y^2}{b^2} = 1.$$

The equivalent parametric equations are

$$x = a\cos\theta, \quad y = b\sin\theta,$$

where θ is a parameter (Fig. 10.8). Hence if $\mathbf{r}$ is the position vector relative to the origin of a point P on the ellipse we may write

$$\mathbf{r} = a\cos\theta\mathbf{i} + b\sin\theta\mathbf{j}.$$

The rectangular hyperbola

The Cartesian form of the equation of a rectangular hyperbola whose centre is at the origin is

$$xy = c^2$$

and in terms of the parameter p this is equivalent to

$$x = cp, \quad y = \frac{c}{p}.$$

Thus the position vector $\mathbf{r}$ relative to the origin of any point on the rectangular hyperbola is given by

$$\mathbf{r} = cp\mathbf{i} + \frac{c}{p}\mathbf{j}.$$

The sphere

Definition. *The locus of a point which moves in space so that its distance from a fixed point is constant is a sphere.*

The terms centre and radius have the same meaning as for a circle.

Let P be any point on the sphere, C the centre and a the radius. If $\mathbf{r}$ and $\mathbf{c}$ are the position vectors of P and C relative to an origin O we have

$$\mathbf{CP} = \mathbf{r} - \mathbf{c}.$$

Since

$$|\mathbf{CP}| = a$$

we have

$$(\mathbf{r} - \mathbf{c})^2 = a^2,$$

i.e.

$$\mathbf{r}^2 - 2\mathbf{r}.\mathbf{c} + \mathbf{c}^2 = a^2. \qquad (10{\cdot}10)$$

This is the vector equation of the sphere with centre at $C(\mathbf{c})$ and radius a. It will be noted that this equation is identical in form to equation (10·3) obtained for the circle. However, its Cartesian equivalent is different since for a sphere we are in three-dimensional space. For the sphere we have

$$\mathbf{r} = (x, y, z), \quad \mathbf{c} = (x', y', z')$$

and equation (10·10) becomes

$$x^2 + y^2 + z^2 - 2(xx' + yy' + zz') + x'^2 + y'^2 + z'^2 = a^2,$$

i.e.

$$(x - x')^2 + (y - y')^2 + (z - z')^2 = a^2. \qquad (10{\cdot}11)$$

This is the standard Cartesian form of the equation of a sphere with centre at the point (x', y', z') and radius a.

In the special case of the origin and centre being the same, equations (10·10) and (10·11) reduce to

$$\mathbf{r}^2 = a^2$$

and $$x^2+y^2+z^2 = a^2.$$

Equation (10·11) can be more generally written as

$$x^2+y^2+z^2+2ux+2vy+2wz+d = 0.$$

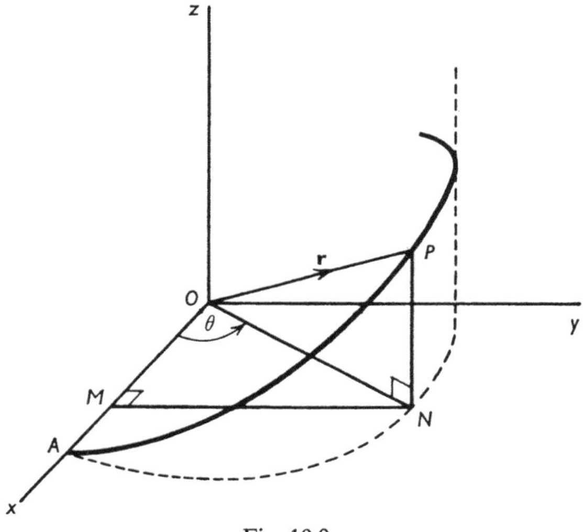

Fig. 10.9

The circular helix

Definition. *A helix is a curve drawn on a cylinder so as to cut the generators at a constant angle. If the cylinder is right-circular then the locus is a circular helix.*

An example of a circular helix is the thread of a bolt. Also, in the theory of the motion of a charged particle in a uniform and constant magnetic field, it may be shown that in general the orbit described by the particle is a helix. (See page 269.)

In Fig. 10.9 the axis of the cylinder is taken as the z-axis. Let A be any point on the helix and take OA as the x-axis. Then the projection of the helix on the x–y plane is a circle. Let P be any other point on the helix and let $\mathbf{r}$ be its position vector relative to the origin O. If

PN is the perpendicular to the x–y plane and *NM* is the perpendicular to the x-axis we have

$$\mathbf{r} = OM\mathbf{i}+MN\mathbf{j}+NP\mathbf{k}.$$

Let angle *MON* be θ and the radius of the cylinder be a. Then

$$ON = a, \quad OM = a\cos\theta \quad \text{and} \quad MN = a\sin\theta.$$

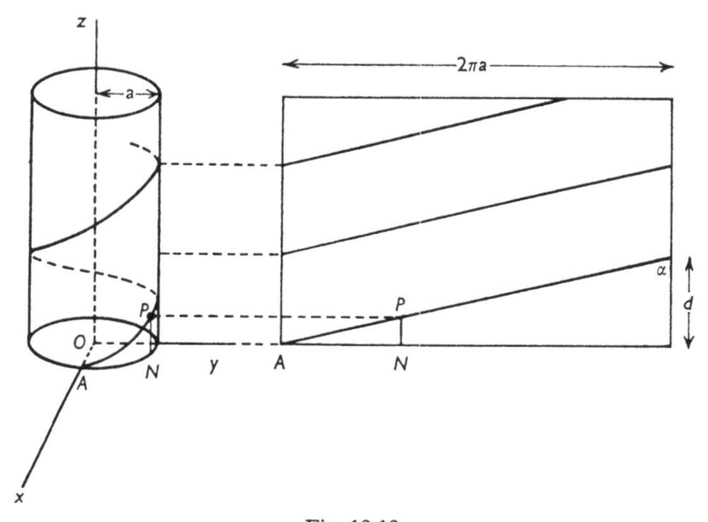

Fig. 10.10

If the cylindrical surface on which a circular helix is drawn is unwrapped on to a plane the helix appears as a set of parallel line segments, each making an angle α with the direction of the generating lines in the plane, where α is the angle of inclination of the helix to the generator (Fig. 10.10). From this it is apparent that *NP* is proportional to θ. We have

$$\operatorname{arc} AN = a\theta.$$

Therefore $$NP = a\theta\cot\alpha.$$

Alternatively $$\frac{NP}{\theta} = \frac{d}{2\pi}$$

$$= \frac{2\pi a\cot\alpha}{2\pi}.$$

Therefore $$NP = a\theta\cot\alpha.$$

Thus the equation of the circular helix is

$$\mathbf{r} = a\cos\theta\mathbf{i}+a\sin\theta\mathbf{j}+(a\theta\cot\alpha)\mathbf{k},$$

where θ is a scalar parameter.

Tangent to a curve

If $$\mathbf{r} = f(t)\mathbf{i}+\phi(t)\mathbf{j}+\psi(t)\mathbf{k}$$

is the vector equation of a curve, then

$$\frac{d\mathbf{r}}{dt} = f'(t)\mathbf{i}+\phi'(t)\mathbf{j}+\psi'(t)\mathbf{k}$$

is a vector in the direction of the tangent to the curve at the point $\mathbf{r}$, parameter t. (See chapter 7, page 88.) Thus the direction-vector of the tangent is given by $d\mathbf{r}/dt$.

Example

Find the equation of the tangent to the curve

$$\mathbf{r} = (6\cos t,\ 6\sin t,\ 5t^2)$$

at the point $t = \frac{1}{3}\pi$.

The direction-vector of the tangent at the point parameter t is given by

$$\frac{d\mathbf{r}}{dt} = (-6\sin t,\ 6\cos t,\ 10t).$$

Hence the direction-vector of the tangent at the point $t = \frac{1}{3}\pi$ is

$$(-6\sin\tfrac{1}{3}\pi,\ 6\cos\tfrac{1}{3}\pi,\ \tfrac{10}{3}\pi),$$

that is, $$(-3\sqrt{3},\ 3,\ \tfrac{10}{3}\pi).$$

The equation of the tangent at the point $t = \frac{1}{3}\pi$ is

$$\frac{x-6\cos\frac{1}{3}\pi}{-3\sqrt{3}} = \frac{y-6\sin\frac{1}{3}\pi}{3} = \frac{z-5(\frac{1}{3}\pi)^2}{\frac{10}{3}\pi},$$

that is $$\frac{x-3}{-\sqrt{3}} = \frac{y-3\sqrt{3}}{1} = \frac{9z-5\pi^2}{10\pi}.$$

Change of co-ordinate system

A co-ordinate system is fixed by (*a*) the choice of origin and (*b*) the choice of the direction and sense of the axes. The latter is specified by three unit vectors. These unit vectors are sometimes called the base-vectors and in the Cartesian rectangular co-ordinate system we have denoted them by **i**, **j**, **k** and their direction and sense give the positive direction and sense of the x-, y-, z-axes. We may change our co-ordinate system in three ways:

(1) a change in the origin from O to O', keeping the base-vectors **i**, **j**, **k** unchanged. In this transformation the axes undergo a translation;

(2) a change in the base-vectors from **i**, **j**, **k** to **i**$'$, **j**$'$, **k**$'$ keeping the origin O unchanged. In this transformation the axes are rotated as a rigid framework about the origin;

(3) a change in both the origin from O to O' and the base-vectors from **i**, **j**, **k** to **i**$'$, **j**$'$, **k**$'$. In this change we have both a translation and a rotation and it may be effected by changing the origin from O to O' keeping the same base-vectors and then changing the base-vectors from **i**, **j**, **k** to **i**$'$, **j**$'$, **k**$'$ keeping O' fixed.

Change of origin

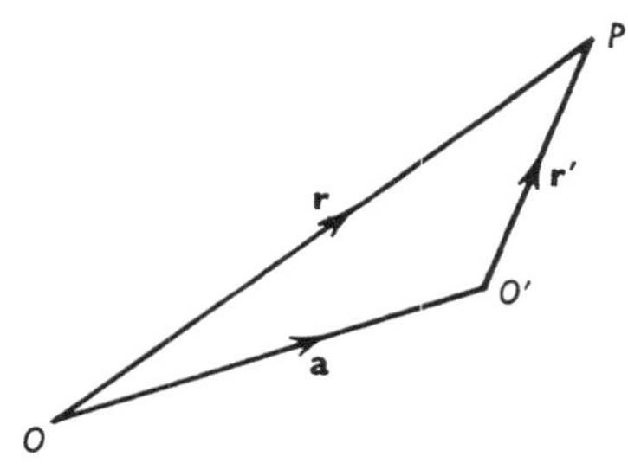

Fig. 10.11

In Fig. 10.11 let O be the original origin and O' be the new origin and the position vector of O' relative to O be $\mathbf{a} = (a, b, c)$. Suppose P is a point whose position vectors relative to O and O' are $\mathbf{r} = (x, y, z)$ and $\mathbf{r}' = (x', y', z')$ respectively. Then

$$\mathbf{r}' = \mathbf{r} - \mathbf{a}. \qquad (10{\cdot}12)$$

This equation expresses the change in position vector of the point P due to the change of origin. In component form (10·12) becomes

$$(x', y', z') = (x, y, z)-(a, b, c),$$

that is, $\quad x' = x-a, \quad y' = y-b, \quad z' = z-c.$

Example

Interpret the locus $\mathbf{r}^2-2\mathbf{r}.\mathbf{c}+\mathbf{c}^2 = a^2$ *where* $\mathbf{r}$ *is the position vector of a variable point* P, $\mathbf{c}$ *is the position vector relative to the origin* O *of a fixed point* C, *and* a *is a constant.*

The equation
$$\mathbf{r}^2-2\mathbf{r}.\mathbf{c}+\mathbf{c}^2 = a^2$$

may be written
$$(\mathbf{r}-\mathbf{c})^2 = a^2.$$

Transfer the origin to C and let $\mathbf{r}'$ be the position vector of P relative to C. Then
$$\mathbf{r}' = \mathbf{r}-\mathbf{c},$$
and the equation becomes
$$(\mathbf{r}')^2 = a^2,$$

that is,
$$|\mathbf{r}'| = a.$$

Thus the locus is a sphere of radius a and centre C.

Rotation of axes

Let $Oxyz$ be the original co-ordinate system and $P(x, y, z)$ be a point whose position vector relative to O is $\mathbf{r}$. Let $\mathbf{i}, \mathbf{j}, \mathbf{k}$ be the unit vectors in the direction of the axes Ox, Oy, Oz respectively. Then

$$\mathbf{r} = x\mathbf{i}+y\mathbf{j}+z\mathbf{k}.$$

But $\quad x = \mathbf{i}.\mathbf{r}, \quad y = \mathbf{j}.\mathbf{r}, \quad z = \mathbf{k}.\mathbf{r},$

Therefore $\quad \mathbf{r} = (\mathbf{i}.\mathbf{r})\mathbf{i}+(\mathbf{j}.\mathbf{r})\mathbf{j}+(\mathbf{k}.\mathbf{r})\mathbf{k}.$

We now rotate the axes as a rigid framework about O so that the new co-ordinate system is $Ox'y'z'$. Let $\mathbf{i}', \mathbf{j}', \mathbf{k}'$ be the unit vectors in the direction of the new axes Ox', Oy', Oz' respectively. Suppose

$$\mathbf{i}' = (l_1, m_1, n_1), \quad \mathbf{j}' = (l_2, m_2, n_2), \quad \mathbf{k}' = (l_3, m_3, n_3)$$

where the components are in terms of the original system. Then the point P will have co-ordinates x', y', z' in the new system given by

$$x' = \mathbf{i}'.\mathbf{r}, \quad y' = \mathbf{j}'.\mathbf{r}, \quad z' = \mathbf{k}'.\mathbf{r}.$$

Therefore

$$x' = l_1x+m_1y+n_1z, \quad y' = l_2x+m_2y+n_2z, \quad z' = l_3x+m_3y+n_3z.$$

These equations give the new co-ordinates x', y', z' in terms of the original co-ordinates x, y, z.

Since $\mathbf{i}'$, $\mathbf{j}'$, $\mathbf{k}'$ are mutually orthogonal vectors we have

$$l_1^2+m_1^2+n_1^2 = l_2^2+m_2^2+n_2^2 = l_3^2+m_3^2+n_3^2 = 1$$

and

$$l_2l_3+m_2m_3+n_2n_3 = l_3l_1+m_3m_1+n_3n_1 = l_1l_2+m_1m_2+n_1n_2 = 0.$$

Hence the nine components of the unit vectors $\mathbf{i}'$, $\mathbf{j}'$, $\mathbf{k}'$ are not independent.

Exercise 10

(1) Show that
$$\mathbf{r} = 5\cos\theta\mathbf{i}+4\sin\theta\mathbf{j},$$
$$\mathbf{r} = 5\cot\theta\mathbf{i}+12\operatorname{cosec}\theta\mathbf{j}$$
and
$$\mathbf{r} = a\sin\theta\mathbf{i}+a\operatorname{cosec}\theta\mathbf{j}$$
are the equations of an ellipse, hyperbola and rectangular hyperbola respectively, where θ is a parameter.

(2) Show that
$$\mathbf{r} = 2a\cos^2\theta\mathbf{i}+a\sin 2\theta\mathbf{j}$$
is the equation of the circle of radius a and centre (a, O), where θ is a parameter.

(3) $P(\mathbf{r})$ is any point on the circle of radius a and centre (O, a). Express $\mathbf{r}$ in terms of the parameter θ, where θ is the angle between $\mathbf{r}$ and the x-axis.

(4) Find the equation of the circle, centre $\mathbf{c} = (1, 2, 3)$, radius 6 units, in the plane containing the perpendicular vectors $\mathbf{u} = (2, 3, -1)$ and $\mathbf{v} = (4, -2, 2)$.

(5) Find the equation of the parabola, vertex $\mathbf{c} = (2, -4, 1)$, latus rectum of length 8 units, in the plane containing the perpendicular vectors $\mathbf{u} = (1, 4, -3)$ and $\mathbf{v} = (3, 0, 1)$, the axis of the parabola being in the direction of $\mathbf{u}$.

(6) Show that the equation of a circle of radius a taking a point O on the circumference as origin is
$$\mathbf{r}.(\mathbf{r}-2\mathbf{a}) = 0,$$
where $\mathbf{a}$ is the position vector of its centre.

(7) Show that the curve

$$\mathbf{r} = \left(1-\frac{16\cos\theta}{\sqrt{14}}+\frac{8\sin\theta}{\sqrt{5}}, \frac{24\cos\theta}{\sqrt{14}}, 2+\frac{8\cos\theta}{\sqrt{14}}+\frac{16\sin\theta}{\sqrt{5}}\right)$$

is a circle. Find its centre, radius and the equation of its plane.

(8) Find the equation of the parabola, vertex (1, 2, 3), focus (2, −1, 4), in the plane $4x+y-z = 3$.

(9) The position vector of a point relative to an origin O is $\mathbf{r}$. Show that

$$\mathbf{r} = a\sin\phi\cos\theta\mathbf{i}+a\sin\phi\sin\theta\mathbf{j}+a\cos\phi\mathbf{k}$$

is the equation of a sphere of radius a and centre at O, where θ and ϕ are parameters.

(10) Show that the equation of the paraboloid formed by rotating the parabola

$$\mathbf{r} = at^2\mathbf{i}+2at\mathbf{j}$$

about the x-axis is

$$\mathbf{r} = at^2\mathbf{i}+2at\cos\theta\mathbf{j}+2at\sin\theta\mathbf{k},$$

where t and θ are parameters.

(11) Show that the equation of the ellipsoid formed by rotating the ellipse

$$\mathbf{r} = a\cos\theta\mathbf{i}+b\sin\theta\mathbf{j}$$

about the x-axis is

$$\mathbf{r} = a\cos\theta\mathbf{i}+b\sin\theta\cos\phi\mathbf{j}+b\sin\theta\sin\phi\mathbf{k},$$

where θ and ϕ are parameters.

(12) Find the equation of the tangent at the point $\mathbf{r} = ap^2\mathbf{i}+2ap\mathbf{j}$ to the parabola $\mathbf{r} = at^2\mathbf{i}+2at\mathbf{j}$.

If the tangents at the points P and Q on the parabola are perpendicular, show that the locus of the mid-point of PQ is

$$\mathbf{r} = a(2t^2-1)\mathbf{i}+2at\mathbf{j},$$

where t is a parameter.

(13) Tangents are drawn from the point (−3, 2) to the parabola $\mathbf{r} = t^2\mathbf{i}+2t\mathbf{j}$, where t is a parameter. Find (i) the equations of the tangents, and (ii) the angle between the tangents.

(14) Find the equations of the tangent and normal to the parabola $\mathbf{r} = at^2\mathbf{i}+2at\mathbf{j}$.

If the tangents to the parabola $\mathbf{r} = 4t^2\mathbf{i}+8t\mathbf{j}$ at the points $t = 2$, $t = -\frac{1}{2}$ intersect at T and the normals at these points intersect at R, prove that $\mathbf{TR}$ is parallel to $\mathbf{i}$.

(15) From a fixed point O a straight line OP is drawn meeting a given straight line AB in P. From O a perpendicular is drawn to AB meeting AB in C. Q is a point in OP such that

$$\mathbf{OP}.\mathbf{OQ} = k^2.$$

Show that the locus of Q is given by

$$\mathbf{r}.\left(\frac{k^2\mathbf{c}}{\mathbf{c}^2}-\mathbf{r}\right) = 0,$$

where $\mathbf{OQ} = \mathbf{r}$, $\mathbf{OC} = \mathbf{c}$. Deduce that the locus is a circle with O at the circumference, with radius $|k^2\mathbf{c}/2\mathbf{c}^2|$ and centre on OC at a distance $|k^2\mathbf{c}/2\mathbf{c}^2|$ from O.

(16) The position vectors of the foci of an ellipse are $\mathbf{c}$ and $-\mathbf{c}$ and the length of the major axis is $2a$. Show that the equation to the ellipse is

$$a^4-a^2(\mathbf{r}^2+\mathbf{c}^2)+(\mathbf{r}.\mathbf{c})^2 = 0.$$

If $\mathbf{r} = (x, y)$ and $\mathbf{c} = (2, 0)$ deduce that the equation to the ellipse is

$$\frac{x^2}{9}+\frac{y^2}{5} = 1.$$

(17) The two helices

$$\mathbf{r}_1 = a\cos\theta\mathbf{i}+a\sin\theta\mathbf{j}+a\theta\mathbf{k} \quad \text{and} \quad \mathbf{r}_2 = a\cos n\theta\mathbf{i}-a\sin n\theta\mathbf{j}+a\theta\mathbf{k},$$

where $n > 0$, intersect at successive points P_1, P_2 and P_3. Prove that

$$P_1P_2 = 2a(\sin^2\phi+\phi^2)^{\frac{1}{2}} \quad \text{where} \quad \phi = \pi/(n+1),$$

and find P_1P_3 in terms of a and ϕ. (L.)

(18) The points A, B and P have coordinates $(-1, 0, 0)$, $(1, 0, 0)$ and (x, y, z) respectively. If

$$2\mathbf{PA}+3\mathbf{PB} = \mathbf{PQ},$$

where Q is the point (x', y', z'), show that

$$(1-4x-x')\mathbf{i}-(4y+y')\mathbf{j}-(4z+z')\mathbf{k} = \mathbf{0}.$$

When the locus of P is a sphere whose equation is

$$x^2+y^2+z^2 = 4,$$

deduce that the locus of Q is a sphere of radius 8 with its centre at the point $(1, 0, 0)$.

(19) If α is a constant such that $0 < \alpha < \frac{1}{2}\pi$, and ϕ is a parameter, prove that the curves

$$\mathbf{r} = b\cos\phi\mathbf{i}+b\sin\phi\mathbf{j}+b\cos\phi\tan\alpha\mathbf{k},$$

$$\mathbf{r} = b\cos\phi\mathbf{i}+b\sin\phi\mathbf{j}+(2h-b\cos\phi\tan\alpha)\mathbf{k}$$

are both ellipses of eccentricity $\sin\alpha$.

Show that, provided $h^2 < b^2\tan^2\alpha$, the ellipses intersect in two points distant $2(b^2-h^2\cot^2\alpha)^{\frac{1}{2}}$ apart. (L.)

(20) Show that $\mathbf{d}.(\mathbf{r}-\mathbf{s}) = 0$ is the equation of the plane passing through the point whose position vector is $\mathbf{s}$ and with the vector $\mathbf{d}$ perpendicular to the plane.

A paraboloid of revolution, of latus rectum $4a$, has its vertex at the origin and its axis along the positive x-axis. P is a point on the paraboloid such that the plane containing P and the x-axis makes an angle θ with the plane $z = 0$. Prove that the position vector of P from the origin is $\mathbf{r} = ap^2\mathbf{i}+2ap\cos\theta\mathbf{j}+2ap\sin\theta\mathbf{k}$, where p is a parameter.

Prove that all normals to the paraboloid at points on the circular section p = constant, pass through a fixed point $\mathbf{r} = (ap^2+2a)\mathbf{i}$.

Deduce that the equation of the tangent plane to the paraboloid at the point p, θ is

$$x-p\cos\theta y-p\sin\theta z+ap^2 = 0. \qquad \text{(L.)}$$

11

THE VECTOR PRODUCT

In this and the following chapters, familiarity with 3×3 determinants is assumed.

Introduction

In the development of the scalar product we considered the condition for two vectors to be perpendicular to one another. We shall define another product of two vectors, in terms of a vector which is perpendicular to each of two other vectors.

Let **a**, **b** and **c** be three non-coplanar and non-zero vectors. Now it is clear that if **c** is perpendicular to both **a** and **b** then the vector $-$**c** is also perpendicular to **a** and **b**. If we decide to take this perpendicular vector as defining a product of **a** and **b** then we shall have to decide whether we take **c** or $-$**c**, so as to obtain a unique vector for our product. To enable us to do this we shall first consider two possible systems formed by a triad of vectors.

Left- and right-handed systems

Referring to Fig. 11.1, O, A and B are points in the plane π. C and D are points on opposite sides of the plane. Let **OA** = **a**, **OB** = **b**, **OC** = **c** and **OD** = **d**. An observer at C will see the angle through which **OA** turns to coincide in direction and sense with **OB** as anticlockwise. In such a case the system [**a**, **b**, **c**] is said to be right-handed or positive. From D the angle sense is clockwise and the system [**a**, **b**, **d**] is left-handed or negative. The order of the vectors is important. The system [**b**, **a**, **c**] is left-handed since in this triad we are considering the angle sense of **b** turning toward **a** as seen from C.

Exercise 11(*a*)

(1) In Fig. 11.2, **a** and **b** are in the plane of the paper and **c** is perpendicular to the plane of the paper. In (*a*) **c** is directed out of the paper (denoted by ⊙) and in (*b*) **c** is directed into the paper (denoted

by ⊗). Consider in each case whether the following systems are left- or right-handed:

(i) [**a**, **b**, **c**], [**b**, **c**, **a**], [**c**, **a**, **b**],

(ii) [**b**, **a**, **c**], [**c**, **b**, **a**], [**a**, **c**, **b**].

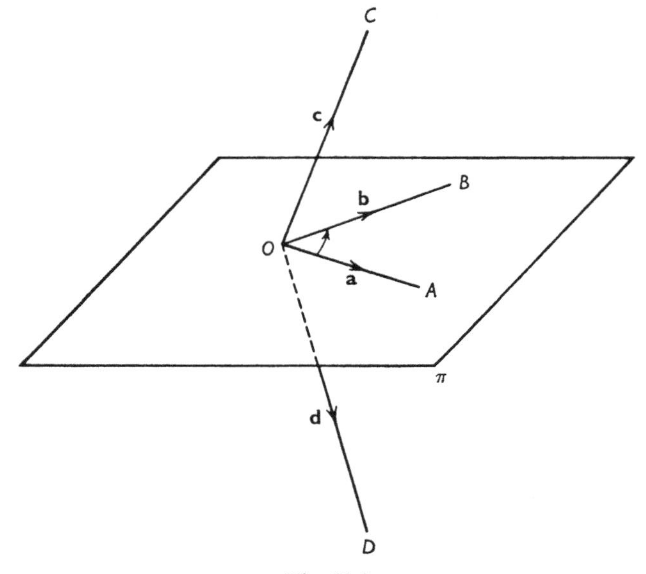

Fig. 11.1

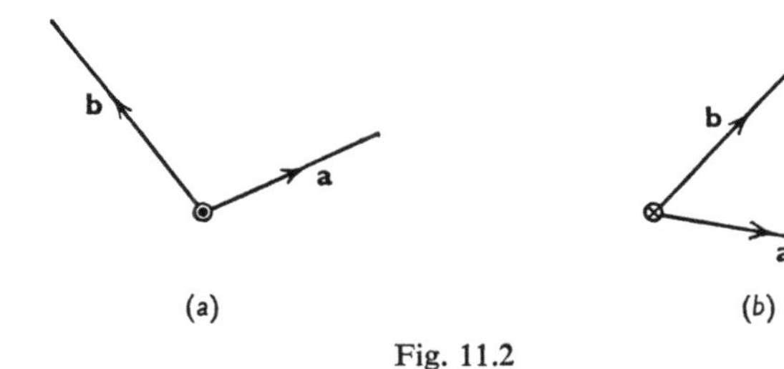

Fig. 11.2

Note. In (i) the systems are obtained from one another by cyclic interchange. In (ii) the systems are obtained from (i) by interchanging pairs of vectors.

(2) In Fig. 11.3, **i**, **j**, **k** are the unit vectors for the right-handed rectangular Cartesian co-ordinate frame. Consider whether the

following systems are left- or right-handed:

$$[\mathbf{i}, \mathbf{j}, \mathbf{k}], \quad [\mathbf{i}, \mathbf{j}, -\mathbf{k}], \quad [\mathbf{j}, \mathbf{i}, \mathbf{k}], \quad [\mathbf{j}, \mathbf{i}, -\mathbf{k}].$$

From Exercise 11(*a*) we see that if the system $[\mathbf{p}, \mathbf{q}, \mathbf{r}]$ is right-handed then the systems obtained from it by cyclic interchange are also right-handed; that is $[\mathbf{q}, \mathbf{r}, \mathbf{p}]$ and $[\mathbf{r}, \mathbf{p}, \mathbf{q}]$ are right-handed.

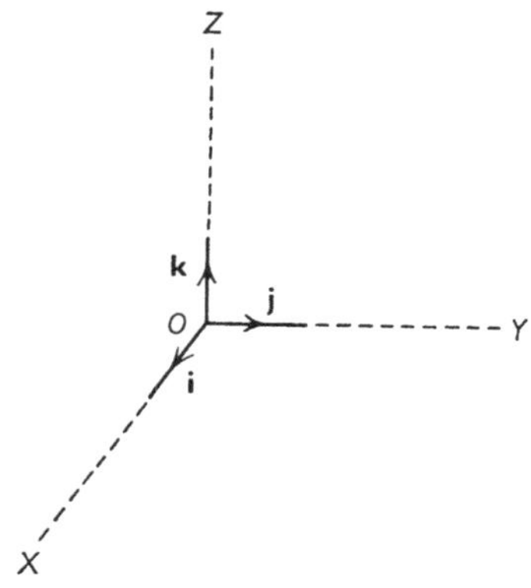

Fig. 11.3

However, the systems obtained by interchanging pairs are opposite in sense; that is $[\mathbf{q}, \mathbf{p}, \mathbf{r}]$, $[\mathbf{p}, \mathbf{r}, \mathbf{q}]$ and $[\mathbf{r}, \mathbf{q}, \mathbf{p}]$ are left-handed. Also changing the sense of a vector changes the sense of the system; that is $[-\mathbf{p}, \mathbf{q}, \mathbf{r}]$, $[\mathbf{p}, -\mathbf{q}, \mathbf{r}]$ and $[\mathbf{p}, \mathbf{q}, -\mathbf{r}]$ are left-handed. Finally if $[\mathbf{p}, \mathbf{q}, \mathbf{r}]$ is right-handed then $[\mathbf{p}, \mathbf{q}, -\mathbf{r}]$ is left-handed. Hence $[\mathbf{q}, \mathbf{p}, -\mathbf{r}]$ is right-handed.

An alternative way of considering left- and right-handed systems is to consider a right-handed screw whose axis lies in the direction of **c** as in Fig. 11.4. Let the screw be turned in the same sense as **OA** turning toward **OB** through the least angle, that is from **a** to **b**. If in doing so the sense of travel of the screw is that of **c** as in the figure then the system $[\mathbf{a}, \mathbf{b}, \mathbf{c}]$ is right-handed; if in the opposite sense then the system $[\mathbf{a}, \mathbf{b}, \mathbf{c}]$ is left-handed.

Vector perpendicular to two vectors

Let $\mathbf{a} = a_1\mathbf{i}+a_2\mathbf{j}+a_3\mathbf{k}$ and $\mathbf{b} = b_1\mathbf{i}+b_2\mathbf{j}+b_3\mathbf{k}$ be two non-parallel and non-zero vectors. We shall now obtain a vector

$$\mathbf{c} = c_1\mathbf{i}+c_2\mathbf{j}+c_3\mathbf{k}$$

which is perpendicular to both **a** and **b**. It is obvious that **a**, **b** and **c** are non-coplanar. Since **a**, **c** and **b**, **c** are perpendicular we have

$$\mathbf{a}.\mathbf{c} = 0 \quad \text{and} \quad \mathbf{b}.\mathbf{c} = 0. \tag{11·1}$$

Since **a** is not parallel to **b**, $\mathbf{a} \neq k\mathbf{b}$ where k is some constant number. This being so at least two of the ratios $a_1{:}b_1$, $a_2{:}b_2$, $a_3{:}b_3$ cannot be equal. Suppose $a_1{:}b_1 \neq a_2{:}b_2$, that is $a_1b_2 - a_2b_1 \neq 0$. From (11·1) we have

$$a_1c_1 + a_2c_2 + a_3c_3 = 0 \tag{11·2}$$

and

$$b_1c_1 + b_2c_2 + b_3c_3 = 0. \tag{11·3}$$

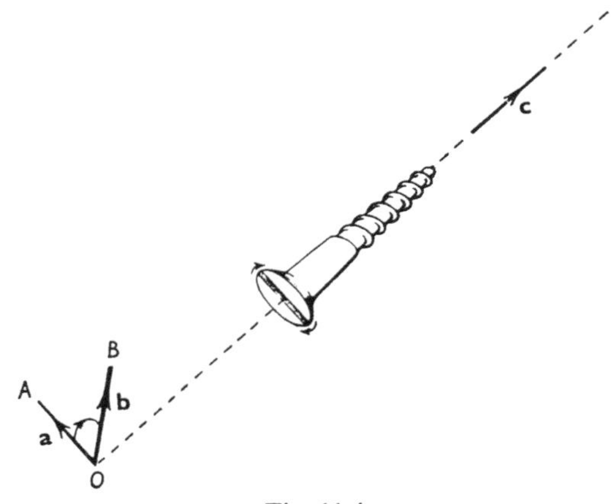

Fig. 11.4

Multiplying (11·2) by b_2 and (11·3) by a_2 we get

$$a_1b_2c_1 + a_2b_2c_2 + a_3b_2c_3 = 0 \tag{11·4}$$

and

$$a_2b_1c_1 + a_2b_2c_2 + a_2b_3c_3 = 0. \tag{11·5}$$

Subtracting (11·4) from (11·5),

$$(a_2b_1 - a_1b_2)c_1 + (a_2b_3 - a_3b_2)c_3 = 0.$$

Therefore

$$c_1 = \frac{a_2b_3 - a_3b_2}{a_1b_2 - a_2b_1}\,c_3,$$

since $a_1b_2 - a_2b_1 \neq 0$. Similarly by multiplying (11·2) by b_1 and (11·3) by a_1 we get

$$c_2 = \frac{a_3b_1 - a_1b_3}{a_1b_2 - a_2b_1}\,c_3.$$

Therefore $$\mathbf{c} = \frac{a_2b_3-a_3b_2}{a_1b_2-a_2b_1}c_3\mathbf{i}+\frac{a_3b_1-a_1b_3}{a_1b_2-a_2b_1}c_3\mathbf{j}+c_3\mathbf{k}.$$

Let $c_3 = \lambda(a_1b_2-a_2b_1)$ where λ is an arbitrary positive or negative number. Then

$$\mathbf{c} = \lambda[(a_2b_3-a_3b_2)\mathbf{i}+(a_3b_1-a_1b_3)\mathbf{j}+(a_1b_2-a_2b_1)\mathbf{k}]. \quad (11{\cdot}6)$$

The solution of the homogeneous equations (11·2) and (11·3) can be immediately written down in terms of determinants, using Cramer's rule:

$$\frac{c_1}{\begin{vmatrix} a_2 & a_3 \\ b_2 & b_3 \end{vmatrix}} = \frac{c_2}{\begin{vmatrix} a_3 & a_1 \\ b_3 & b_1 \end{vmatrix}} = \frac{c_3}{\begin{vmatrix} a_1 & a_2 \\ b_1 & b_2 \end{vmatrix}} = \lambda \quad \text{(say)},$$

i.e. $$\frac{c_1}{a_2b_3-a_3b_2} = \frac{c_2}{a_3b_1-a_1b_3} = \frac{c_3}{a_1b_2-a_2b_1} = \lambda,$$

where if a particular denominator is zero the corresponding numerator is to be interpreted as zero.

By taking different values of λ in (11·6) we obtain various parallel vectors all perpendicular to **a** and **b**. In particular when $\lambda = 1$ we have the vector

$$\mathbf{c} = (a_2b_3-a_3b_2)\mathbf{i}+(a_3b_1-a_1b_3)\mathbf{j}+(a_1b_2-a_2b_1)\mathbf{k}. \quad (11{\cdot}7)$$

Definition of the vector product

The vector **c** as given by (11·7) is not only perpendicular to **a** and **b** but also has one of two senses such that either [**a**, **b**, **c**] or [**b**, **a**, **c**] forms a right-handed system. If [**a**, **b**, **c**] forms a right-handed system then the vector **c** as given by (11·7) is known as the vector product of **a** and **b**, written $\mathbf{a}\times\mathbf{b}$ (pronounced '**a** cross **b**') or $\mathbf{a}\wedge\mathbf{b}$ (pronounced '**a** vec **b**'). However if [**b**, **a**, **c**] forms a right-handed system then (11·7) defines the vector product of **b** and **a**, written $\mathbf{b}\times\mathbf{a}$. The vector product is also known as the cross-product or the outer product.

We shall now determine whether **c** as defined by (11·7) is to be taken as the vector product $\mathbf{a}\times\mathbf{b}$ or $\mathbf{b}\times\mathbf{a}$. A simple case will help us to decide.

Consider the vectors $\mathbf{a} = a_1\mathbf{i}$ $(a_1 > 0)$ and $\mathbf{b} = b_1\mathbf{i}+b_2\mathbf{j}$ $(b_2 > 0)$. Since we require the system $[\mathbf{a}, \mathbf{b}, \mathbf{a}\times\mathbf{b}]$ to be right-handed we see from Fig. 11.5(*a*) that $\mathbf{a}\times\mathbf{b}$ is a vector having the same direction and sense as $\mathbf{k}$. We can therefore write

$$\mathbf{a}\times\mathbf{b} = p\mathbf{k} \quad (p > 0).$$

Similarly since the system $[\mathbf{b}, \mathbf{a}, \mathbf{b}\times\mathbf{a}]$ has to be right-handed we see from Fig. 11.5(*b*) that $\mathbf{b}\times\mathbf{a}$ is a vector having the same direction and sense as $-\mathbf{k}$ and therefore

$$\mathbf{b}\times\mathbf{a} = -q\mathbf{k} \quad (q > 0).$$

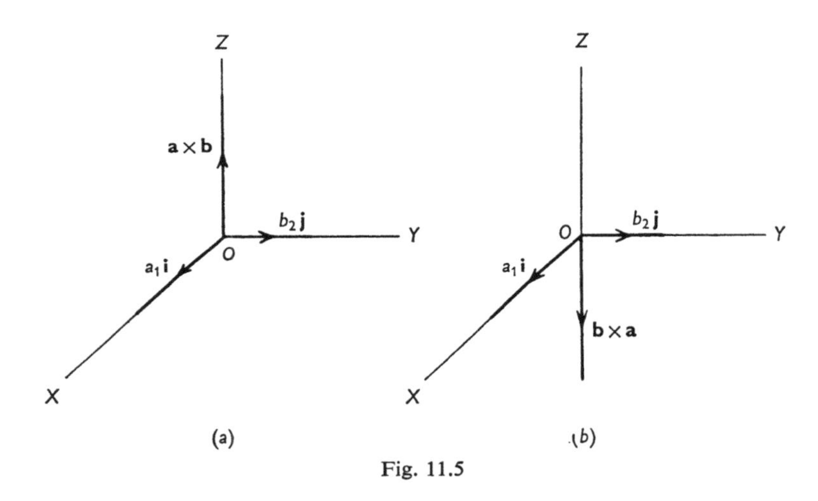

Fig. 11.5

Now substituting in (11·7) we get

$$\mathbf{c} = a_1 b_2 \mathbf{k}.$$

This is a vector having the same direction and sense as $\mathbf{k}$ since $a_1 > 0$ and $b_2 > 0$. Therefore $\mathbf{a}\times\mathbf{b}$ and $\mathbf{c}$ have the same direction and sense, and hence $[\mathbf{a}, \mathbf{b}, \mathbf{c}]$ is a right-handed system. It follows that $\mathbf{c}$ defines the vector product of $\mathbf{a}$ and $\mathbf{b}$, that is

$$\mathbf{a}\times\mathbf{b} = (a_2 b_3 - a_3 b_2)\mathbf{i} + (a_3 b_1 - a_1 b_3)\mathbf{j} + (a_1 b_2 - a_2 b_1)\mathbf{k}. \quad (11\cdot 8)$$

Interchanging $\mathbf{b}$ and $\mathbf{a}$ we get

$$\mathbf{b}\times\mathbf{a} = (b_2 a_3 - b_3 a_2)\mathbf{i} + (b_3 a_1 - b_1 a_3)\mathbf{j} + (b_1 a_2 - b_2 a_1)\mathbf{k}. \quad (11\cdot 9)$$

We have defined the vector product in terms of a co-ordinate system. This definition is itself incomplete until it is shown that it is invariant under a change of axes. In fact, as we shall see later, the vector product defined in terms of co-ordinates does not depend on the frame of reference, that is, its magnitude, direction and sense are independent of the choice of origin and of the orientation of the axes.

The definition given by (11·8) or (11·9) can be easily remembered by expressing it in a determinantal form bringing out a symmetrical pattern. Thus

$$\mathbf{a}\times\mathbf{b} = \begin{vmatrix} \mathbf{i} & \mathbf{j} & \mathbf{k} \\ a_1 & a_2 & a_3 \\ b_1 & b_2 & b_3 \end{vmatrix}, \quad \mathbf{b}\times\mathbf{a} = \begin{vmatrix} \mathbf{i} & \mathbf{j} & \mathbf{k} \\ b_1 & b_2 & b_3 \\ a_1 & a_2 & a_3 \end{vmatrix}. \qquad (11\cdot10)$$

It must be stressed that these are not true determinants since the elements are not all numbers. They are to be regarded as an aid to writing out the components of the vector product. Thus on expanding (11·10) we get

$$\begin{aligned} \mathbf{a}\times\mathbf{b} &= \begin{vmatrix} a_2 & a_3 \\ b_2 & b_3 \end{vmatrix}\mathbf{i} - \begin{vmatrix} a_1 & a_3 \\ b_1 & b_3 \end{vmatrix}\mathbf{j} + \begin{vmatrix} a_1 & a_2 \\ b_1 & b_2 \end{vmatrix}\mathbf{k} \\ &= (a_2b_3 - a_3b_2)\mathbf{i} - (a_1b_3 - a_3b_1)\mathbf{j} + (a_1b_2 - a_2b_1)\mathbf{k} \\ &= (a_2b_3 - a_3b_2)\mathbf{i} + (a_3b_1 - a_1b_3)\mathbf{j} + (a_1b_2 - a_2b_1)\mathbf{k}. \end{aligned}$$

Example

If $\mathbf{a} = 3\mathbf{i} - 5\mathbf{j} + \mathbf{k}$, $\mathbf{b} = 2\mathbf{j} - 4\mathbf{k}$ *find* $\mathbf{a}\times\mathbf{b}$.

$$\begin{aligned} \mathbf{a}\times\mathbf{b} &= \begin{vmatrix} \mathbf{i} & \mathbf{j} & \mathbf{k} \\ 3 & -5 & 1 \\ 0 & 2 & -4 \end{vmatrix} \\ &= \begin{vmatrix} -5 & 1 \\ 2 & -4 \end{vmatrix}\mathbf{i} - \begin{vmatrix} 3 & 1 \\ 0 & -4 \end{vmatrix}\mathbf{j} + \begin{vmatrix} 3 & -5 \\ 0 & 2 \end{vmatrix}\mathbf{k} \\ &= (20-2)\mathbf{i} - (-12-0)\mathbf{j} + (6-0)\mathbf{k} \\ &= 18\mathbf{i} + 12\mathbf{j} + 6\mathbf{k}. \end{aligned}$$

Alternatively the vector product $\mathbf{a}\times\mathbf{b}$ can be obtained by writing down the components of $\mathbf{a}$ and $\mathbf{b}$ with those of $\mathbf{a}$ above those of $\mathbf{b}$ and starting and ending with the second component in each case.

$$\begin{array}{ccccccc} & \mathbf{i} & & \mathbf{j} & & \mathbf{k} & \\ a_2 & & a_3 & & a_1 & & a_2 \\ & \times & & \times & & \times & \\ b_2 & & b_3 & & b_1 & & b_2 \end{array}$$

The product obtained along a downward arrow has a plus sign and that along an upward arrow has a minus sign. We have therefore

$$\mathbf{a}\times\mathbf{b} = (a_2b_3-a_3b_2)\mathbf{i}+(a_3b_1-a_1b_3)\mathbf{j}+(a_1b_2-a_2b_1)\mathbf{k}.$$

Example

If $\mathbf{p} = 4\mathbf{i}-2\mathbf{j}-3\mathbf{k}$, $\mathbf{q} = 2\mathbf{i}+\mathbf{j}-5\mathbf{k}$ *find* $\mathbf{q}\times\mathbf{p}$.

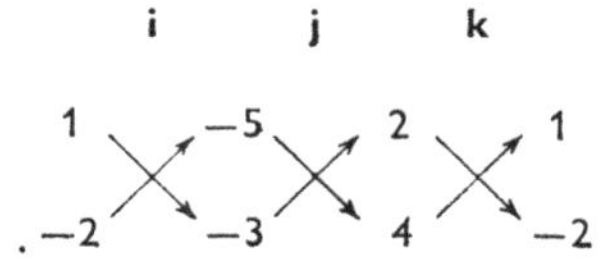

i j k

1 −5 2 1

−2 −3 4 −2

$$\begin{aligned}\mathbf{q}\times\mathbf{p} &= (-3-10)\mathbf{i}+(-20+6)\mathbf{j}+(-4-4)\mathbf{k}\\ &= -13\mathbf{i}-14\mathbf{j}-8\mathbf{k}\\ &= -(13\mathbf{i}+14\mathbf{j}+8\mathbf{k}).\end{aligned}$$

The vector product involving the zero vector or parallel vectors

In obtaining the definition of the vector product we excluded the case of either **a** or **b** being the zero vector, or **a** and **b** being parallel.

Let us now extend our definition of the vector product

$$\mathbf{a}\times\mathbf{b} = (a_2b_3-a_3b_2)\mathbf{i}+(a_3b_1-a_1b_3)\mathbf{j}+(a_1b_2-a_2b_1)\mathbf{k}$$

to cover all cases of **a** and **b**. Then if either **a** or **b** is the zero vector we have either $a_1 = a_2 = a_3 = 0$ or $b_1 = b_2 = b_3 = 0$ and on substitution we get $\mathbf{a}\times\mathbf{b} = \mathbf{0}$. If **a** and **b** are parallel we have $a_1 = kb_1$, $a_2 = kb_2$, $a_3 = kb_3$ $(k \neq 0)$ and again we get $\mathbf{a}\times\mathbf{b} = \mathbf{0}$. Therefore we have

$$\mathbf{a}\times\mathbf{0} = \mathbf{0}\times\mathbf{a} = \mathbf{0} \quad \text{for all } \mathbf{a}$$

and

$$\mathbf{a}\times\mathbf{b} = \mathbf{0} \quad \text{for } \mathbf{a} \text{ parallel to } \mathbf{b}.$$

Consequences of the vector product definition

In the following, we take $\mathbf{a} = a_1\mathbf{i}+a_2\mathbf{j}+a_3\mathbf{k}$, $\mathbf{b} = b_1\mathbf{i}+b_2\mathbf{j}+b_3\mathbf{k}, \ldots$. We shall investigate particularly whether the commutative, associative, and distributive laws hold for the vector product.

(1) $\mathbf{a}\times\mathbf{b} = -(\mathbf{b}\times\mathbf{a})$

Proof. Since

$$\mathbf{a}\times\mathbf{b} = (a_2b_3-a_3b_2)\mathbf{i}+(a_3b_1-a_1b_3)\mathbf{j}+(a_1b_2-a_2b_1)\mathbf{k}$$

and $$\mathbf{b}\times\mathbf{a} = (b_2a_3-b_3a_2)\mathbf{i}+(b_3a_1-b_1a_3)\mathbf{j}+(b_1a_2-b_2a_1)\mathbf{k}$$

we have immediately $\quad \mathbf{a}\times\mathbf{b} = -(\mathbf{b}\times\mathbf{a})$.

Thus the vector product is not commutative.

(2) $(\mathbf{a}\times\mathbf{b})\times\mathbf{c} \neq \mathbf{a}\times(\mathbf{b}\times\mathbf{c})$ ***in general***

Proof. (*a*) Suppose $\mathbf{a} = \mathbf{0}$ or $\mathbf{b} = \mathbf{0}$ or $\mathbf{c} = \mathbf{0}$. Then $(\mathbf{a}\times\mathbf{b})\times\mathbf{c} = \mathbf{0}$, $\mathbf{a}\times(\mathbf{b}\times\mathbf{c}) = \mathbf{0}$, and so we have

$$(\mathbf{a}\times\mathbf{b})\times\mathbf{c} = \mathbf{a}\times(\mathbf{b}\times\mathbf{c}) \quad \text{when} \quad \mathbf{a} = \mathbf{0} \text{ or } \mathbf{b} = \mathbf{0} \text{ or } \mathbf{c} = \mathbf{0}.$$

(*b*) Generally however the associative law does not hold for the vector product. For $(\mathbf{a}\times\mathbf{b})\times\mathbf{c}$ is in the plane containing $\mathbf{a}$, $\mathbf{b}$ and $\mathbf{a}\times(\mathbf{b}\times\mathbf{c})$ is in the plane containing $\mathbf{b}$, $\mathbf{c}$. These planes are, in general, different. Hence

$$(\mathbf{a}\times\mathbf{b})\times\mathbf{c} \neq \mathbf{a}\times(\mathbf{b}\times\mathbf{c}) \text{ in general.}$$

The reader may like to consider when $(\mathbf{a}\times\mathbf{b})\times\mathbf{c} = \mathbf{a}\times(\mathbf{b}\times\mathbf{c})$ apart from the case of $\mathbf{a} = \mathbf{0}$ or $\mathbf{b} = \mathbf{0}$ or $\mathbf{c} = \mathbf{0}$. (See Exercise 11(*b*), Question 6.)

(3) $(p\mathbf{a})\times(q\mathbf{b}) = pq(\mathbf{a}\times\mathbf{b})$

Proof.

$$p\mathbf{a} = pa_1\mathbf{i}+pa_2\mathbf{j}+pa_3\mathbf{k},$$
$$q\mathbf{b} = qb_1\mathbf{i}+qb_2\mathbf{j}+qb_3\mathbf{k}.$$

Therefore

$$\begin{aligned}(p\mathbf{a})\times(q\mathbf{b}) &= (pa_2qb_3-pa_3qb_2)\mathbf{i}+(pa_3qb_1-pa_1qb_3)\mathbf{j}\\ &\qquad +(pa_1qb_2-pa_2qb_1)\mathbf{k}\\ &= pq\{(a_2b_3-a_3b_2)\mathbf{i}+(a_3b_1-a_1b_3)\mathbf{j}+(a_1b_2-a_2b_1)\mathbf{k}\}\\ &= pq(\mathbf{a}\times\mathbf{b}).\end{aligned}$$

In particular when $p = 1$, $q = n$,

$$\mathbf{a}\times(n\mathbf{b}) = n(\mathbf{a}\times\mathbf{b}),$$

and when $q = 1$, $p = n$,

$$(n\mathbf{a})\times\mathbf{b} = n(\mathbf{a}\times\mathbf{b}),$$

i.e. $$\mathbf{a}\times(n\mathbf{b}) = (n\mathbf{a})\times\mathbf{b} = n(\mathbf{a}\times\mathbf{b}).$$

Again in particular when $n = -1$,

$$\mathbf{a}\times(-\mathbf{b}) = (-\mathbf{a})\times\mathbf{b} = -(\mathbf{a}\times\mathbf{b}).$$

(4) $\mathbf{a}\times(\mathbf{b}+\mathbf{c}) = \mathbf{a}\times\mathbf{b}+\mathbf{a}\times\mathbf{c}$

Proof. $\mathbf{b}+\mathbf{c} = (b_1+c_1)\mathbf{i}+(b_2+c_2)\mathbf{j}+(b_3+c_3)\mathbf{k}.$

Therefore

$$\begin{aligned}\mathbf{a}\times(\mathbf{b}+\mathbf{c}) &= \{a_2(b_3+c_3)-a_3(b_2+c_2)\}\mathbf{i}+\{a_3(b_1+c_1)\\ &\qquad -a_1(b_3+c_3)\}\mathbf{j}+\{a_1(b_2+c_2)-a_2(b_1+c_1)\}\mathbf{k}\\ &= (a_2b_3+a_2c_3-a_3b_2-a_3c_2)\mathbf{i}\\ &\qquad +(a_3b_1+a_3c_1-a_1b_3-a_1c_3)\mathbf{j}\\ &\qquad +(a_1b_2+a_1c_2-a_2b_1-a_2c_1)\mathbf{k}.\\ \mathbf{a}\times\mathbf{b} &= (a_2b_3-a_3b_2)\mathbf{i}+(a_3b_1-a_1b_3)\mathbf{j}+(a_1b_2-a_2b_1)\mathbf{k},\\ \mathbf{a}\times\mathbf{c} &= (a_2c_3-a_3c_2)\mathbf{i}+(a_3c_1-a_1c_3)\mathbf{j}+(a_1c_2-a_2c_1)\mathbf{k}.\end{aligned}$$

Therefore

$$\begin{aligned}\mathbf{a}\times\mathbf{b}+\mathbf{a}\times\mathbf{c} &= (a_2b_3+a_2c_3-a_3b_2-a_3c_2)\mathbf{i}\\ &\qquad +(a_3b_1+a_3c_1-a_1b_3-a_1c_3)\mathbf{j}\\ &\qquad +(a_1b_2+a_1c_2-a_2b_1-a_2c_1)\mathbf{k}.\end{aligned}$$

Hence

$$\mathbf{a}\times(\mathbf{b}+\mathbf{c}) = \mathbf{a}\times\mathbf{b}+\mathbf{a}\times\mathbf{c}.$$

Multiplying throughout by -1 we get

$$-\mathbf{a}\times(\mathbf{b}+\mathbf{c}) = -(\mathbf{a}\times\mathbf{b})-(\mathbf{a}\times\mathbf{c}),$$

which on application of (1) gives

$$(\mathbf{b}+\mathbf{c})\times\mathbf{a} = \mathbf{b}\times\mathbf{a}+\mathbf{c}\times\mathbf{a}.$$

Extending the argument we have

$$\mathbf{a}\times(\mathbf{b}+\mathbf{c}+\mathbf{d}+\ldots) = \mathbf{a}\times\mathbf{b}+\mathbf{a}\times\mathbf{c}+\mathbf{a}\times\mathbf{d}+\ldots.$$

Thus the vector product is distributive over addition and this enables us to expand expressions such as $(\mathbf{a}+\mathbf{b})\times(\mathbf{c}+\mathbf{d})$:

$$\begin{aligned}(\mathbf{a}+\mathbf{b})\times(\mathbf{c}+\mathbf{d}) &= (\mathbf{a}+\mathbf{b})\times\mathbf{c}+(\mathbf{a}\times\mathbf{b})\times\mathbf{d}\\ &= \mathbf{a}\times\mathbf{c}+\mathbf{b}\times\mathbf{c}+\mathbf{a}\times\mathbf{d}+\mathbf{b}\times\mathbf{d}.\end{aligned}$$

It is of course important to maintain the correct order of the factors in each of the vector products.

Note. The truth of (1) and (3) is easily seen if either $\mathbf{a} = \mathbf{0}$ or $\mathbf{b} = \mathbf{0}$ or $\mathbf{a}$ is parallel to $\mathbf{b}$ since then each side reduces to the zero vector. Again in (4) if $\mathbf{a}$ is the zero vector then each side reduces to the zero vector.

(5) ***The unit vectors* i, j, k**

Since $\mathbf{i} = (1, 0, 0)$, $\mathbf{j} = (0, 1, 0)$, $\mathbf{k} = (0, 0, 1)$ it immediately follows by substitution in the definition of the vector product that

$$\mathbf{i}\times\mathbf{i} = \mathbf{j}\times\mathbf{j} = \mathbf{k}\times\mathbf{k} = \mathbf{0},$$

$$\mathbf{j}\times\mathbf{k} = \mathbf{i}, \quad \mathbf{k}\times\mathbf{i} = \mathbf{j}, \quad \mathbf{i}\times\mathbf{j} = \mathbf{k}, \tag{11·11}$$

$$\mathbf{k}\times\mathbf{j} = -\mathbf{i}, \quad \mathbf{i}\times\mathbf{k} = -\mathbf{j}, \quad \mathbf{j}\times\mathbf{i} = -\mathbf{k}. \tag{11·12}$$

Note that the results (11·11) are obtained by cyclic interchange of the right-handed system [**i**, **j**, **k**] and the results (11·12) are obtained by upsetting the cyclic arrangement.

Using (3), (4) and (5) the numerical evaluation of the vector product is easily obtained without substituting in the definition; for example consider $\mathbf{a}\times\mathbf{b}$ when $\mathbf{a} = 2\mathbf{i}-5\mathbf{j}+\mathbf{k}$, $\mathbf{b} = \mathbf{i}+2\mathbf{j}-3\mathbf{k}$.

$$\begin{aligned}\mathbf{a}\times\mathbf{b} &= (2\mathbf{i}-5\mathbf{j}+\mathbf{k})\times(\mathbf{i}+2\mathbf{j}-3\mathbf{k})\\ &= 2(\mathbf{i}\times\mathbf{i})+4(\mathbf{i}\times\mathbf{j})-6(\mathbf{i}\times\mathbf{k})-5(\mathbf{j}\times\mathbf{i})-10(\mathbf{j}\times\mathbf{j})\\ &\qquad +15(\mathbf{j}\times\mathbf{k})+(\mathbf{k}\times\mathbf{i})+2(\mathbf{k}\times\mathbf{j})-3(\mathbf{k}\times\mathbf{k})\\ &= 4\mathbf{k}+6\mathbf{j}+5\mathbf{k}+15\mathbf{i}+\mathbf{j}-2\mathbf{i}\\ &= 13\mathbf{i}+7\mathbf{j}+9\mathbf{k}.\end{aligned}$$

The modulus of the vector product

Since

$$\mathbf{a}\times\mathbf{b} = (a_2b_3-a_3b_2)\mathbf{i}+(a_3b_1-a_1b_3)\mathbf{j}+(a_1b_2-a_2b_1)\mathbf{k}$$

we have immediately

$$|\mathbf{a}\times\mathbf{b}| = \{(a_2b_3-a_3b_2)^2+(a_3b_1-a_1b_3)^2+(a_1b_2-a_2b_1)^2\}^{\frac{1}{2}}.$$

Suppose we are given the vectors **a**, **b** and the angle θ $(0 \leqslant \theta \leqslant \pi)$ between them. Can we obtain the modulus of $\mathbf{a}\times\mathbf{b}$ without first obtaining $\mathbf{a}\times\mathbf{b}$? On consideration it seems reasonable to assume that the modulus of the vector product depends on the product of the moduli of **a** and **b** and on the angle between them. To see if such a relation exists we shall calculate $|\mathbf{a}\times\mathbf{b}|$, $|\mathbf{a}|$, $|\mathbf{b}|$ and $|\mathbf{a}\times\mathbf{b}|/(|\mathbf{a}|\,|\mathbf{b}|)$ for the simple cases of Fig. 11.6.

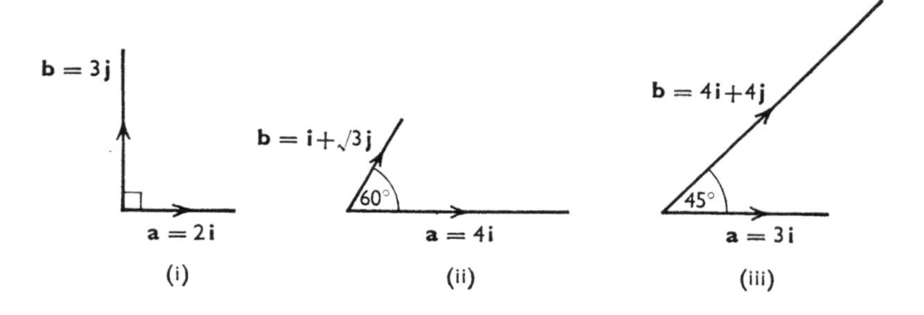

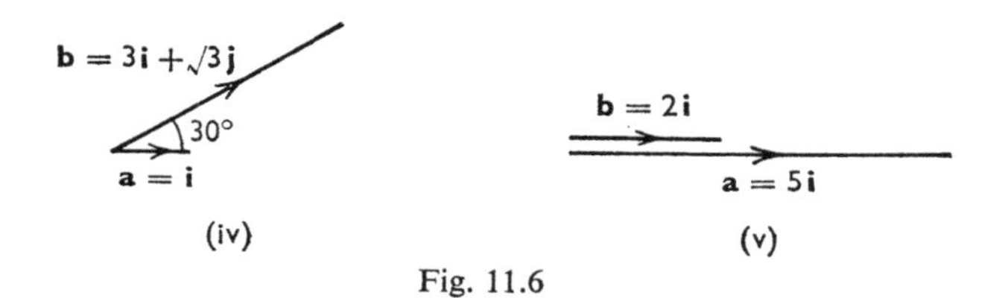

Fig. 11.6

Case	Angle θ	$\mathbf{a}\times\mathbf{b}$	$\lvert\mathbf{a}\times\mathbf{b}\rvert$	$\lvert\mathbf{a}\rvert$	$\lvert\mathbf{b}\rvert$	$\lvert\mathbf{a}\rvert\,\lvert\mathbf{b}\rvert$	$\dfrac{\lvert\mathbf{a}\times\mathbf{b}\rvert}{\lvert\mathbf{a}\rvert\,\lvert\mathbf{b}\rvert}$
(i)	90°	$6\mathbf{k}$	6	2	3	6	1
(ii)	60°	$4\sqrt{3}\mathbf{k}$	$4\sqrt{3}$	4	2	8	$\sqrt{3}/2$
(iii)	45°	$12\mathbf{k}$	12	3	$4\sqrt{2}$	$12\sqrt{2}$	$1/\sqrt{2}$
(iv)	30°	$\sqrt{3}\mathbf{k}$	$\sqrt{3}$	1	$2\sqrt{3}$	$2\sqrt{3}$	1/2
(v)	0°	$\mathbf{0}$	0	5	2	10	0

The values of $|\mathbf{a}\times\mathbf{b}|/(|\mathbf{a}|\,|\mathbf{b}|)$ correspond to those of $\sin\theta$. This leads us to expect that in general $|\mathbf{a}\times\mathbf{b}| = |\mathbf{a}|\,|\mathbf{b}|\,|\sin\theta|$. We now investigate this generally.

We have

$$|\mathbf{a}\times\mathbf{b}|^2 = (a_2b_3-a_3b_2)^2+(a_3b_1-a_1b_3)^2+(a_1b_2-a_2b_1)^2,$$

which on expansion and simplication, gives

$$|\mathbf{a}\times\mathbf{b}|^2 = (a_1^2+a_2^2+a_3^2)(b_1^2+b_2^2+b_3^2)-(a_1b_1+a_2b_2+a_3b_3)^2.$$

Now

$$(a_1^2+a_2^2+a_3^2) = |\mathbf{a}|^2,\ (b_1^2+b_2^2+b_3^2) = |\mathbf{b}|^2,\ a_1b_1+a_2b_2+a_3b_3 = \mathbf{a}.\mathbf{b}.$$

Therefore
$$|\mathbf{a}\times\mathbf{b}|^2 = |\mathbf{a}|^2|\mathbf{b}|^2-(\mathbf{a}.\mathbf{b})^2$$
$$= |\mathbf{a}|^2|\mathbf{b}|^2-|\mathbf{a}|^2|\mathbf{b}|^2\cos^2\theta$$
$$= |\mathbf{a}|^2|\mathbf{b}|^2\sin^2\theta.$$

Hence
$$|\mathbf{a}\times\mathbf{b}| = |\mathbf{a}|\,|\mathbf{b}|\sin\theta,\ (0 \leqslant \theta \leqslant \pi).$$

If $\hat{\mathbf{n}}$ is the unit vector perpendicular to **a** and **b** such that $[\mathbf{a}, \mathbf{b}, \hat{\mathbf{n}}]$ forms a right-handed system, then we can write

$$\mathbf{a}\times\mathbf{b} = |\mathbf{a}|\,|\mathbf{b}|\sin\theta\hat{\mathbf{n}}.$$

Vector product involving the projection of one of its factors

We shall now show that $\mathbf{a}\times\mathbf{b}$ is unaltered if we replace one factor by its projection on a plane perpendicular to the other.

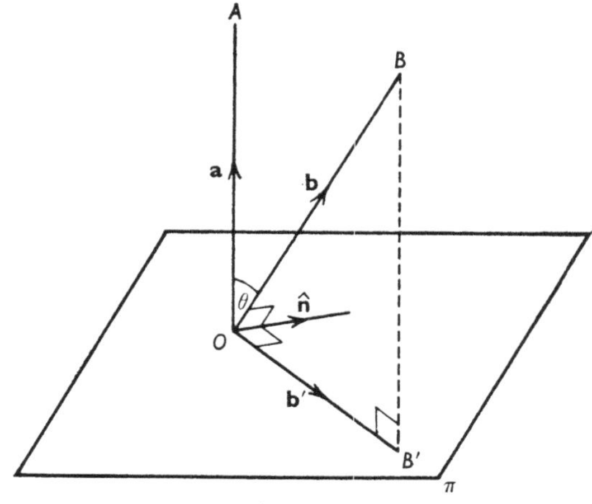

Fig. 11.7

Referring to Fig. 11.7, $\mathbf{OA} = \mathbf{a}$, $\mathbf{OB} = \mathbf{b}$, π is the plane through O perpendicular to **a**. Let $\mathbf{OB'} = \mathbf{b'}$ be the projection of **b** on π and $\hat{\mathbf{n}}$ the unit vector perpendicular to both **a** and **b**, hence perpendicular to both **a** and $\mathbf{b'}$, $[\mathbf{a}, \mathbf{b}, \hat{\mathbf{n}}]$ forming a right-handed system. If θ is the angle between **a** and **b** we have

$$\mathbf{a}\times\mathbf{b} = |\mathbf{a}|\,|\mathbf{b}|\,|\sin\theta|\,\hat{\mathbf{n}}.$$

But since $$|\mathbf{b}'| = |\mathbf{b}|\,|\sin\theta|,$$

$$\mathbf{a}\times\mathbf{b} = |\mathbf{a}|\,|\mathbf{b}'|\,\hat{\mathbf{n}}.$$

Now $$\mathbf{a}\times\mathbf{b}' = |\mathbf{a}|\,|\mathbf{b}'|\,|\sin\tfrac{1}{2}\pi|\,\hat{\mathbf{n}}$$

$$= |\mathbf{a}|\,|\mathbf{b}'|\,\hat{\mathbf{n}}.$$

Therefore $$\mathbf{a}\times\mathbf{b} = \mathbf{a}\times\mathbf{b}'.$$

Thus the vector product $\mathbf{a}\times\mathbf{b}$ is equivalent to the vector product $\mathbf{a}\times\mathbf{b}'$ where $\mathbf{b}'$ is the projection of **b** on the plane through O perpendicular to **a**.

Also $\mathbf{a}\times\mathbf{b} = |\mathbf{a}|\,|\mathbf{b}'|\,\hat{\mathbf{n}}$ and since $|\mathbf{b}'|\,\hat{\mathbf{n}}$ is the vector obtained by rotating $\mathbf{b}'$ through a right angle we have $\mathbf{a}\times\mathbf{b} = |\mathbf{a}|\,\mathbf{b}''$ where $\mathbf{b}'' = |\mathbf{b}'|\,\hat{\mathbf{n}}$ and hence

$$\mathbf{a}\times\mathbf{b} = \mathbf{a}\times\mathbf{b}' = |\mathbf{a}|\,|\mathbf{b}'|\,\hat{\mathbf{n}} = |\mathbf{a}|\,\mathbf{b}''.$$

From this we see that the vector product can be obtained in three stages regarding one of the vectors, say **a**, as fixed:

(1) Obtain the projection $\mathbf{b}'$ of **b** on the plane through O perpendicular to **a**.

(2) Rotate $\mathbf{b}'$ through a right angle in the plane obtaining the vector $\mathbf{b}''$ such that $[\mathbf{a}, \mathbf{b}', \mathbf{b}'']$ is a right-handed system.

(3) Multiply $\mathbf{b}''$ by $|\mathbf{a}|$.

We shall (page 198) make use of the above in another proof of the distributive law for the vector product.

Alternative treatment of the vector product

We now give another method of defining the vector product and of developing its algebra. The sections indicated by * may be omitted without any loss of continuity, but the reader is advised not to do so in order to gain further insight into the technique involved in creating the necessary algebra, starting from a different definition.

* ***Definition of the vector product.*** *If* **a**, **b** *are non-zero, non-parallel vectors and* θ *is the angle between them then the vector product of* **a** *and* **b** *written* $\mathbf{a}\times\mathbf{b}$ *is the vector whose*

(i) *magnitude is* $|\mathbf{a}|\,|\mathbf{b}|\,|\sin\theta|$,

(ii) *direction is perpendicular to both* $\mathbf{a}$ *and* $\mathbf{b}$, *and*

(iii) *sense is such that* $[\mathbf{a}, \mathbf{b}, \mathbf{a}\times\mathbf{b}]$ *forms a right-handed system.*

This definition can be put also into the following equivalent form:

If $\mathbf{a}$, $\mathbf{b}$ *are non-zero, non-parallel vectors,* θ *the angle between them and* $\hat{\mathbf{n}}$ *the unit vector perpendicular to both* $\mathbf{a}$ *and* $\mathbf{b}$ *such that* $[\mathbf{a}, \mathbf{b}, \hat{\mathbf{n}}]$ *forms a right-handed system, the vector product of* $\mathbf{a}$ *and* $\mathbf{b}$ *written* $\mathbf{a}\times\mathbf{b}$ *is the vector given by*

$$\mathbf{a}\times\mathbf{b} = |\mathbf{a}|\,|\mathbf{b}|\,|\sin\theta|\,\hat{\mathbf{n}}.$$

If either $\mathbf{a}$ or $\mathbf{b}$ is the zero vector or if $\mathbf{a}$ is parallel to $\mathbf{b}$ then the above definition is amended since conditions (ii) and (iii) cannot be satisfied. In this case we define the vector product to be the zero vector $\mathbf{0}$ which agrees with (i) since either $|\mathbf{a}|$ or $|\mathbf{b}|$ is zero when either $\mathbf{a}$ or $\mathbf{b}$ is the zero vector, or $\sin\theta = 0$ when $\mathbf{a}$ is parallel to $\mathbf{b}$. Therefore we have

$$\mathbf{0}\times\mathbf{a} = \mathbf{a}\times\mathbf{0} = \mathbf{0} \quad \text{for all } \mathbf{a}$$

and

$$\mathbf{a}\times\mathbf{b} = \mathbf{0} \quad \text{for } \mathbf{a} \text{ parallel to } \mathbf{b}.$$

* *Consequences of the definition*

(1) $\mathbf{a}\times\mathbf{b} = -(\mathbf{b}\times\mathbf{a})$

Proof. This follows directly from the definition since $\mathbf{a}\times\mathbf{b}$ and $\mathbf{b}\times\mathbf{a}$ have the same magnitude and direction but opposite sense, apart from the special cases of $\mathbf{a} = \mathbf{0}$ or $\mathbf{b} = \mathbf{0}$ or $\mathbf{a}$ parallel to $\mathbf{b}$ when each side of (1) reduces to the zero vector.

Alternatively,

$$\mathbf{a}\times\mathbf{b} = |\mathbf{a}|\,|\mathbf{b}|\,|\sin\theta|\,\hat{\mathbf{n}}$$

and

$$\begin{aligned}\mathbf{b}\times\mathbf{a} &= |\mathbf{b}|\,|\mathbf{a}|\,|\sin\theta|\,(-\hat{\mathbf{n}})\\ &= -|\mathbf{a}|\,|\mathbf{b}|\,|\sin\theta|\,\hat{\mathbf{n}}\\ &= -(\mathbf{a}\times\mathbf{b}),\end{aligned}$$

i.e.

$$\mathbf{a}\times\mathbf{b} = -(\mathbf{b}\times\mathbf{a}).$$

Thus the vector product is not commutative.

(2) $(\mathbf{a}\times\mathbf{b})\times\mathbf{c} \neq \mathbf{a}\times(\mathbf{b}\times\mathbf{c})$ ***in general***

Proof. If $\mathbf{a} = \mathbf{0}$ or $\mathbf{b} = \mathbf{0}$ or $\mathbf{c} = \mathbf{0}$ then each side of (2) reduces to the zero vector and in this case

$$(\mathbf{a}\times\mathbf{b})\times\mathbf{c} = \mathbf{a}\times(\mathbf{b}\times\mathbf{c}).$$

Generally however the associative law does not hold for the vector product. For $(\mathbf{a}\times\mathbf{b})\times\mathbf{c}$ is in the plane containing **a**, **b** and $\mathbf{a}\times(\mathbf{b}\times\mathbf{c})$ is in the plane containing **b**, **c**. These planes are, in general, different. Hence

$$(\mathbf{a}\times\mathbf{b})\times\mathbf{c} \neq \mathbf{a}\times(\mathbf{b}\times\mathbf{c}) \text{ in general.}$$

(3) $(p\mathbf{a})\times(q\mathbf{b}) = pq(\mathbf{a}\times\mathbf{b})$ ***where*** p, q ***are numbers***

Proof. (*a*) $\mathbf{a} = \mathbf{0}$ or $\mathbf{b} = \mathbf{0}$ or **a** parallel to **b**.

Each side of (3) reduces to the zero vector and thus (3) holds for these special cases.

(*b*) $\mathbf{a} \neq \mathbf{0}$, or $\mathbf{b} \neq \mathbf{0}$, or **a** not parallel to **b**.

The magnitude of both $(p\mathbf{a})\times(q\mathbf{b})$ and $pq(\mathbf{a}\times\mathbf{b})$ is

$$|p|\,|q|\,|\mathbf{a}|\,|\mathbf{b}|\,|\sin\theta|.$$

The direction of both $(p\mathbf{a})\times(q\mathbf{b})$ and $pq(\mathbf{a}\times\mathbf{b})$ is perpendicular to both **a** and **b**.

The sense of both $(p\mathbf{a})\times(q\mathbf{b})$ and $pq(\mathbf{a}\times\mathbf{b})$ is that of $\mathbf{a}\times\mathbf{b}$ if $p \gtrless 0$, $q \gtrless 0$ but is that of $-(\mathbf{a}\times\mathbf{b})$ if $p \gtrless 0$, $q \lessgtr 0$.

Thus $(p\mathbf{a})\times(q\mathbf{b})$ and $pq(\mathbf{a}\times\mathbf{b})$ have the same magnitude, direction and sense. Therefore

$$(p\mathbf{a})\times(q\mathbf{b}) = pq(\mathbf{a}\times\mathbf{b}).$$

In particular
$$\mathbf{a}\times(n\mathbf{b}) = (n\mathbf{a})\times\mathbf{b} = n(\mathbf{a}\times\mathbf{b})$$
and
$$\mathbf{a}\times(-\mathbf{b}) = (-\mathbf{a})\times\mathbf{b} = -(\mathbf{a}\times\mathbf{b}).$$

(4) $\mathbf{a}\times(\mathbf{b}+\mathbf{c}) = \mathbf{a}\times\mathbf{b}+\mathbf{a}\times\mathbf{c}$.

Proof. If $\mathbf{a} = \mathbf{0}$ each side of (4) is the zero vector and the distributive law is obviously true. We shall now consider the case when $\mathbf{a} \neq \mathbf{0}$.

In Fig. 11.8 π is the plane through O perpendicular to $\mathbf{OA} = \mathbf{a}$. $\mathbf{OB}' = \mathbf{b}'$, $\mathbf{OC}' = \mathbf{c}'$, $\mathbf{OD}' = (\mathbf{b}+\mathbf{c})'$ are the projections of the vectors $\mathbf{OB} = \mathbf{b}$, $\mathbf{OC} = \mathbf{c}$, $\mathbf{OD} = \mathbf{b}+\mathbf{c}$ respectively on π. Since the projection of a sum of vectors on a plane is equal to the sum of the projections of the separate vectors on the plane, $OB'D'C'$ is a parallelogram.

The parallelogram $OB'D'C'$ is now rotated in the plane π about O through a right angle resulting in the parallelogram $OB''D''C''$, in which $\mathbf{OB}'' = \mathbf{b}''$, $\mathbf{OC}'' = \mathbf{c}''$, $\mathbf{OD}'' = (\mathbf{b}+\mathbf{c})''$, and the sense of rotation such that $[\mathbf{a}, \mathbf{b}', \mathbf{b}'']$, $[\mathbf{a}, \mathbf{c}', \mathbf{c}'']$ and $[\mathbf{a}, (\mathbf{b}+\mathbf{c})', (\mathbf{b}+\mathbf{c})'']$ are right-handed. Fig. 11.9 shows these vectors, the plane π taken to be

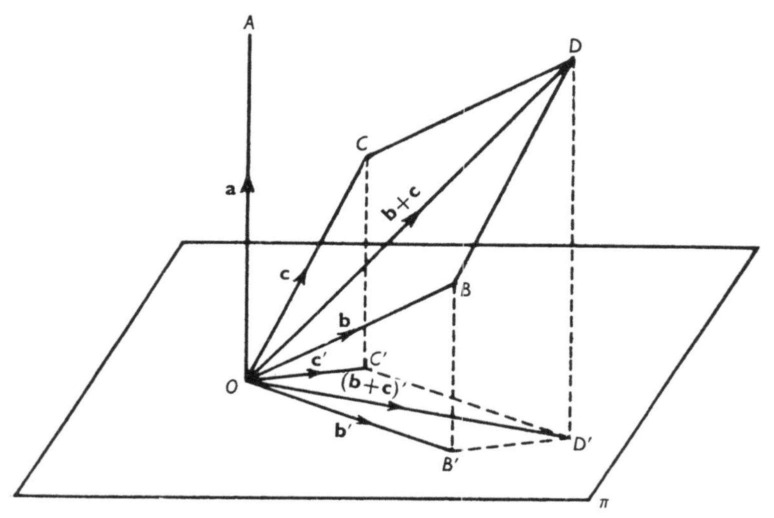

Fig. 11.8

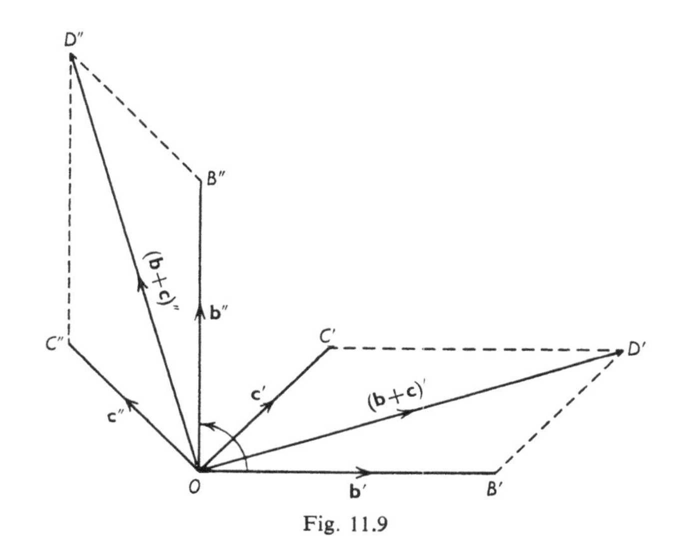

Fig. 11.9

that of the paper, with $\mathbf{OA} = \mathbf{a}$ perpendicular to and out of the plane of the paper. Then

$$(\mathbf{b}+\mathbf{c})'' = \mathbf{b}''+\mathbf{c}''. \tag{11·13}$$

The effect of the rotation is the same as obtaining the vectors $\mathbf{b}''$, $\mathbf{c}''$, $(\mathbf{b}+\mathbf{c})''$ by rotating the vectors $\mathbf{b}'$, $\mathbf{c}'$, $(\mathbf{b}+\mathbf{c})'$ each through a right angle. Now we have shown (page 195) that the vector $\mathbf{a}\times\mathbf{b}$ can be obtained by multiplying $\mathbf{b}''$ by $|\mathbf{a}|$. Thus we have

$$\mathbf{a}\times\mathbf{b} = |\mathbf{a}|\mathbf{b}'',$$

$$\mathbf{a}\times\mathbf{c} = |\mathbf{a}|\mathbf{c}''$$

and

$$\mathbf{a}\times(\mathbf{b}+\mathbf{c}) = |\mathbf{a}|(\mathbf{b}+\mathbf{c})''.$$

Multiplying (11·13) by $|\mathbf{a}|$ gives

$$|\mathbf{a}|(\mathbf{b}+\mathbf{c})'' = |\mathbf{a}|\mathbf{b}''+|\mathbf{a}|\mathbf{c}''.$$

Therefore

$$\mathbf{a}\times(\mathbf{b}+\mathbf{c}) = \mathbf{a}\times\mathbf{b}+\mathbf{a}\times\mathbf{c}. \tag{11·14}$$

Extending this result we have

$$\mathbf{a}\times(\mathbf{b}+\mathbf{c}+\mathbf{d}+\ldots) = \mathbf{a}\times\mathbf{b}+\mathbf{a}\times\mathbf{c}+\mathbf{a}\times\mathbf{d}+\ldots,$$

and thus the vector product is distributive over addition.

Multiplying (11·14) by -1 we get

$$(\mathbf{b}+\mathbf{c})\times\mathbf{a} = \mathbf{b}\times\mathbf{a}+\mathbf{c}\times\mathbf{a}.$$

(5) *The unit vectors* **i**, **j**, **k**. Since by definition $\mathbf{a}\times\mathbf{b} = \mathbf{0}$ when $\mathbf{a} = \mathbf{0}$ or $\mathbf{b} = \mathbf{0}$ or **a** is parallel to **b** we have

$$\mathbf{i}\times\mathbf{i} = \mathbf{j}\times\mathbf{j} = \mathbf{k}\times\mathbf{k} = \mathbf{0}.$$

Using the general definition, if **a** and **b** are unit vectors perpendicular to one another then $\mathbf{a}\times\mathbf{b}$ is the vector whose modulus is unity, direction is perpendicular to both **a** and **b** and whose sense is such that $[\mathbf{a}, \mathbf{b}, \mathbf{a}\times\mathbf{b}]$ forms a right-handed system. Thus for the mutually perpendicular unit vectors **i**, **j**, **k** which form a right-handed system we have

$$\mathbf{j}\times\mathbf{k} = \mathbf{i}, \quad \mathbf{k}\times\mathbf{i} = \mathbf{j}, \quad \mathbf{i}\times\mathbf{j} = \mathbf{k}$$

and

$$\mathbf{k}\times\mathbf{j} = -\mathbf{i}, \quad \mathbf{i}\times\mathbf{k} = -\mathbf{j}, \quad \mathbf{j}\times\mathbf{i} = -\mathbf{k}.$$

* *Vector product in terms of components of the vectors*

By making use of the results of (3), (4) and (5) of the previous section, the formula for the vector product in terms of the components of the vectors may be obtained.

Let
$$\mathbf{a} = a_1\mathbf{i}+a_2\mathbf{j}+a_3\mathbf{k},$$
$$\mathbf{b} = b_1\mathbf{i}+b_2\mathbf{j}+b_3\mathbf{k}.$$

Therefore

$$\begin{aligned}\mathbf{a}\times\mathbf{b} &= (a_1\mathbf{i}+a_2\mathbf{j}+a_3\mathbf{k})\times(b_1\mathbf{i}+b_2\mathbf{j}+b_3\mathbf{k})\\ &= a_1b_1(\mathbf{i}\times\mathbf{i})+a_1b_2(\mathbf{i}\times\mathbf{j})+a_1b_3(\mathbf{i}\times\mathbf{k})\\ &\quad +a_2b_1(\mathbf{j}\times\mathbf{i})+a_2b_2(\mathbf{j}\times\mathbf{j})+a_2b_3(\mathbf{j}\times\mathbf{k})\\ &\quad +a_3b_1(\mathbf{k}\times\mathbf{i})+a_3b_2(\mathbf{k}\times\mathbf{j})+a_3b_3(\mathbf{k}\times\mathbf{k})\\ &= (a_2b_3-a_3b_2)\mathbf{i}+(a_3b_1-a_1b_3)\mathbf{j}+(a_1b_2-a_2b_1)\mathbf{k}.\end{aligned}$$

For ways of remembering the components and for worked examples see page 189.

Parallel vectors

At this point it is convenient to note the following four statements which are equivalent if $\mathbf{a}$ and $\mathbf{b}$ are non-zero vectors.

(i) $\mathbf{a}$ is parallel to $\mathbf{b}$,

(ii) $\mathbf{a} = k\mathbf{b}$ $(k \neq 0)$,

(iii) $\mathbf{a}.\mathbf{b} = |\mathbf{a}|\,|\mathbf{b}|$,

(iv) $\mathbf{a}\times\mathbf{b} = \mathbf{0}$.

Geometrical meaning of the vector product

If θ is the angle between $\mathbf{a}$ and $\mathbf{b}$ then

$$|\mathbf{a}\times\mathbf{b}| = |\mathbf{a}|\,|\mathbf{b}|\,|\sin\theta|.$$

Now the area of the parallelogram $OACB$ in which $\mathbf{OA} = \mathbf{a}$, $\mathbf{OB} = \mathbf{b}$ is given by $|\mathbf{a}|\,|\mathbf{b}|\,|\sin\theta|$. Therefore the modulus of the vector product is equal to the area of the parallelogram with adjacent sides formed by the two vectors.

Also since the area of the triangle OAB is half that of the parallelogram $OACB$ we have
$$\triangle OAB = \tfrac{1}{2}|\mathbf{a}\times\mathbf{b}|.$$

Comparison of scalar and vector products

The scalar and vector products enable us to decide whether two non-zero vectors $\mathbf{a}$, $\mathbf{b}$ are perpendicular or parallel; for if $\mathbf{a}.\mathbf{b} = 0$ the vectors are perpendicular and if $\mathbf{a}\times\mathbf{b} = \mathbf{0}$ the vectors are parallel.

Whenever the sine of the angle between two vectors is required the vector product is used. However if the angle is required the scalar product is used since θ is given uniquely by $\cos\theta = \mathbf{a}.\mathbf{b}/(|\mathbf{a}|\,|\mathbf{b}|)$, which is either positive or negative depending on the value of $\mathbf{a}.\mathbf{b}$. Using the vector product we have $|\sin\theta| = |\mathbf{a}\times\mathbf{b}|/(|\mathbf{a}|\,|\mathbf{b}|)$, which is always positive and thus it is not possible to distinguish between an acute angle and its obtuse supplement. It will be remembered that the same difficulty arises when finding the angle of a triangle by using the sine rule, whereas the cosine rule gives the angle unambiguously.

General examples

(1) *Find the unit vectors perpendicular to both* $\mathbf{a} = 3\mathbf{i}+\mathbf{j}-2\mathbf{k}$, $\mathbf{b} = 2\mathbf{i}-3\mathbf{j}-\mathbf{k}$.

We first find the vector $\mathbf{a}\times\mathbf{b}$ since this is a vector perpendicular to $\mathbf{a}$ and $\mathbf{b}$.

$$\mathbf{a}\times\mathbf{b} = (-1-6)\mathbf{i}+(-4+3)\mathbf{j}+(-9-2)\mathbf{k}$$

$$= -7\mathbf{i}-\mathbf{j}-11\mathbf{k}.$$

But
$$|\mathbf{a}\times\mathbf{b}| = (49+1+121)^{\frac{1}{2}} = 171^{\frac{1}{2}}.$$

The unit vectors are therefore $\pm(7\mathbf{i}+\mathbf{j}+11\mathbf{k})/\sqrt{171}$.

(2) *The position vectors of the points* A, B, C *are* $(8, 4, -3)$, $(6, 3, -4)$, $(7, 5, -5)$ *respectively. Find the area of triangle* ABC.

Since $\triangle ABC = \frac{1}{2}BA.BC\sin B$ and $|\mathbf{BA}\times\mathbf{BC}| = |\mathbf{BA}|\,|\mathbf{BC}|\,|\sin B|$ we have

$$\triangle ABC = \tfrac{1}{2}|\mathbf{BA}\times\mathbf{BC}|.$$

Now $\mathbf{BA} = \mathbf{a}-\mathbf{b} = 2\mathbf{i}+\mathbf{j}+\mathbf{k}$ and $\mathbf{BC} = \mathbf{c}-\mathbf{b} = \mathbf{i}+2\mathbf{j}-\mathbf{k}$.

Therefore
$$\triangle ABC = \tfrac{1}{2}|(2\mathbf{i}+\mathbf{j}+\mathbf{k})\times(\mathbf{i}+2\mathbf{j}-\mathbf{k})|$$

$$= \tfrac{1}{2}|(-1-2)\mathbf{i}+(1+2)\mathbf{j}+(4-1)\mathbf{k}|$$

$$= \tfrac{1}{2}|-3\mathbf{i}+3\mathbf{j}+3\mathbf{k}|$$

$$= \tfrac{3}{2}|-\mathbf{i}+\mathbf{j}+\mathbf{k}|$$

$$= \tfrac{3}{2}\sqrt{3}.$$

The area can also be found by finding the angle B using the scalar product $\mathbf{BA}.\mathbf{BC}$ but this method is not as straight-forward as using the vector product.

(3) *Find the perpendicular distance of a corner of a unit cube from a diagonal which does not pass through it.*

Referring to Fig. 11.10 the required distance is OH where

$$OH = FO \sin O\hat{F}H.$$

Now $$|\mathbf{FO} \times \mathbf{FC}| = |\mathbf{FO}|\,|\mathbf{FC}| \sin O\hat{F}H.$$

Therefore $$OH = \frac{|\mathbf{FO} \times \mathbf{FC}|}{|\mathbf{FC}|}.$$

$$\mathbf{FO} = -\mathbf{OF} = -\mathbf{i}-\mathbf{j},$$

$$\mathbf{FC} = \mathbf{OC}-\mathbf{OF} = \mathbf{k}-(\mathbf{i}+\mathbf{j}) = -\mathbf{i}-\mathbf{j}+\mathbf{k}.$$

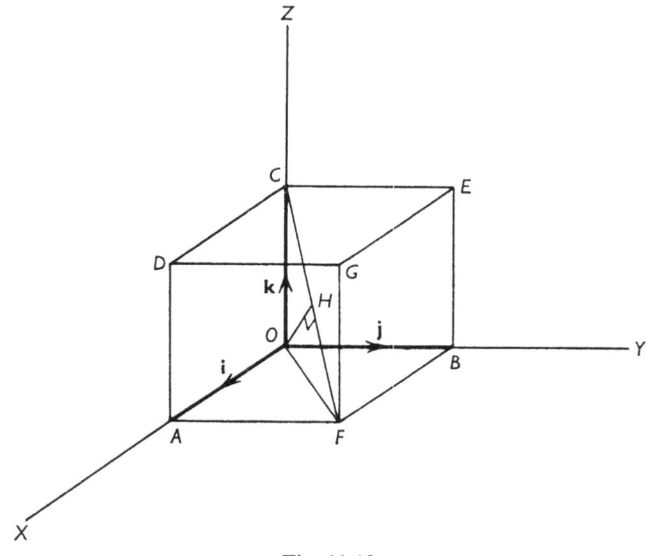

Fig. 11.10

Therefore $$|\mathbf{FO} \times \mathbf{FC}| = |(-\mathbf{i}-\mathbf{j}) \times (-\mathbf{i}-\mathbf{j}+\mathbf{k})|$$
$$= |-\mathbf{i}+\mathbf{j}|$$
$$= \sqrt{2}$$

and $$|\mathbf{FC}| = \sqrt{3}.$$

Hence $$OH = \frac{\sqrt{2}}{\sqrt{3}} = \frac{\sqrt{6}}{3}.$$

(4) *Prove the sine rule* $a/\sin A = b/\sin B = c/\sin C$ *for any triangle* ABC *by using vector products.*

Let $\mathbf{BC} = \mathbf{a}$, $\mathbf{CA} = \mathbf{b}$, $\mathbf{AB} = \mathbf{c}$ (Fig. 11.11).

Now

$$\mathbf{a}+\mathbf{b}+\mathbf{c} = \mathbf{0}.$$

By multiplying each side vectorially by $\mathbf{a}$ we obtain

$$\mathbf{a}\times(\mathbf{a}+\mathbf{b}+\mathbf{c}) = \mathbf{a}\times\mathbf{0}.$$
$$= \mathbf{0}.$$

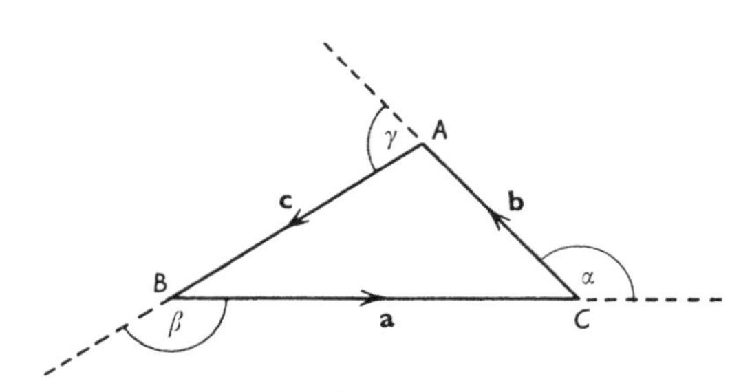

Fig. 11.11

Expanding $\qquad \mathbf{a}\times\mathbf{a}+\mathbf{a}\times\mathbf{b}+\mathbf{a}\times\mathbf{c} = \mathbf{0}.$

Since $\qquad \mathbf{a}\times\mathbf{a} = \mathbf{0},$

$$\mathbf{a}\times\mathbf{b}+\mathbf{a}\times\mathbf{c} = \mathbf{0},$$

that is, $\qquad \mathbf{a}\times\mathbf{b} = -(\mathbf{a}\times\mathbf{c}).$

Therefore $\qquad |\mathbf{a}\times\mathbf{b}| = |\mathbf{a}\times\mathbf{c}|,$

giving $\qquad |\mathbf{a}|\,|\mathbf{b}|\,|\sin\alpha| = |\mathbf{a}|\,|\mathbf{c}|\,|\sin\beta|,$

that is, $\qquad ab\sin C = ac\sin B.$

Therefore $\qquad \dfrac{b}{\sin B} = \dfrac{c}{\sin C} \quad (a \neq 0).$

Similarly by multiplying vectorially by $\mathbf{b}$ we get

$$\frac{a}{\sin A} = \frac{c}{\sin C}.$$

Therefore $\qquad \dfrac{a}{\sin A} = \dfrac{b}{\sin B} = \dfrac{c}{\sin C}.$

(5) *Prove that* $\mathbf{b} = \mathbf{c}$ when $\mathbf{a}.\mathbf{b} = \mathbf{a}.\mathbf{c}$ *and* $\mathbf{a}\times\mathbf{b} = \mathbf{a}\times\mathbf{c}$ $(\mathbf{a} \neq \mathbf{0})$, *but* $\mathbf{b}$ *is not necessarily equal to* $\mathbf{c}$ *when only* $\mathbf{a}.\mathbf{b} = \mathbf{a}.\mathbf{c}$ *or* $\mathbf{a}\times\mathbf{b} = \mathbf{a}\times\mathbf{c}$.

$$\mathbf{a}.\mathbf{b} = \mathbf{a}.\mathbf{c}.$$

This may be written $\mathbf{a}.(\mathbf{b}-\mathbf{c}) = 0.$ (11·15)

$$\mathbf{a}\times\mathbf{b} = \mathbf{a}\times\mathbf{c}.$$

This may be written $\mathbf{a}\times(\mathbf{b}-\mathbf{c}) = \mathbf{0}.$ (11·16)

From (11·15) since $\mathbf{a} \neq \mathbf{0}$, either $\mathbf{b}-\mathbf{c} = \mathbf{0}$, that is, $\mathbf{b} = \mathbf{c}$ or $(\mathbf{b}-\mathbf{c})$ is perpendicular to $\mathbf{a}$. From (11·16) since $\mathbf{a} \neq \mathbf{0}$, either $\mathbf{b}-\mathbf{c} = \mathbf{0}$ or $(\mathbf{b}-\mathbf{c})$ is parallel to $\mathbf{a}$, that is $\mathbf{b}-\mathbf{c} = k\mathbf{a}$ (k a constant number). Thus when $\mathbf{a}.\mathbf{b} = \mathbf{a}.\mathbf{c}$ and $\mathbf{a}\times\mathbf{b} = \mathbf{a}\times\mathbf{c}$ we have $\mathbf{b} = \mathbf{c}$. Otherwise, when either $\mathbf{a}.\mathbf{b} = \mathbf{a}.\mathbf{c}$ or $\mathbf{a}\times\mathbf{b} = \mathbf{a}\times\mathbf{c}$, $\mathbf{b}$ may not be equal to $\mathbf{c}$, that is, $(\mathbf{b}-\mathbf{c})$ may not be the zero vector, but $(\mathbf{b}-\mathbf{c})$ may be perpendicular to $\mathbf{a}$ in the former case, and $(\mathbf{b}-\mathbf{c})$ may be parallel to $\mathbf{a}$ (i.e. $\mathbf{b} = \mathbf{c}+k\mathbf{a}$) in the latter case.

(6) *Moment about the origin of the force* $\mathbf{F} = F_x\mathbf{i}+F_y\mathbf{j}$ *acting at the point* $P(x, y)$.

The magnitude of the moment of the force $\mathbf{F}$ about the origin O is given by pF where p is the perpendicular distance of the origin from the line of action of the force. We have (Fig. 11.12)

$$\begin{aligned} pF &= OP\sin\theta F \\ &= |\mathbf{r}|\,|\mathbf{F}|\sin\theta \\ &= |\mathbf{r}\times\mathbf{F}| \\ &= |(x\mathbf{i}+y\mathbf{j})\times(F_x\mathbf{i}+F_y\mathbf{j})| \\ &= |(F_y x-F_x y)\mathbf{k}| \\ &= |F_y x-F_x y|, \end{aligned}$$

which is the sum of the moments of $F_x\mathbf{i}$ and $F_y\mathbf{j}$ about O, the moment being reckoned positive or negative according as the sense of rotation about O implied by the direction of the force is anticlockwise or clockwise.

If we define the vector moment of a force $\mathbf{F}$ about the origin O to be the vector $\mathbf{r}\times\mathbf{F}$, where $\mathbf{r}$ is the position vector of a point on its line of action, we see that all coplanar forces will have vector moments perpendicular to the plane containing the forces, the senses of their moments being either into or out of the plane. The direction of each vector moment may be described by the motion of the tip of a right-handed screw as it is rotated either anticlockwise or clockwise in accordance with the rotation of $\mathbf{r}$ on to $\mathbf{F}$. It follows from the definition of the vector moment that for coplanar forces the sum of their moments is the algebraic sum of their several moments. Hence the definition is in agreement with the scalar way of finding the sum of the moments of coplanar forces about a point in their plane. It is when we come to deal with the moments of forces in three-dimensional space that we find the necessity of defining the moment as a vector (page 216).

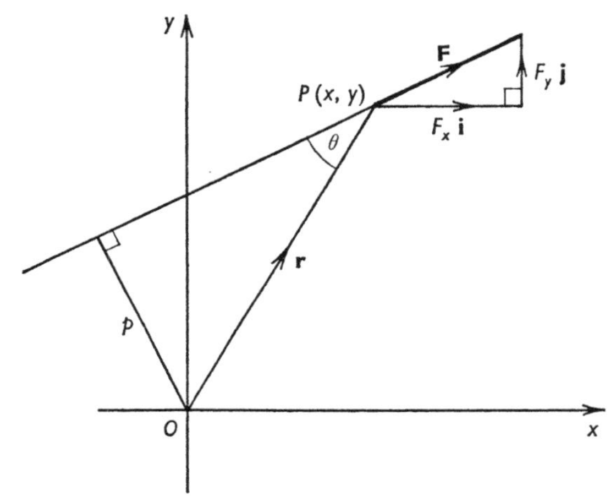

Fig. 11.12

Exercise 11(*b*)

(1) If $\mathbf{a} = 3\mathbf{i}+\mathbf{j}-2\mathbf{k}$, $\mathbf{b} = -2\mathbf{i}+3\mathbf{j}-\mathbf{k}$ find $\mathbf{a}\times\mathbf{b}$, and deduce the sine of the angle between **a** and **b**.

(2) Find the unit vectors perpendicular to both $\mathbf{p} = 3\mathbf{i}-\mathbf{j}+\mathbf{k}$, $\mathbf{q} = \mathbf{i}+2\mathbf{j}+2\mathbf{k}$.

(3) If $\mathbf{a} = (2, -2, 1)$, $\mathbf{b} = (-1, -1, 2)$, $\mathbf{c} = (1, -1, 1)$ evaluate (i) $\mathbf{a}.(\mathbf{b}\times\mathbf{c})$, (ii) $\mathbf{a}\times(\mathbf{b}\times\mathbf{c})$, (iii) $(\mathbf{a}\times\mathbf{b})\times\mathbf{c}$, (iv) $(\mathbf{a}\times\mathbf{b}).(\mathbf{a}\times\mathbf{c})$, (v) $(\mathbf{a}\times\mathbf{b})\times(\mathbf{a}\times\mathbf{c})$.

(4) If $\mathbf{a} = (2, 0, 3)$, $\mathbf{b} = (1, -1, 2)$, $\mathbf{c} = (4, 3, -1)$ verify that

$$\mathbf{a}\times(\mathbf{b}+\mathbf{c}) = \mathbf{a}\times\mathbf{b}+\mathbf{a}\times\mathbf{c}$$

and

$$\mathbf{a}\times(\mathbf{b}\times\mathbf{c}) = (\mathbf{a}.\mathbf{c})\mathbf{b}-(\mathbf{a}.\mathbf{b})\mathbf{c}.$$

(5) (i) If $|\mathbf{a}+\mathbf{b}| = |\mathbf{a}|+|\mathbf{b}|$ prove $\mathbf{a}\times\mathbf{b} = \mathbf{0}$.
(ii) If $\mathbf{c} = \mathbf{a}+\mathbf{b}$ prove $\mathbf{b}\times\mathbf{a} = \mathbf{b}\times\mathbf{c} = \mathbf{c}\times\mathbf{a}$.

(6) By expressing in components show that

$$\mathbf{a}\times(\mathbf{b}\times\mathbf{c}) = (\mathbf{a}\times\mathbf{b})\times\mathbf{c} \quad \text{when} \quad (\mathbf{a}.\mathbf{b})\mathbf{c} = (\mathbf{b}.\mathbf{c})\mathbf{a}.$$

(7) P, Q, R are the points $(0, 7, 3)$, $(-1, 2, 1)$, $(3, 0, -2)$ respectively. Find the area of the triangle PQR. Also find the unit vectors normal to the plane of the triangle.

(8) If **a**, **b**, **c** are the position vectors of the points A, B, C respectively, show that the area of the triangle ABC is

$$\tfrac{1}{2}|\mathbf{b}\times\mathbf{c}+\mathbf{c}\times\mathbf{a}+\mathbf{a}\times\mathbf{b}|.$$

(9) If **a** and **b** are unit vectors such that $\mathbf{a} = \mathbf{i}\cos\theta+\mathbf{j}\sin\theta$ and $\mathbf{b} = \mathbf{i}\cos\phi+\mathbf{j}\sin\phi$, obtain $\mathbf{a}.\mathbf{b}$ and $\mathbf{a}\times\mathbf{b}$. Hence prove that

$$\cos(\theta-\phi) = \cos\theta\cos\phi+\sin\theta\sin\phi,$$

$$\sin(\theta-\phi) = \sin\theta\cos\phi-\cos\theta\sin\phi.$$

(10) Prove that the moment about the origin O of the resultant of several forces in the x–y plane and acting through the point (x, y) is equal to the sum of the moments of the forces about O. This is Varignon's theorem in two dimensions for intersecting forces. (*Hint*. Use the result of page 205.)

Application of the vector product to three-dimensional coordinate geometry

The vector product is useful in finding the direction-vector of a line perpendicular to two given lines. Hence the normal-vector of the plane containing two given lines, or parallel to two given lines, can be found, by using the vector product.

Examples

(1) *A plane passes through the points* $A(1, 2, 3)$, $B(-1, 2, 0)$ *and* $C(2, -1, -1)$. *Find its equation.*

A solution of this problem involves the normal-vector of the plane together with a point on it, of which we have a choice of three. Another solution has been given on page 151.

The normal-vector is perpendicular to both **AB** and **AC**. Therefore the vector product of **AB** and **AC** may be taken as the normal-vector **n**. We have

$$\mathbf{n} = (-2, 0, -3)\times(1, -3, -4)$$
$$= (-9, -11, 6).$$

The equation of the plane is given by

$$\mathbf{n}.\mathbf{r} = k, \quad \text{i.e.} \quad -9x-11y+6z = k,$$

where k is a constant to be determined. Since $A(1, 2, 3)$ is a point on the plane we have

$$-9-22+18 = k,$$

i.e.

$$k = -13.$$

Thus the equation of the plane is

$$9x+11y-6z = 13.$$

(2) *A tetrahedron has its vertices at* $O(0, 0, 0)$, $A(1, 2, 1)$, $B(2, 1, 3)$ *and* $C(-1, 1, 2)$. *Calculate the angle between the faces OAB and ABC.*

We have

$$\mathbf{OA} = \mathbf{i}+2\mathbf{j}+\mathbf{k}, \quad \mathbf{OB} = 2\mathbf{i}+\mathbf{j}+3\mathbf{k}, \quad \mathbf{OC} = -\mathbf{i}+\mathbf{j}+2\mathbf{k}.$$

The angle between the faces OAB and ABC is the angle between any two vectors normal to these faces. Since the vector product $\mathbf{OA}\times\mathbf{OB}$ is perpendicular to both $\mathbf{OA}$ and $\mathbf{OB}$, it is perpendicular to the face OAB, i.e. $\mathbf{OA}\times\mathbf{OB}$ is a normal-vector of the face OAB. Similarly $\mathbf{AB}\times\mathbf{AC}$ is a normal-vector of the face ABC. Denoting these normal-vectors by $\mathbf{n}_1$, $\mathbf{n}_2$ we have

$$\begin{aligned}\mathbf{n}_1 &= \mathbf{OA}\times\mathbf{OB}\\ &= (\mathbf{i}+2\mathbf{j}+\mathbf{k})\times(2\mathbf{i}+\mathbf{j}+3\mathbf{k})\\ &= 5\mathbf{i}-\mathbf{j}-3\mathbf{k}\end{aligned}$$

and

$$\begin{aligned}\mathbf{n}_2 &= \mathbf{AB}\times\mathbf{AC}\\ &= (\mathbf{i}-\mathbf{j}+2\mathbf{k})\times(-2\mathbf{i}-\mathbf{j}+\mathbf{k})\\ &= \mathbf{i}-5\mathbf{j}-3\mathbf{k}.\end{aligned}$$

The angle θ between the two faces is given by

$$\begin{aligned}\cos\theta &= \frac{\mathbf{n}_1.\mathbf{n}_2}{|\mathbf{n}_1|\,|\mathbf{n}_2|}\\ &= \frac{5+5+9}{\sqrt{35}.\sqrt{35}}.\end{aligned}$$

Hence the required angle is $\cos^{-1}(19/35)$.

(3) *The shortest distance between two skew lines.*

The shortest distance between two skew lines is the length of the common perpendicular P_1P_2 of the lines. Let L_1, L_2 be two lines, $\mathbf{l}_1$, $\mathbf{l}_2$ their direction-vectors and $A_1(\mathbf{r}_1)$, $A_2(\mathbf{r}_2)$ be two points on L_1, L_2 respectively (Fig. 11.13). The common perpendicular has direction-

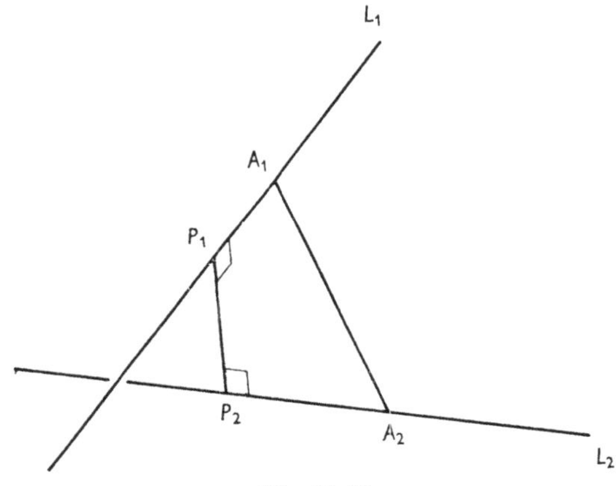

Fig. 11.13

vector $\mathbf{l}_1 \times \mathbf{l}_2$. The shortest distance is given by the projection of $\mathbf{A}_1\mathbf{A}_2$ on the unit vector in the direction of P_1P_2, that is, by

$$\mathbf{A}_1\mathbf{A}_2 . \frac{\mathbf{l}_1 \times \mathbf{l}_2}{|\mathbf{l}_1 \times \mathbf{l}_2|} = (\mathbf{r}_2 - \mathbf{r}_1) . \frac{(\mathbf{l}_1 \times \mathbf{l}_2)}{|\mathbf{l}_1 \times \mathbf{l}_2|}.$$

Denoting the numerical value of this distance by p we have

$$p = \frac{|(\mathbf{r}_1 - \mathbf{r}_2) . (\mathbf{l}_1 \times \mathbf{l}_2)|}{|\mathbf{l}_1 \times \mathbf{l}_2|}.$$

It is interesting to compare this method of finding the shortest distance with the method of the example on page 148. We have from this example

$$\mathbf{r}_1 = (-7, 5, 4), \quad \mathbf{r}_2 = (-4, 0, 19),$$

$$\mathbf{l}_1 = (-8, 3, 1), \quad \mathbf{l}_2 = (4, 3, -2).$$

Hence $\mathbf{r}_1 - \mathbf{r}_2 = (-3, 5, -15)$, $\mathbf{l}_1 \times \mathbf{l}_2 = (-9, -12, -36)$.

Therefore

$$p = \frac{|(-3, 5, -15) . (-9, -12, -36)|}{|(-9, -12, -36)|} = \frac{507}{39} = 13.$$

Exercise 11(*c*)

(1) By finding the normal-vector of the required plane, repeat Questions 1, 2 and 3 of Exercise 9(*c*).

(2) Show that the points

$$(0, 1, 6), \quad (3, 4, 3), \quad (-2, -1, 8), \quad (-1, -8, -1)$$

are coplanar. Find the equation of the plane containing these four points.

(3) Determine the nature of the following straight lines in relation to the plane $3x-2y+z = 2$.

(i) $\dfrac{x}{2} = \dfrac{y-2}{-3} = \dfrac{z-6}{-12}$,

(ii) $\dfrac{x+7}{-1} = \dfrac{y-3}{4} = \dfrac{z-4}{11}$,

(iii) $\dfrac{x-2}{5} = \dfrac{y+6}{6} = \dfrac{z+16}{3}$,

(iv) $\dfrac{x-1}{3} = \dfrac{y-2}{-2} = \dfrac{z+11}{1}$.

(4) Prove that the length of the shortest distance from the point $\mathbf{a}$ to the line joining the points $\mathbf{b}$ and $\mathbf{c}$ is given by

$$\frac{|\mathbf{a}\times\mathbf{b}+\mathbf{b}\times\mathbf{c}+\mathbf{c}\times\mathbf{a}|}{|\mathbf{b}-\mathbf{c}|}.$$

(5) The rectangular Cartesian co-ordinates of points A, B, C, D are $(1, 0, 3)$, $(2, 2, 2)$, $(4, 1, 0)$, $(6, -2, 5)$ respectively. Find a unit vector $\mathbf{n}$ which is perpendicular to both AB and CD.

Show that the shortest distance between the lines AB and CD is $|\mathbf{n}.\overrightarrow{AC}|$ and evaluate this.

Show also that the equation of the plane through AB parallel to CD may be written $\mathbf{n}.\mathbf{r} = \mathbf{n}.\overrightarrow{OA}$ and this is equivalent to

$$x-y-z+2 = 0. \qquad \text{(M.A.)}$$

(6) A tetrahedron has its vertices at the points $O(0, 0, 0)$, $P(2, 1, 1)$, $Q(1, 2, 2)$, $R(0, 0, 3)$. Find (i) the angle between the planes OPR, OQR, (ii) the angle between the line OP and the plane PQR, (iii) the area of the face PQR, (iv) the shortest distance between OP and QR.

(7) Prove that the shortest distance between two opposite edges of a regular tetrahedron is $\sqrt{2}a$, where $2a$ is the length of an edge.

Differentiation of the vector product

By definition, if $\mathbf{a} = \mathbf{f}(t)$, $\mathbf{b} = \boldsymbol{\phi}(t)$ then

$$\frac{d}{dt}(\mathbf{a}\times\mathbf{b}) = \underset{\delta t\to 0}{\mathrm{Lt}} \frac{(\mathbf{a}+\delta\mathbf{a})\times(\mathbf{b}+\delta\mathbf{b})-(\mathbf{a}\times\mathbf{b})}{\delta t}$$

$$= \underset{\delta t\to 0}{\mathrm{Lt}} \frac{(\mathbf{a}\times\mathbf{b})+(\mathbf{a}\times\delta\mathbf{b})+(\delta\mathbf{a}\times\mathbf{b})+(\delta\mathbf{a}\times\delta\mathbf{b})-(\mathbf{a}\times\mathbf{b})}{\delta t}$$

$$= \underset{\delta t\to 0}{\mathrm{Lt}} \left(\mathbf{a}\times\frac{\delta\mathbf{b}}{\delta t}+\frac{\delta\mathbf{a}}{\delta t}\times\mathbf{b}+\frac{\delta\mathbf{a}}{\delta t}\times\delta\mathbf{b}\right)$$

$$= \mathbf{a}\times\frac{d\mathbf{b}}{dt}+\frac{d\mathbf{a}}{dt}\times\mathbf{b}.$$

Thus the derivatives of the vector and scalar products are formed in the same way as the derivative of the product of two scalar functions. However in the case of the derivative of the vector product the order of **a** and **b** must remain unaltered.

The following special case is of note:

By putting $\mathbf{b} = d\mathbf{a}/dt$ we have

$$\frac{d}{dt}\left(\mathbf{a}\times\frac{d\mathbf{a}}{dt}\right) = \mathbf{a}\times\frac{d}{dt}\left(\frac{d\mathbf{a}}{dt}\right)+\frac{d\mathbf{a}}{dt}\times\frac{d\mathbf{a}}{dt}$$

$$= \mathbf{a}\times\frac{d^2\mathbf{a}}{dt^2}$$

since
$$\frac{d\mathbf{a}}{dt}\times\frac{d\mathbf{a}}{dt} = \mathbf{0}.$$

Example

If $\mathbf{a} = 6t^2\mathbf{i}+4t\mathbf{j}-3\mathbf{k}$, $\mathbf{b} = 2t^2\mathbf{i}-t\mathbf{j}+\mathbf{k}$ *find* $d(\mathbf{a}\times\mathbf{b})/dt$.

$$\frac{d}{dt}(\mathbf{a}\times\mathbf{b}) = \mathbf{a}\times\frac{d\mathbf{b}}{dt}+\frac{d\mathbf{a}}{dt}\times\mathbf{b}$$

$$= (6t^2\mathbf{i}+4t\mathbf{j}-3\mathbf{k})\times(4t\mathbf{i}-\mathbf{j})$$

$$+(12t\mathbf{i}+4\mathbf{j})\times(2t^2\mathbf{i}-t\mathbf{j}+\mathbf{k})$$

$$= -3\mathbf{i}-12\mathbf{j}-22t^2\mathbf{k}+4\mathbf{i}-12t\mathbf{j}-20t^2\mathbf{k}$$

$$= \mathbf{i}-24t\mathbf{j}-42t^2\mathbf{k}.$$

Alternatively, it is quicker to evaluate $(\mathbf{a}\times\mathbf{b})$ first and then to differentiate the result:

$$\begin{aligned}(\mathbf{a}\times\mathbf{b}) &= (6t^2\mathbf{i}+4t\mathbf{j}-3\mathbf{k})\times(2t^2\mathbf{i}-t\mathbf{j}+\mathbf{k})\\ &= t\mathbf{i}-12t^2\mathbf{j}-14t^3\mathbf{k}.\end{aligned}$$

Therefore $$\frac{d}{dt}(\mathbf{a}\times\mathbf{b}) = \mathbf{i}-24t\mathbf{j}-42t^2\mathbf{k}.$$

Integration

Suppose $\mathbf{a}$, $\mathbf{b}$ are vector functions of the scalar t. The indefinite integral of $\mathbf{a}\times\mathbf{b}$ with respect to t, written $\int(\mathbf{a}\times\mathbf{b})dt$ is defined as the general solution for $\mathbf{u}$ of the equation $d\mathbf{u}/dt = \mathbf{a}\times\mathbf{b}$, where $\mathbf{u}$ is another vector function of the scalar t. Suppose $\mathbf{u}$ has been found such that $d\mathbf{u}/dt = \mathbf{a}\times\mathbf{b}$. Let $\mathbf{c}$ be any constant vector. Then $d(\mathbf{u}+\mathbf{c})/dt$ is also equal to $\mathbf{a}\times\mathbf{b}$. Thus, in general, if $d\mathbf{u}/dt = \mathbf{a}\times\mathbf{b}$, we write

$$\int(\mathbf{a}\times\mathbf{b})dt = \mathbf{u}+\mathbf{c},$$

where $\mathbf{c}$ is the arbitrary constant of integration.

From the previous section, we have immediately

(i) $\int\{\mathbf{a}\times(d\mathbf{b}/dt)\}dt = \mathbf{a}\times\mathbf{b}+\mathbf{c}$ ($\mathbf{c}$ = constant of integration), when $\mathbf{a}$ is a constant vector, and

(ii) $\int\{\mathbf{a}\times(d^2\mathbf{a}/dt^2)\}dt = \mathbf{a}\times(d\mathbf{a}/dt)+\mathbf{c}$ ($\mathbf{c}$ = constant of integration), as a consequence of the special case on page 211 already noted.

The following examples indicate the use of components in order to evaluate integrals involving the vector product.

Examples

(1) *Evaluate* $\int(\mathbf{a}\times\mathbf{b})dt$ *if* $\mathbf{a} = \mathbf{p}\sin t+\mathbf{q}\cos t$, $\mathbf{b} = \mathbf{p}\cos t+\mathbf{q}\sin t$ *where* $\mathbf{p}$, $\mathbf{q}$ *are constant vectors and t is a parameter.*

$$\begin{aligned}\mathbf{a}\times\mathbf{b} &= (\mathbf{p}\sin t+\mathbf{q}\cos t)\times(\mathbf{p}\cos t+\mathbf{q}\sin t)\\ &= \sin t\cos t(\mathbf{p}\times\mathbf{p})+\sin^2 t(\mathbf{p}\times\mathbf{q})+\cos^2 t(\mathbf{q}\times\mathbf{p})\\ &\quad+\sin t\cos t(\mathbf{q}\times\mathbf{q})\\ &= (\cos^2 t-\sin^2 t)(\mathbf{q}\times\mathbf{p})\\ &= \cos 2t(\mathbf{q}\times\mathbf{p}).\end{aligned}$$

Therefore

$$\begin{aligned}\int(\mathbf{a}\times\mathbf{b})\,dt &= \int\cos 2t(\mathbf{q}\times\mathbf{p})\,dt\\ &= \tfrac{1}{2}\sin 2t(\mathbf{q}\times\mathbf{p})+\mathbf{c}\quad(\mathbf{c}\text{ arbitrary constant})\\ &= -\tfrac{1}{2}\sin 2t(\mathbf{p}\times\mathbf{q})+\mathbf{c}.\end{aligned}$$

(2) *If* $\mathbf{P} = y\mathbf{i}-z\mathbf{j}+x\mathbf{k}$, $\mathbf{r} = x\mathbf{i}+y\mathbf{j}+z\mathbf{k}$ *evaluate* $\int\mathbf{P}\times d\mathbf{r}$ *for* $x = 2t$, $y = t^2$, $z = 1$ *where t is a parameter.*

This integral corresponds to the type $\int\mathbf{P}.d\mathbf{r}$ discussed in chapter 8 on page 129.

$$\begin{aligned}\mathbf{P} &= y\mathbf{i}-z\mathbf{j}+x\mathbf{k}\\ &= t^2\mathbf{i}-\mathbf{j}+2t\mathbf{k}.\\ \mathbf{r} &= x\mathbf{i}+y\mathbf{j}+z\mathbf{k}\\ &= 2t\mathbf{i}+t^2\mathbf{j}+\mathbf{k}.\\ \int\mathbf{P}\times d\mathbf{r} &= \int\{(t^2\mathbf{i}-\mathbf{j}+2t\mathbf{k})\times(2\mathbf{i}+2t\mathbf{j})\}\,dt\\ &= \int\{-4t^2\mathbf{i}+4t\mathbf{j}+(2t^3+2)\mathbf{k}\}\,dt\\ &= -\frac{4t^3}{3}\mathbf{i}+2t^2\mathbf{j}+\left(\frac{t^4}{2}+2t\right)\mathbf{k}+\mathbf{c}.\end{aligned}$$

Exercise 11(*d*)

(1) If $\mathbf{a} = (1, -t, 2t^2)$, $\mathbf{b} = (2, 4t, t^2)$ obtain $d(\mathbf{a}\times\mathbf{b})/dt$.

(2) Differentiate with respect to t, $\mathbf{r}\times(d\mathbf{r}/dt)$.

(3) If $\mathbf{r} = \mathbf{a}\cos\omega t+\mathbf{b}\sin\omega t$, where ω, $\mathbf{a}$, $\mathbf{b}$ are constants, verify that $\mathbf{r}\times(d\mathbf{r}/dt) = \omega\mathbf{a}\times\mathbf{b}$.

(4) Show that if $\mathbf{a} = t\mathbf{p}+(1/t)\mathbf{q}$ and $\mathbf{b} = (1/t)\mathbf{p}+t\mathbf{q}$, where t, $\mathbf{p}$, $\mathbf{q}$ are constants, then $d(\mathbf{a}\times\mathbf{b})/dt = 2\{t+(1/t^3)\}\mathbf{p}\times\mathbf{q}$.

(5) Evaluate $\int(\mathbf{a}\times\mathbf{b})\,dt$ where $\mathbf{a} = 2t^2\mathbf{i}+t\mathbf{j}-\mathbf{k}$, $\mathbf{b} = t^2\mathbf{i}-2t\mathbf{j}+\mathbf{k}$.

(6) If $\mathbf{r}\times(d^2\mathbf{r}/dt^2) = \mathbf{0}$, show that $\mathbf{r}\times(d\mathbf{r}/dt) = \mathbf{c}$, where $\mathbf{c}$ is a constant vector.

(7) If $\mathbf{r}.(d\mathbf{r}/dt) = 0$ and $\mathbf{r}\times(d\mathbf{r}/dt) = \mathbf{0}$, show $\mathbf{r}$ is constant.

(8) If $(d\mathbf{r}/dt)\times\mathbf{s} = (d\mathbf{s}/dt)\times\mathbf{r}$, prove $\mathbf{r}\times\mathbf{s}$ is constant.

(9) If $(d\mathbf{r}/dt)\times(d^2\mathbf{r}/dt^2) = \mathbf{0}$ prove that $\mathbf{r} = \mathbf{a}+s\mathbf{b}$, where s is a scalar function of t.

(10) If $\mathbf{r}$ is the position vector of a variable point, relative to the origin, find the magnitude and direction of the vector $\int\mathbf{r}\times d\mathbf{r}$, where the integral is taken round the circle $x = a\cos\theta$, $y = a\sin\theta$, $z = 0$.

Applications of the vector product in mechanics

There are two applications of the vector product in mechanics which are of fundamental importance. They are the velocity of a particle of a rigid body rotating with a given angular velocity about an axis and the moment of a force about a point.

The angular velocity vector

It can be shown that the motion of a rigid body about a fixed point is one of rotation about an axis through the fixed point. Every point of the body on this axis is instantaneously at rest. Every other point of the body will describe a circle whose plane is perpendicular to this axis and whose centre is on this axis.

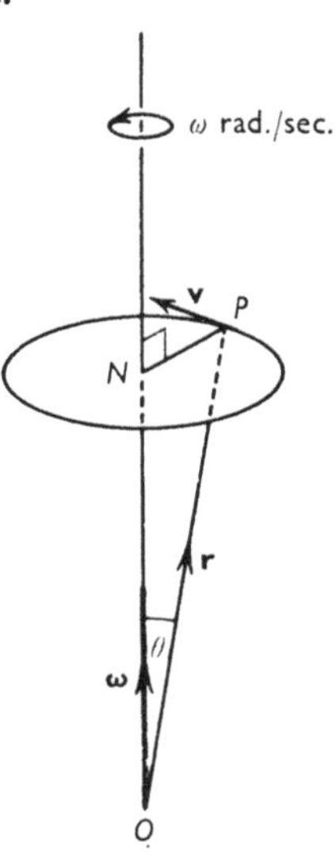

Fig. 11.14

The angular velocity of a rigid body with one point fixed can be specified by a vector $\boldsymbol{\omega}$ whose magnitude $|\boldsymbol{\omega}| = \omega$ is the measure of the angular velocity, whose direction indicates the axis through the fixed point about which the body rotates and whose sense is the same as that moved by a right-handed screw placed along the axis and driven by the rotation. If $\boldsymbol{\omega}$ is not constant in time, the direction of $\boldsymbol{\omega}$ is called the instantaneous axis.

We remark here that we are assuming the existence of the angular velocity vector, that is, we have not shown how such a vector arises in the consideration of the rotation of a rigid body.

Referring to Fig. 11.14 let O be any point on the axis, P be any point of the body and PN the perpendicular from P to the axis. The point P will describe a circle centre N and radius PN about the axis. At any instant its linear velocity $\mathbf{v}$ is tangential to this circle and hence perpendicular to the plane OPN. Since the magnitude of the linear velocity is given by the product of the magnitude of the angular velocity and the radius we have

$$|\mathbf{v}| = |\boldsymbol{\omega}|\,|\mathbf{OP}|\sin P\hat{O}N.$$

If $\mathbf{OP} = \mathbf{r}$ and θ is the angle between $\boldsymbol{\omega}$ and $\mathbf{r}$

$$|\mathbf{v}| = |\boldsymbol{\omega}|\,|\mathbf{r}|\sin\theta.$$

Now the vector $\boldsymbol{\omega}\times\mathbf{r}$ has magnitude $|\boldsymbol{\omega}|\,|\mathbf{r}|\sin\theta$ and its direction and sense are the same as those of $\mathbf{v}$. Therefore

$$\mathbf{v} = \dot{\mathbf{r}} = \boldsymbol{\omega}\times\mathbf{r}. \tag{11·17}$$

The velocity $\mathbf{v}$ given by (11·17) is independent of the choice of O on the axis. For suppose O' is another point on the axis and $\mathbf{r}'$ the position vector of P relative to O'. Then

$$\mathbf{v} = \boldsymbol{\omega}\times\mathbf{r}'.$$

Since $$\mathbf{OO}'+\mathbf{r}' = \mathbf{r}$$

we have
$$\begin{aligned}\mathbf{v} &= \boldsymbol{\omega}\times(\mathbf{r}-\mathbf{OO}')\\ &= \boldsymbol{\omega}\times\mathbf{r}-\boldsymbol{\omega}\times\mathbf{OO}'\\ &= \boldsymbol{\omega}\times\mathbf{r}.\end{aligned}$$

The angular velocity vector $\boldsymbol{\omega}$ is a unique vector for all points of the body. For suppose there is another vector $\boldsymbol{\omega}'$ for which we can find two non-parallel vectors $\mathbf{r}_1$, $\mathbf{r}_2$ such that

$$\boldsymbol{\omega}\times\mathbf{r}_1 = \boldsymbol{\omega}'\times\mathbf{r}_1, \quad \text{i.e.} \quad (\boldsymbol{\omega}-\boldsymbol{\omega}')\times\mathbf{r}_1 = \mathbf{0},$$

and $$\boldsymbol{\omega}\times\mathbf{r}_2 = \boldsymbol{\omega}'\times\mathbf{r}_2, \quad \text{i.e.} \quad (\boldsymbol{\omega}-\boldsymbol{\omega}')\times\mathbf{r}_2 = \mathbf{0}.$$

These imply $\boldsymbol{\omega}-\boldsymbol{\omega}' = \mathbf{0}$ or $\boldsymbol{\omega}-\boldsymbol{\omega}'$ is parallel to both $\mathbf{r}_1$ and $\mathbf{r}_2$ which cannot be so since $\mathbf{r}_1$ and $\mathbf{r}_2$ are non-parallel. Hence $\boldsymbol{\omega} = \boldsymbol{\omega}'$.

Example

A rigid body is rotating with angular velocity 14 *radians per second about an axis parallel to* $2\mathbf{i}-6\mathbf{j}+3\mathbf{k}$ *passing through the point* $A\,(1, 1, -2)$. *Calculate the velocity of the point* $P\,(1, 1, -3)$, *the unit of length being* 1 *metre.*

The unit vector parallel to the axis is $\frac{1}{7}(2\mathbf{i}-6\mathbf{j}+3\mathbf{k})$. Hence

$$\begin{aligned}\boldsymbol{\omega} &= \tfrac{14}{7}(2\mathbf{i}-6\mathbf{j}+3\mathbf{k})\\ &= 4\mathbf{i}-12\mathbf{j}+6\mathbf{k}.\\ \mathbf{r} &= \mathbf{AP} = -\mathbf{k}.\end{aligned}$$

Since $$\mathbf{v} = \boldsymbol{\omega}\times\mathbf{r},$$

$$\begin{aligned}\mathbf{v} &= (4\mathbf{i}-12\mathbf{j}+6\mathbf{k})\times(-\mathbf{k})\\ &= 12\mathbf{i}+4\mathbf{j}\\ &= 4(3\mathbf{i}+\mathbf{j}).\end{aligned}$$

Hence P moves with a speed of $4\sqrt{10}$ m/s in the direction and sense of the vector $(3\mathbf{i}+\mathbf{j})$.

Moment of a force about a point

Suppose a rigid body free to rotate about a fixed point O is acted on by a force **F**. Since force is a vector quantity localized in a straight line, the free vector **F** represents only the magnitude, direction and sense of the force but not its line of action. Suppose **F** is localized in the line AB, i.e. AB is the line of action of the force. Let P be any point on AB, the position vector of P relative to O being **r** (Fig. 11.15).

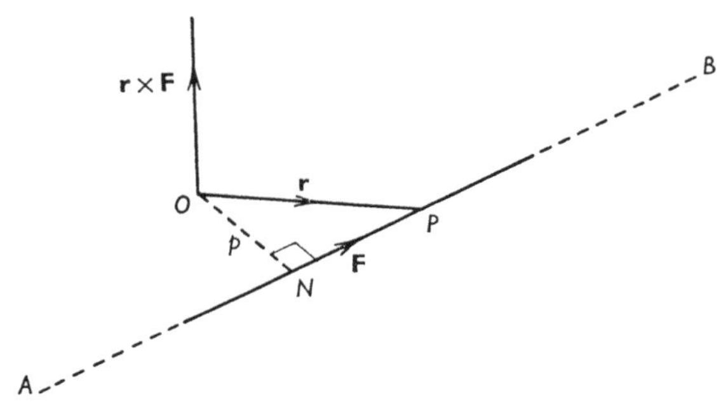

Fig. 11.15

Definition. *The moment of the force about the point O is the vector*

$$\mathbf{m} = \mathbf{r} \times \mathbf{F}.$$

From this we see that the moment is a vector perpendicular to the plane containing O and the line of action of the force. Its magnitude is $|\mathbf{r}|\,|\mathbf{F}|\sin\theta$, i.e. $p|\mathbf{F}|$ where p is the perpendicular from O to the line of action. Its sense is determined by the right-hand rule.

This moment is independent of the choice of P. For suppose P' is another point on the line of action. Then

$$\mathbf{m} = \mathbf{OP}' \times \mathbf{F}.$$

But
$$\mathbf{OP}' = \mathbf{OP} + \mathbf{PP}'.$$

Therefore
$$\begin{aligned}\mathbf{m} &= (\mathbf{OP} + \mathbf{PP}') \times \mathbf{F} \\ &= \mathbf{OP} \times \mathbf{F} + \mathbf{PP}' \times \mathbf{F} \\ &= \mathbf{r} \times \mathbf{F}\end{aligned}$$

since $\mathbf{PP}' \times \mathbf{F} = \mathbf{0}$, as both $\mathbf{PP}'$ and **F** have the same direction.

The two free vectors **F** and **m** completely specify the effect of the force acting on a rigid body. Its magnitude, direction and sense are given by $\mathbf{F}$. Its line of action lies in the plane through O perpendicular to $\mathbf{m}$, that is in the plane OPN, at a distance p given by $p|\mathbf{F}| = |\mathbf{m}|$, and on the side of O such that $[\mathbf{r}, \mathbf{F}, \mathbf{m}]$ is a right-handed system. In general any localized vector, i.e. a line vector, requires two free vectors for its complete specification.

Suppose we require the moment about the point A whose position vector relative to O is $\mathbf{a}$. Since

$$\mathbf{AP} = \mathbf{OP} - \mathbf{OA} = \mathbf{r} - \mathbf{a}$$

we have

$$\mathbf{m} = \mathbf{AP} \times \mathbf{F} = (\mathbf{r} - \mathbf{a}) \times \mathbf{F}.$$

Let $\mathbf{F}_1, \mathbf{F}_2, \mathbf{F}_3, \ldots, \mathbf{F}_n$ be forces acting through the point P and $\mathbf{R}$ their resultant. The moment of $\mathbf{R}$ about O is

$$\begin{aligned}\mathbf{r} \times \mathbf{R} &= \mathbf{r} \times (\mathbf{F}_1 + \mathbf{F}_2 + \mathbf{F}_3 + \ldots + \mathbf{F}_n) \\ &= \mathbf{r} \times \mathbf{F}_1 + \mathbf{r} \times \mathbf{F}_2 + \mathbf{r} \times \mathbf{F}_3 + \ldots + \mathbf{r} \times \mathbf{F}_n.\end{aligned}$$

Thus the moment about O of the resultant of the several forces through P is equal to the sum of the moments of the forces about the same point O (Varignon's theorem for concurrent forces).

Example

A force of 6 *newtons acts along the line AB where A, B are the points* $(2, -1, 3)$, $(4, -2, 5)$ *respectively. Find its moment about the point* $C\,(-1, 4, 2)$, *the unit of length being* 1 *metre.*

$$\mathbf{AB} = 2\mathbf{i} - \mathbf{j} + 2\mathbf{k}.$$

Let $\mathbf{F}$ be the force of 6 newtons acting along $\mathbf{AB}$.

$$\begin{aligned}\mathbf{F} &= \tfrac{6}{3}(2\mathbf{i} - \mathbf{j} + 2\mathbf{k}) \\ &= 4\mathbf{i} - 2\mathbf{j} + 4\mathbf{k}.\end{aligned}$$

Any point on the line of action of $\mathbf{F}$ may be taken as the point of application. Taking A as the point we have

$$\mathbf{CA} = 3\mathbf{i} - 5\mathbf{j} + \mathbf{k}.$$

Then the moment $\mathbf{m}$ of the force about C is given by

$$\begin{aligned}\mathbf{m} &= \mathbf{CA} \times \mathbf{F} \\ &= (3\mathbf{i} - 5\mathbf{j} + \mathbf{k}) \times (4\mathbf{i} - 2\mathbf{j} + 4\mathbf{k}) \\ &= -18\mathbf{i} - 8\mathbf{j} + 14\mathbf{k} \\ &= 2(-9\mathbf{i} - 4\mathbf{j} + 7\mathbf{k}).\end{aligned}$$

Hence the moment is $2\sqrt{146}$ Nm in the direction and sense of the vector $(-9\mathbf{i}-4\mathbf{j}+7\mathbf{k})$.

Vector equations

The following examples show how by forming vector or scalar products the solutions of some vector equations are found. Vector equations are considered in greater detail in the next chapter after the development of the triple products.

In keeping with the usual practice in number algebra we denote the unknown vector by $\mathbf{x}$. In solving for $\mathbf{x}$ we must find all values of $\mathbf{x}$ for which the equation is true.

(1) *Solve for* $\mathbf{x}$ *the equation* $\mathbf{a}\times\mathbf{x} = \mathbf{a}\times\mathbf{b}$, $\mathbf{a} \neq \mathbf{0}$.

Method I. Since
$$\mathbf{a}\times\mathbf{x} = \mathbf{a}\times\mathbf{b}$$
we have
$$\mathbf{a}\times(\mathbf{x}-\mathbf{b}) = \mathbf{0}.$$
Hence $\mathbf{a}$ is parallel to $(\mathbf{x}-\mathbf{b})$, that is,
$$\mathbf{x}-\mathbf{b} = \lambda\mathbf{a},$$
where λ is a parameter. Thus we see that the solution is not unique, the general solution being
$$\mathbf{x} = \mathbf{b}+\lambda\mathbf{a}.$$

Method II. Let $\mathbf{p} = \mathbf{a}\times\mathbf{x}$. Hence $\mathbf{p}$ is perpendicular to both $\mathbf{a}$, $\mathbf{x}$. Also since $\mathbf{p} = \mathbf{a}\times\mathbf{b}$, $\mathbf{p}$ is also perpendicular to both $\mathbf{a}$, $\mathbf{b}$. It follows that $\mathbf{x}$, $\mathbf{a}$, $\mathbf{b}$ are coplanar. We may therefore express $\mathbf{x}$ in the form
$$\mathbf{x} = \lambda\mathbf{a}+\mu\mathbf{b}.$$
Multiplying vectorially by $\mathbf{a}$ we have
$$\begin{aligned}\mathbf{a}\times\mathbf{x} &= \mathbf{a}\times(\lambda\mathbf{a}+\mu\mathbf{b})\\ &= \mu(\mathbf{a}\times\mathbf{b}).\end{aligned}$$
Since $\mathbf{a}\times\mathbf{x} = \mathbf{a}\times\mathbf{b}$ it follows that $\mu = 1$. Thus the general solution is
$$\mathbf{x} = \mathbf{b}+\lambda\mathbf{a}.$$

(2) *Find the vector* $\mathbf{x}$ *satisfying the equations* $\mathbf{a}\times\mathbf{x} = \mathbf{a}\times\mathbf{b}$, $\mathbf{a}.\mathbf{x} = k$. *Hence solve for* $\mathbf{x}$ *when* $\mathbf{a} = (-1, 1, -2)$, $\mathbf{b} = (2, -1, 3)$, $k = 3$.

From the previous example, the solution of
$$\mathbf{a}\times\mathbf{x} = \mathbf{a}\times\mathbf{b}$$
is
$$\mathbf{x} = \mathbf{b}+\lambda\mathbf{a},$$
where λ is a parameter.

Multiplying scalarly by $\mathbf{a}$ we have

$$\mathbf{a}.\mathbf{x} = \mathbf{a}.\mathbf{b}+\lambda\mathbf{a}^2.$$

Hence $$k = \mathbf{a}.\mathbf{b}+\lambda\mathbf{a}^2.$$

The general solution is therefore

$$\mathbf{x} = \mathbf{b}+\frac{k-\mathbf{a}.\mathbf{b}}{a^2}\mathbf{a} \quad (\mathbf{a} \neq \mathbf{0}).$$

Evaluating $$\mathbf{x} = (2, -1, 3)+\frac{3-(-9)}{6}(-1, 1, -2)$$
$$= (0, 1, -1).$$

(3) *Solve the equation* $p\mathbf{x}+(\mathbf{x}.\mathbf{c})\mathbf{a} = \mathbf{b}\ (p \neq 0)$.
Forming the scalar product with $\mathbf{c}$ we have

$$p(\mathbf{x}.\mathbf{c})+(\mathbf{x}.\mathbf{c})(\mathbf{a}.\mathbf{c}) = (\mathbf{b}.\mathbf{c}),$$

that is, $$(\mathbf{x}.\mathbf{c})(p+\mathbf{a}.\mathbf{c}) = \mathbf{b}.\mathbf{c}.$$

Therefore $$\mathbf{x}.\mathbf{c} = \frac{\mathbf{b}.\mathbf{c}}{p+\mathbf{a}.\mathbf{c}} \quad (p+\mathbf{a}.\mathbf{c} \neq 0). \qquad (11\cdot18)$$

Eliminating $\mathbf{x}.\mathbf{c}$ from the given equation we have

$$p\mathbf{x}+\frac{\mathbf{b}.\mathbf{c}}{p+\mathbf{a}.\mathbf{c}}\mathbf{a} = \mathbf{b}.$$

Hence $$\mathbf{x} = \frac{\mathbf{b}}{p}-\frac{\mathbf{b}.\mathbf{c}}{p(p+\mathbf{a}.\mathbf{c})}\mathbf{a},$$

provided $p+\mathbf{a}.\mathbf{c} \neq 0$. Hence if $p+\mathbf{a}.\mathbf{c} \neq 0$ we have a unique solution.

If $p+\mathbf{a}.\mathbf{c} = 0$ we see from the given equation that if a solution $\mathbf{x}$ exists it is a linear sum of $\mathbf{a}$ and $\mathbf{b}$. Therefore we write

$$\mathbf{x} = \lambda\mathbf{a}+\mu\mathbf{b}.$$

Also if $p+\mathbf{a}.\mathbf{c} = 0$ it follows from (11·18) that $\mathbf{b}.\mathbf{c} = 0$.

Substituting for $\mathbf{x}$ in the given equation we obtain

$$p(\lambda\mathbf{a}+\mu\mathbf{b})+(\lambda\mathbf{a}.\mathbf{c}+\mu\mathbf{b}.\mathbf{c})\mathbf{a} = \mathbf{b}.$$

But $$p = -\mathbf{a}.\mathbf{c} \quad \text{and} \quad \mathbf{b}.\mathbf{c} = 0.$$

Therefore $$-(\mathbf{a}.\mathbf{c})(\lambda\mathbf{a}+\mu\mathbf{b})+\lambda(\mathbf{a}.\mathbf{c})\mathbf{a} = \mathbf{b},$$

giving $$-\mu(\mathbf{a}.\mathbf{c})\mathbf{b} = \mathbf{b}.$$

Therefore $$\mu = -\frac{1}{\mathbf{a}.\mathbf{c}}$$

provided $\mathbf{a}.\mathbf{c} \neq 0$. Thus the given equation is satisfied by

$$\mathbf{x} = \lambda\mathbf{a} - \frac{\mathbf{b}}{\mathbf{a}.\mathbf{c}}$$

for any value of λ. We see that the solution is not unique when $p = -\mathbf{a}.\mathbf{c}$ and $\mathbf{b}.\mathbf{c} = 0$.

Furthermore if $\mathbf{a}.\mathbf{c} = 0$, $p = 0$ and therefore $\mathbf{b} = \mathbf{0}$. The given equation then reduces to

$$(\mathbf{x}.\mathbf{c})\mathbf{a} = \mathbf{0}.$$

Therefore $$\mathbf{x}.\mathbf{c} = 0 \quad (\mathbf{a} \neq \mathbf{0}),$$

that is, $\mathbf{x}$ is any vector perpendicular to $\mathbf{c}$.

Vector algebra and frames of reference

Consider a vector $\mathbf{OA} = \mathbf{a}$. By drawing axes OX, OY, OZ the vector may be expressed in terms of components a_1, a_2, a_3 parallel to OX, OY, OZ. The values of these components will depend on the orientation of the axes relative to $\mathbf{a}$, that is on the orientation of the frame of reference. However the vector itself is independent of its reference frame and, in this sense, is more fundamental than its components.

The scalar and vector products are also frame-invariant. For if we consider the scalar product $|\mathbf{a}|\,|\mathbf{b}|\cos\theta$, we see that it depends only on the magnitudes of $\mathbf{a}$, $\mathbf{b}$ and on the angle between them. Again the vector product $|\mathbf{a}|\,|\mathbf{b}|\sin\theta\hat{\mathbf{n}}$ is also frame-invariant since its magnitude depends on the magnitudes of $\mathbf{a}$, $\mathbf{b}$ and on the angle between them, its direction is perpendicular to both $\mathbf{a}$, $\mathbf{b}$ and its sense is given by the right-handed system $[\mathbf{a}, \mathbf{b}, \hat{\mathbf{n}}]$.

As we have seen, certain physical quantities are vectors or can be obtained by vectors entering into combination, e.g. the work done by a force and the moment of a force about a point. Since the physical laws of nature when formulated with suitable generality must hold independently of the frame of reference we may expect a relation between physical quantities to be capable of being expressed mathematically by a relation between vectors which is itself independent of a frame of reference, i.e. it is frame-invariant. This being so there is no need to use a frame of reference involving writing down components. It is because vector algebra can deal directly with frame-invariant quantities and also because it offers economy of expression in three-dimensional work that it is of importance particularly in mechanics and physics.

Exercise 11(*e*) Miscellaneous

(1) Find the vector moment about the point (1, 1, 2) of a force acting through the point (2, −1, 3) and having the direction and sense of the vector (3, −6, 2), given that the magnitude of the force is 8 newtons and that the unit of length is 1 metre.

(2) Prove $(\mathbf{a}+\mathbf{b})\times(\mathbf{a}-\mathbf{b}) = 2\mathbf{b}\times\mathbf{a}$. Interpret geometrically this result.

(3) A rigid body is rotating at the rate of 11 revolutions per second about an axis through the origin whose direction ratios are 2, −6, 9. Find the velocity (in metres per second) of the point of the body whose co-ordinates are (1, −4, 1), the unit of length being 1 metre.

(4) A tetrahedron has its vertices at A (0, 0, 1), B (3, 0, 1), C (2, 3, 1) and D (1, 1, 2). Find the angles between the faces ABC and BCD, between the edges AB, AC, and between the edge BC and the face ADC. (L.U.)

(5) Evaluate

(i) $\mathbf{a}\times\mathbf{b}$ if $\mathbf{a}.\mathbf{b} = |\mathbf{a}|\,|\mathbf{b}|$,

(ii) $\mathbf{a}.(\mathbf{b}\times\mathbf{c})$ if $\mathbf{a}$, $\mathbf{b}$, $\mathbf{c}$ are coplanar.

(6) The position vectors of the points A, B, C relative to the origin O are $\mathbf{a}$, $\mathbf{b}$, $\mathbf{c}$ respectively. Give the geometrical interpretation of the following:

(i) $\mathbf{a}.\mathbf{b} = \mathbf{a}.\mathbf{c}$,

(ii) $\mathbf{a}\times\mathbf{b} = \mathbf{a}\times\mathbf{c}$.

(7) Prove

(i) $\mathbf{a}.\{\mathbf{a}\times(\mathbf{b}\times\mathbf{c})\} = 0$,

(ii) $|\mathbf{a}\times\mathbf{b}|^2 = |\mathbf{a}|^2\,|\mathbf{b}|^2-(\mathbf{a}.\mathbf{b})^2$.

(8) Given that $\mathbf{a}$ and $\mathbf{b}$ are perpendicular vectors, show that the solution of the vector equation

$$\mathbf{x}\times\mathbf{a} = \mathbf{b}$$

is

$$\mathbf{x} = \frac{1}{a^2}(\mathbf{a}\times\mathbf{b})+\lambda\mathbf{a}$$

where λ is a parameter.

(*Hint*. Since $\mathbf{a}$, $\mathbf{b}$, $\mathbf{a}\times\mathbf{b}$ are mutually perpendicular vectors we can write

$$\mathbf{x} = \lambda\mathbf{a}+\mu\mathbf{b}+\nu(\mathbf{a}\times\mathbf{b}),$$

where λ, μ, ν are numbers. Choose unit vectors $\mathbf{i}$ and $\mathbf{j}$ such that $\mathbf{a} = a\mathbf{i}$ and $\mathbf{b} = b\mathbf{j}$.)

(9) A vector $\mathbf{x}$ makes an angle $\cos^{-1}(\frac{1}{3})$ with the vector $\mathbf{i}-\mathbf{j}+\mathbf{k}$ and $\mathbf{x}\times(\mathbf{i}+2\mathbf{j}) = -2\mathbf{i}+\mathbf{j}+\mathbf{k}$. Show that there are two such vectors $\mathbf{x}$ and find the angle between them.

(10) Prove that the lines joining the mid-points of opposite sides of a skew quadrilateral bisect each other.

If $\mathbf{a}$, $\mathbf{b}$, $\mathbf{c}$, $\mathbf{d}$ are the sides of a skew quadrilateral taken in order, show that the area of the parallelogram formed by the mid-points of the sides is equal to $\frac{1}{4}|\mathbf{b}\times\mathbf{c}+\mathbf{c}\times\mathbf{d}+\mathbf{d}\times\mathbf{b}|$.

(11) Prove that parallelograms on the same base and between the same parallels are equal in area.

(12) The edges OP, OQ, OR of a tetrahedron $OPQR$ are the vectors $\mathbf{a}$, $\mathbf{b}$, $\mathbf{c}$ respectively, where

$$\mathbf{a} = 2\mathbf{i}+4\mathbf{j}, \quad \mathbf{b} = 2\mathbf{i}-\mathbf{j}+3\mathbf{k}, \quad \mathbf{c} = 4\mathbf{i}-2\mathbf{j}+5\mathbf{k}.$$

Evaluate $\mathbf{b}\times\mathbf{c}$ and deduce that OP is perpendicular to the plane OQR. Hence obtain the volume of the tetrahedron.

(13) The unit vectors $\mathbf{u}$, $\mathbf{v}$ are perpendicular and the unit vector $\mathbf{w}$ is inclined at an angle θ to each of $\mathbf{u}$ and $\mathbf{v}$. Prove that

$$\mathbf{w} = (\mathbf{u}+\mathbf{v})\cos\theta+\mathbf{u}\times\mathbf{v}\sqrt{(-\cos 2\theta)}.$$

(14) A rigid body is rotated in a right-handed sense about an axis through the origin, the direction and sense of the axis being given by the unit vector $\mathbf{u}$. If $\mathbf{r}$ is the position vector of a point of the body before rotation, show that its position vector after rotation through an angle of 90° is

$$\mathbf{u}\times\mathbf{r}+(\mathbf{r}.\mathbf{u})\mathbf{u},$$

and that its position vector after rotation through an angle θ is

$$(1-\cos\theta)(\mathbf{r}.\mathbf{u})\mathbf{u}+\cos\theta\mathbf{r}+\sin\theta(\mathbf{u}\times\mathbf{r}).$$

Hence find the position vector, after rotation through an angle of 60°, of the point (1, 2, 1) if the point (3, 0, 4) is on the axis.

(15) A particle of mass m has position vector $\mathbf{r}$ relative to a fixed point O. It is acted upon by a force $\lambda\mathbf{k}\times\dot{\mathbf{r}}-\mu\mathbf{r}$, where λ and μ are positive constants, and $\mathbf{k}$ is one of the right-handed set $\mathbf{i}$, $\mathbf{j}$, $\mathbf{k}$, of

mutually orthogonal unit vectors. Write down the vector equation of motion of the particle.

Show that if $\mathbf{r} = x\mathbf{i}+y\mathbf{j}+z\mathbf{k}$ then z satisfies the equation

$$m\ddot{z}+\mu z = 0.$$

Deduce that if $\mathbf{r}$ and $\dot{\mathbf{r}}$ are initially perpendicular to $\mathbf{k}$ the particle moves in a plane.

Show also that the vector equation of motion is satisfied by

$$\mathbf{r} = a(\mathbf{i}\cos\omega t+\mathbf{j}\sin\omega t)$$

where a is an arbitrary constant and ω is either of the two real roots of a certain quadratic equation. Deduce that the particle can describe a circle about O under this force with either of two constant angular speeds. (J.M.B.)

12

PRODUCT OF THREE VECTORS

Introduction

Since $(\mathbf{b}\times\mathbf{c})$ is a vector we may combine it with the vector $\mathbf{a}$ to form two distinct products:

(i) $\mathbf{a}.(\mathbf{b}\times\mathbf{c})$ which is a number and known as the scalar triple product.

(ii) $\mathbf{a}\times(\mathbf{b}\times\mathbf{c})$ which is a vector and known as the vector triple product.

The scalar triple product

Let
$$\mathbf{a} = a_1\mathbf{i}+a_2\mathbf{j}+a_3\mathbf{k}$$
and
$$\mathbf{b}\times\mathbf{c} = (b_2c_3-b_3c_2)\mathbf{i}+(b_3c_1-b_1c_3)\mathbf{j}+(b_1c_2-b_2c_1)\mathbf{k}.$$
Then
$$\mathbf{a}.(\mathbf{b}\times\mathbf{c}) = a_1(b_2c_3-b_3c_2)+a_2(b_3c_1-b_1c_3)+a_3(b_1c_2-b_2c_1).$$

This may be written as a determinant:

$$\mathbf{a}.(\mathbf{b}\times\mathbf{c}) = \begin{vmatrix} a_1 & a_2 & a_3 \\ b_1 & b_2 & b_3 \\ c_1 & c_2 & c_3 \end{vmatrix}.$$

Alternatively by writing $\mathbf{b}\times\mathbf{c}$ in the symbolic determinant form we have

$$\mathbf{a}.(\mathbf{b}\times\mathbf{c}) = (a_1\mathbf{i}+a_2\mathbf{j}+a_3\mathbf{k}).\begin{vmatrix} \mathbf{i} & \mathbf{j} & \mathbf{k} \\ b_1 & b_2 & b_3 \\ c_1 & c_2 & c_3 \end{vmatrix} = \begin{vmatrix} a_1 & a_2 & a_3 \\ b_1 & b_2 & b_3 \\ c_1 & c_2 & c_3 \end{vmatrix}.$$

Properties of the scalar triple product

An important property of determinants is that the value of a determinant is unaltered in magnitude, but changed in sign, when two rows are interchanged.

Therefore
$$\begin{vmatrix} a_1 & a_2 & a_3 \\ b_1 & b_2 & b_3 \\ c_1 & c_2 & c_3 \end{vmatrix} = -\begin{vmatrix} c_1 & c_2 & c_3 \\ b_1 & b_2 & b_3 \\ a_1 & a_2 & a_3 \end{vmatrix},$$

i.e.
$$\mathbf{a}.(\mathbf{b}\times\mathbf{c}) = -\mathbf{c}.(\mathbf{b}\times\mathbf{a}).$$

Now $\mathbf{c}.(\mathbf{b}\times\mathbf{a}) = -\mathbf{c}.(\mathbf{a}\times\mathbf{b})$ and $\mathbf{c}.(\mathbf{a}\times\mathbf{b}) = (\mathbf{a}\times\mathbf{b}).\mathbf{c}$.

Hence $\quad \mathbf{a}.(\mathbf{b}\times\mathbf{c}) = -\mathbf{c}.(\mathbf{b}\times\mathbf{a}) = \mathbf{c}.(\mathbf{a}\times\mathbf{b}) = (\mathbf{a}\times\mathbf{b}).\mathbf{c}.$

Similarly $\quad \mathbf{c}.(\mathbf{a}\times\mathbf{b}) = -\mathbf{b}.(\mathbf{a}\times\mathbf{c}) = \mathbf{b}.(\mathbf{c}\times\mathbf{a}) = (\mathbf{c}\times\mathbf{a}).\mathbf{b},$

$$\mathbf{b}.(\mathbf{c}\times\mathbf{a}) = -\mathbf{a}.(\mathbf{c}\times\mathbf{b}) = \mathbf{a}.(\mathbf{b}\times\mathbf{c}) = (\mathbf{b}\times\mathbf{c}).\mathbf{a}.$$

(i) Since
$$\mathbf{a}.(\mathbf{b}\times\mathbf{c}) = (\mathbf{a}\times\mathbf{b}).\mathbf{c},$$
$$\mathbf{b}.(\mathbf{c}\times\mathbf{a}) = (\mathbf{b}\times\mathbf{c}).\mathbf{a},$$
$$\mathbf{c}.(\mathbf{a}\times\mathbf{b}) = (\mathbf{c}\times\mathbf{a}).\mathbf{b},$$

the scalar triple product is unaltered when the dot and cross are interchanged provided the factors remain in the same order.

(ii) Since
$$\mathbf{a}.(\mathbf{b}\times\mathbf{c}) = \mathbf{b}.(\mathbf{c}\times\mathbf{a}) = \mathbf{c}.(\mathbf{a}\times\mathbf{b})$$

the scalar triple product is unaltered by a cyclic interchange of the factors.

(iii) Since
$$\mathbf{a}.(\mathbf{b}\times\mathbf{c}) = -\mathbf{c}.(\mathbf{b}\times\mathbf{a}) = -\mathbf{b}.(\mathbf{a}\times\mathbf{c}) = -\mathbf{a}.(\mathbf{c}\times\mathbf{b})$$

the scalar triple product is altered in sign by a non-cyclic interchange of the factors.

Summary

The value of the scalar triple product depends on the order of the factors but is independent of the position of the dot and cross which may be interchanged at will.

In particular, we note that if any two of **a**, **b**, **c** are equal, then $\mathbf{a}.(\mathbf{b}\times\mathbf{c}) = 0$; for $\mathbf{a}.(\mathbf{a}\times\mathbf{b}) = (\mathbf{a}\times\mathbf{a}).\mathbf{b} = 0$.

Since the dot and cross are interchangeable the symbol

$$[\mathbf{a}, \mathbf{b}, \mathbf{c}] \quad \text{or} \quad [\mathbf{a}\ \mathbf{b}\ \mathbf{c}] \quad \text{or} \quad (\mathbf{a}, \mathbf{b}, \mathbf{c}) \quad \text{or} \quad (\mathbf{a}\ \mathbf{b}\ \mathbf{c})$$

is used to denote the scalar triple product of **a**, **b**, **c** in that order. Using this notation we may write

$$[\mathbf{a}, \mathbf{b}, \mathbf{c}] = [\mathbf{b}, \mathbf{c}, \mathbf{a}] = [\mathbf{c}, \mathbf{a}, \mathbf{b}]$$

expressing the fact that the scalar triple product is unaltered by a cyclic interchange of its factors and

$$[\mathbf{a}, \mathbf{b}, \mathbf{c}] = -[\mathbf{c}, \mathbf{b}, \mathbf{a}]$$

expressing the fact that the scalar triple product is altered in sign by a non-cyclic interchange of its factors.

Since $[\mathbf{i}, \mathbf{j}, \mathbf{k}] = 1$ the scalar triple product is positive for a right-handed system and negative for a left-handed system. This gives us a way of determining whether the system $[\mathbf{a}, \mathbf{b}, \mathbf{c}]$ is right- or left-handed, that is, positive or negative.

Since the distributive law holds for both the scalar and vector products it holds for the scalar triple product. Thus

$$(m\mathbf{a}).(\mathbf{b}\times\mathbf{c}) = m\mathbf{a}.(\mathbf{b}\times\mathbf{c}),$$

i.e.

$$[m\mathbf{a}, \mathbf{b}, \mathbf{c}] = m[\mathbf{a}, \mathbf{b}, \mathbf{c}].$$

Also

$$\mathbf{a}.\{(\mathbf{b}+\mathbf{p})\times(\mathbf{c}+\mathbf{q})\} = \mathbf{a}.(\mathbf{b}\times\mathbf{c})+\mathbf{a}.(\mathbf{b}\times\mathbf{q})+\mathbf{a}.(\mathbf{p}\times\mathbf{c})+\mathbf{a}.(\mathbf{p}\times\mathbf{q}),$$

i.e.

$$[\mathbf{a}, \mathbf{b}+\mathbf{p}, \mathbf{c}+\mathbf{q}] = [\mathbf{a}, \mathbf{b}, \mathbf{c}]+[\mathbf{a}, \mathbf{b}, \mathbf{q}]+[\mathbf{a}, \mathbf{p}, \mathbf{c}]+[\mathbf{a}, \mathbf{p}, \mathbf{q}].$$

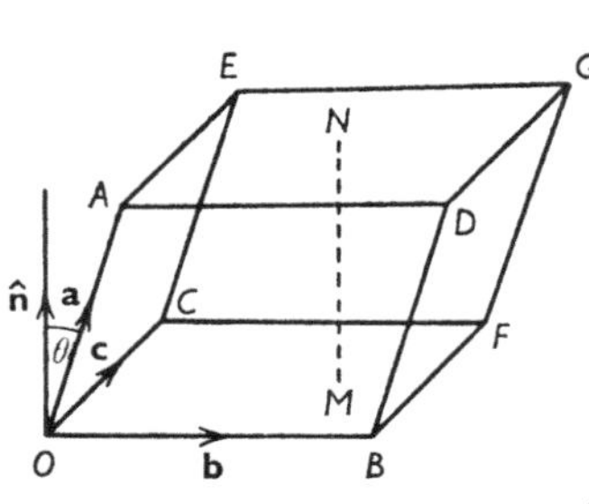

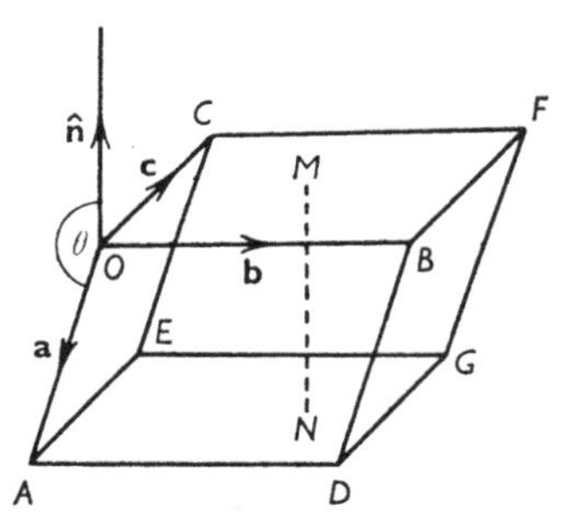

Fig. 12.1

Geometrical interpretation of the scalar triple product

Consider the parallelepiped with edges OA, OB, OC represented by the vectors $\mathbf{a}$, $\mathbf{b}$, $\mathbf{c}$ (Fig. 12.1). Let MN be the perpendicular distance between the faces $OBFC$ and $ADGE$. Let $\hat{\mathbf{n}}$ be the unit vector perpendicular to the base and having the same sense as $(\mathbf{b}\times\mathbf{c})$. If θ is the angle between $\mathbf{a}$ and $\hat{\mathbf{n}}$ we have

$$MN = |\mathbf{a}.\hat{\mathbf{n}}| = |\mathbf{a}|\,|\cos\theta|.$$

The area of the base is the magnitude of the vector product $(\mathbf{b}\times\mathbf{c})$, and since the volume V of a parallelepiped is given by the product of the base and the height we have

$$V = |\mathbf{b}\times\mathbf{c}|\,MN = |\mathbf{a}|\,|\mathbf{b}\times\mathbf{c}|\,|\cos\theta|.$$

Now $\cos\theta$ is positive when θ is acute and this is so when **a**, **b**, **c** form a right-handed system. Then

$$V = \mathbf{a}.(\mathbf{b}\times\mathbf{c}).$$

However, if **a**, **b**, **c** form a left-handed system we have

$$V = -\mathbf{a}.(\mathbf{b}\times\mathbf{c}).$$

Thus we see that

(i) the numerical value of the scalar triple product [**a**, **b**, **c**] gives the volume of the parallelepiped with coterminous edges **a**, **b**, **c** and

(ii) the scalar triple product [**a**, **b**, **c**] is positive or negative depending on whether **a**, **b**, **c** form a right- or left-handed system respectively.

In the same way we can show that for the right-handed system [**a**, **b**, **c**]

$$V = \mathbf{b}.(\mathbf{c}\times\mathbf{a}) = \mathbf{c}.(\mathbf{a}\times\mathbf{b}).$$

Hence $$V = \mathbf{a}.(\mathbf{b}\times\mathbf{c}) = \mathbf{b}.(\mathbf{c}\times\mathbf{a}) = \mathbf{c}.(\mathbf{a}\times\mathbf{b}).$$

Since the commutative law holds for the scalar product

$$V = (\mathbf{b}\times\mathbf{c}).\mathbf{a} = (\mathbf{c}\times\mathbf{a}).\mathbf{b} = (\mathbf{a}\times\mathbf{b}).\mathbf{c},$$

showing again that the dot and cross may be interchanged.

Remembering that $\mathbf{a}\times\mathbf{b} = -(\mathbf{b}\times\mathbf{a})$ and that interchanging the order of two vectors in a right-handed triad results in a left-handed triad, we have

$$\mathbf{V} = -\mathbf{a}.(\mathbf{c}\times\mathbf{b}) = -\mathbf{b}.(\mathbf{a}\times\mathbf{c}) = -\mathbf{c}.(\mathbf{b}\times\mathbf{a})$$
$$= -(\mathbf{c}\times\mathbf{b}).\mathbf{a} = -(\mathbf{a}\times\mathbf{c}).\mathbf{b} = -(\mathbf{b}\times\mathbf{a}).\mathbf{c},$$

showing again that the value of the scalar triple product is altered in sign by a non-cyclic interchange of the factors.

Finally if **a**, **b**, **c** are coplanar the volume vanishes. Therefore when **a**, **b**, **c** are coplanar we have

$$\mathbf{a}.(\mathbf{b}\times\mathbf{c}) = 0.$$

Also the volume vanishes when one of **a**, **b**, **c** is the zero vector, or any two of **a**, **b**, **c** are parallel or equal.

Condition for three vectors to be coplanar

Three vectors are coplanar if they are all parallel to the one and same plane. If **a, b, c** are coplanar then $(\mathbf{b}\times\mathbf{c})$ is perpendicular to **a**. Therefore

$$\mathbf{a}.(\mathbf{b}\times\mathbf{c}) = 0.$$

Thus when three vectors are coplanar, their scalar triple product vanishes.

Alternatively, if **a, b, c** are coplanar we can express **a** in terms of components parallel to **b** and **c**. Let

$$\mathbf{a} = m\mathbf{b}+n\mathbf{c},$$

where m, n are some numbers. Therefore

$$\mathbf{a} = (mb_1+nc_1)\mathbf{i}+(mb_2+nc_2)\mathbf{j}+(mb_3+nc_3)\mathbf{k}.$$

Hence

$$\mathbf{a}.(\mathbf{b}\times\mathbf{c}) = \begin{vmatrix} mb_1+nc_1 & mb_2+nc_2 & mb_3+nc_3 \\ b_1 & b_2 & b_3 \\ c_1 & c_2 & c_3 \end{vmatrix}$$

$$= 0 \text{ (on expansion).}$$

Now suppose $\mathbf{a}.(\mathbf{b}\times\mathbf{c}) = 0$. Then either (i) $\mathbf{a} = \mathbf{0}$, or (ii) $\mathbf{b} = \mathbf{0}$, or $\mathbf{c} = \mathbf{0}$, or **b** is parallel or equal to **c**, or (iii) **a** is perpendicular to $\mathbf{b}\times\mathbf{c}$, that is, **a, b, c** are coplanar. Hence if the scalar triple product vanishes either

(i) one of the vectors is the zero vector, or

(ii) two of the vectors are parallel or equal, or

(iii) the three vectors are coplanar.

Thus a necessary condition for **a, b, c** to be coplanar is that $[\mathbf{a}, \mathbf{b}, \mathbf{c}] = 0$, that is, the scalar triple product vanishes.

The following examples show how the scalar triple product is used in obtaining the equation of a plane and in obtaining four coplanar points, in each case by finding three coplanar vectors.

Examples

(1) *A plane passes through the points* $A\,(1, 2, 3)$, $B\,(-1, 2, 0)$ *and* $C\,(2, -1, -1)$. *Find its equation.* (See example on page 207.)

Let $P\,(x, y, z)$ be any point on the plane. The vectors **AP, AB, AC** are coplanar. We have

$$\mathbf{AP} = (x-1, y-2, z-3), \quad \mathbf{AB} = (-2, 0, 3), \quad \mathbf{AC} = (1, -3, -4).$$

The equation of the plane is given by

$$\mathbf{AP}.(\mathbf{AB}\times\mathbf{AC}) = 0,$$

that is,

$$\begin{vmatrix} x-1 & y-2 & z-3 \\ -2 & 0 & -3 \\ 1 & -3 & -4 \end{vmatrix} = 0.$$

On expansion, the equation of the plane is

$$9x+11y-6z = 13.$$

(2) *Find the value of* λ *such that the points* $A\,(-\lambda, 3, 4)$, $B\,(\lambda, -1, 8)$, $C\,(-2, 2, -1)$, $D\,(-\lambda, 2, 3)$ *are coplanar.*

$$\mathbf{AB} = (2\lambda, -4, 4),$$

$$\mathbf{AC} = (\lambda-2, -1, -5),$$

$$\mathbf{AD} = (0, -1, -1).$$

If the points are coplanar then the vectors **AB**, **AC**, **AD** are coplanar. This is so when

$$\mathbf{AB}.(\mathbf{AC}\times\mathbf{AD}) = 0,$$

that is,

$$\begin{vmatrix} 2\lambda & -4 & 4 \\ \lambda-2 & -1 & -5 \\ 0 & -1 & -1 \end{vmatrix} = 0.$$

The solution is $\lambda = 1.$

The tetrahedron

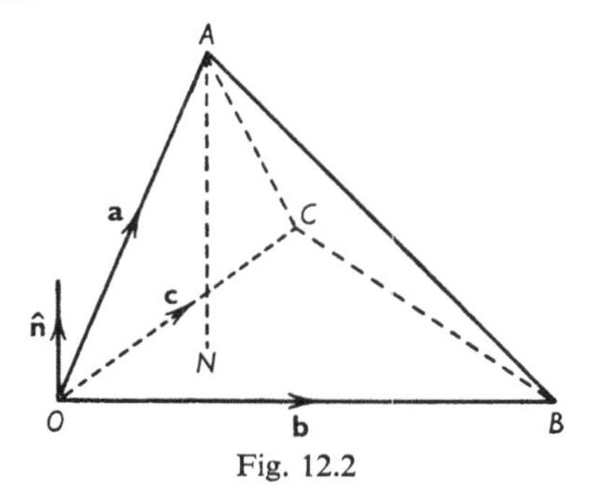

Fig. 12.2

Consider the tetrahedron with vertices at O, A, B, C (Fig. 12.2). Let $\mathbf{OA} = \mathbf{a}$, $\mathbf{OB} = \mathbf{b}$, $\mathbf{OC} = \mathbf{c}$ and $\hat{\mathbf{n}}$ be the unit vector having the same direction and sense as $\mathbf{b}\times\mathbf{c}$. We have

$$\text{area of } \triangle OBC = \tfrac{1}{2}|\mathbf{b}\times\mathbf{c}|,$$

and
$$\text{height } AN = |\mathbf{a}.\hat{\mathbf{n}}|.$$

The volume V of the tetrahedron is given by

$$\begin{aligned} V &= \tfrac{1}{3}\text{ height}\times\text{area of base} \\ &= \tfrac{1}{3}|\mathbf{a}.\hat{\mathbf{n}}|\,|\tfrac{1}{2}(\mathbf{b}\times\mathbf{c})| \\ &= \tfrac{1}{6}|\mathbf{a}.(\mathbf{b}\times\mathbf{c})| \\ &= \tfrac{1}{6}|[\mathbf{a}, \mathbf{b}, \mathbf{c}]|. \end{aligned}$$

Thus the volume of a tetrahedron is $\frac{1}{6}$ of the value of the scalar triple product formed by the three vectors representing coterminous edges of the tetrahedron.

An interesting relation occurs between the volume of a tetrahedron and the shortest distance between two opposite edges of the tetrahedron. In Example (3) on page 209, the formula

$$p = \frac{|(\mathbf{r}_1-\mathbf{r}_2).(\mathbf{l}_1\times\mathbf{l}_2)|}{|\mathbf{l}_1\times\mathbf{l}_2|}$$

was obtained for the shortest distance p between two skew lines having direction-vectors $\mathbf{l}_1$, $\mathbf{l}_2$ and passing through the points $\mathbf{r}_1$, $\mathbf{r}_2$ respectively.

Consider the tetrahedron $OABC$. $A(\mathbf{a})$ is a point on the line OA whose direction-vector is $\mathbf{a}$ and $B(\mathbf{b})$ is a point on the line BC whose direction-vector is $(\mathbf{b}-\mathbf{c})$. Hence the shortest distance p between OA and BC is given by

$$p = \frac{|(\mathbf{a}-\mathbf{b}).\{\mathbf{a}\times(\mathbf{b}-\mathbf{c})\}|}{|\mathbf{a}\times(\mathbf{b}-\mathbf{c})|}.$$

On expansion of the numerator we have

$$\begin{aligned} p &= \frac{|\mathbf{b}.(\mathbf{a}\times\mathbf{c})|}{|\mathbf{a}\times(\mathbf{b}-\mathbf{c})|} \\ &= \frac{|[\mathbf{a}, \mathbf{b}, \mathbf{c}]|}{|\mathbf{a}\times(\mathbf{b}-\mathbf{c})|}. \end{aligned}$$

Now
$$V = \tfrac{1}{6}|[\mathbf{a}, \mathbf{b}, \mathbf{c}]|.$$

Therefore
$$\begin{aligned} p &= \frac{6V}{|\mathbf{a}\times(\mathbf{b}-\mathbf{c})|} \\ &= \frac{6V}{OA.BC\sin\theta}, \end{aligned}$$

where θ is the angle between $\mathbf{OA}$ and $\mathbf{BC}$.

Hence the shortest distance between two opposite edges of a tetrahedron is equal to six times the volume of the tetrahedron divided by the product of the lengths of the edges and the sine of the angle between them.

Example

Calculate the volume of a tetrahedron whose vertices are at the points $A\,(1, 2, -1)$, $B\,(2, 0, 1)$, $C\,(-1, 1, 2)$, $D\,(3, 2, 4)$.

The volume V is given by

$$V = \tfrac{1}{6}|[\mathbf{AB}, \mathbf{AC}, \mathbf{AD}]|$$

$$= \tfrac{1}{6}|[\mathbf{b}-\mathbf{a}, \mathbf{c}-\mathbf{a}, \mathbf{d}-\mathbf{a}]|.$$

Now

$$\mathbf{b}-\mathbf{a} = (1, -2, 2), \quad \mathbf{c}-\mathbf{a} = (-2, -1, 3), \quad \mathbf{d}-\mathbf{a} = (2, 0, 5).$$

Therefore

$$[\mathbf{b}-\mathbf{a}, \mathbf{c}-\mathbf{a}, \mathbf{d}-\mathbf{a}] = \begin{vmatrix} 1 & -2 & 2 \\ -2 & -1 & 3 \\ 2 & 0 & 5 \end{vmatrix}.$$

Hence

$$V = \tfrac{1}{6}|(-5-0)+2(-10-6)+2(0+2)|$$

$$= \tfrac{1}{6}|-5-32+4|$$

$$= 5\tfrac{1}{2}.$$

Moment of a force about a straight line

We shall show that the moment of a force $\mathbf{F}$ about any straight line through a point O is equal to the scalar component along this line of the moment $\mathbf{m}$ of $\mathbf{F}$ about O.

Take the line through O as the Ox axis. Let $\mathbf{F}$ act through the point $P\,(x, y, z)$ and X, Y, Z be its components (Fig. 12.3). Since the moment of $\mathbf{F}$ about Ox is given by the sum of the moments of its components about Ox we have

$$\text{moment of } \mathbf{F} \text{ about } Ox = yZ - zY$$

since neither x nor X affects this moment. We have shown

$$\text{moment of } \mathbf{F} \text{ about } O = \mathbf{OP} \times \mathbf{F}.$$

Therefore

$$\mathbf{m} = (x\mathbf{i}+y\mathbf{j}+z\mathbf{k}) \times (X\mathbf{i}+Y\mathbf{j}+Z\mathbf{k})$$

$$= (yZ-zY)\mathbf{i}+(zX-xZ)\mathbf{j}+(xY-yX)\mathbf{k}.$$

Thus we see that the moment of **F** about Ox is the scalar component along Ox of the vector moment **m** of **F** about O. Since there is no special significance in the choice of the direction of the axes we have the following:

Definition. *The moment of a force* **F** *about any straight line through O is the component along this line of the moment of* **F** *about O.*

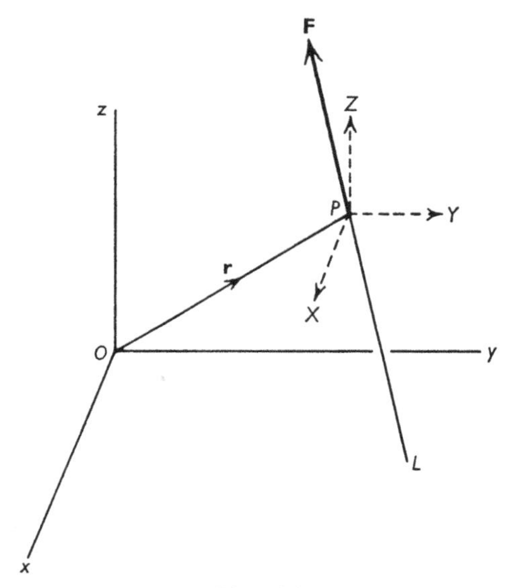

Fig. 12.3

This moment may be expressed as a scalar triple product. Suppose we require the moment of **F** about the line l. Let O be a point on l and let P be any point on the line of action L of **F**. If $\hat{\mathbf{l}}$ is the unit vector parallel to l and **r** the position vector of P relative to O, we have

$$\begin{aligned} \text{moment of } \mathbf{F} \text{ about } l &= \hat{\mathbf{l}}.\mathbf{m} \\ &= \hat{\mathbf{l}}.(\mathbf{r}\times\mathbf{F}). \end{aligned}$$

This result is independent of the choice of O and P. For suppose O', P' are any other points on l and L. Then

$$\begin{aligned} \text{moment of } \mathbf{F} \text{ about } l &= \hat{\mathbf{l}}.(\mathbf{O'P'}\times\mathbf{F}) \\ &= \hat{\mathbf{l}}.\{(\mathbf{O'O}+\mathbf{OP}+\mathbf{PP'})\}\times\mathbf{F} \\ &= \hat{\mathbf{l}}.(\mathbf{r}\times\mathbf{F}) \end{aligned}$$

since $\mathbf{OO'}$ is parallel to $\hat{\mathbf{l}}$ and $\mathbf{PP'}$ is parallel to **F**.

In particular the moments of $\mathbf{F}$ about the axes Ox, Oy, Oz are $\mathbf{i}.(\mathbf{r}\times\mathbf{F})$, $\mathbf{j}.(\mathbf{r}\times\mathbf{F})$, $\mathbf{k}.(\mathbf{r}\times\mathbf{F})$ respectively.

It should be noted that the moment of a force about a point is a vector whereas the moment about a line is a scalar.

Exercise 12(*a*)

(1) If $\mathbf{a} = 2\mathbf{i}+\mathbf{j}-\mathbf{k}$, $\mathbf{b} = -\mathbf{i}+2\mathbf{j}$, $\mathbf{c} = \mathbf{i}-\mathbf{j}+\mathbf{k}$ evaluate $\mathbf{a}.(\mathbf{b}\times\mathbf{c})$, by expressing it as a determinant. Hence deduce the values of $\mathbf{c}.(\mathbf{a}\times\mathbf{b})$ and $\mathbf{b}.(\mathbf{a}\times\mathbf{c})$.

(2) Given $\mathbf{a} = (1, 0, 2)$, $\mathbf{b} = (2, -1, 3)$, $\mathbf{c} = (3, 1, -1)$ determine whether $\mathbf{a}$, $\mathbf{b}$, $\mathbf{c}$ form a positive or negative triad.

(3) Simplify $(\mathbf{a}+\mathbf{b}).(\mathbf{a}+\mathbf{a}\times\mathbf{c})+\mathbf{a}.(\mathbf{b}\times\mathbf{c})$.

(4) Find the volume of the tetrahedron whose vertices are the points $(1, 3, -1)$, $(2, 2, 3)$, $(3, 7, 4)$, $(4, 2, -2)$.

(5) (*a*) Show that the points $(1, 1, -2)$, $(1, 0, -3)$, $(2, 1, -1)$, $(5, -1, 0)$ are coplanar.

(*b*) The position vectors of the points A, B, C and D are $\mathbf{a}$, $\mathbf{b}$, $\mathbf{c}$ and $\mathbf{d}$ respectively. Show that the condition for the four points to be coplanar is

$$[\mathbf{a}, \mathbf{b}, \mathbf{c}]+[\mathbf{a}, \mathbf{c}, \mathbf{d}]+[\mathbf{a}, \mathbf{d}, \mathbf{b}] = [\mathbf{b}, \mathbf{c}, \mathbf{d}].$$

(6) Find the volume of the parallelepiped whose edges are parallel to, and of the same magnitude as, the vectors $(8, -3, 3)$, $(4, 2, 3)$, $(-3, 4, 1)$.

(7) Given that A, B, C, D are the points $(0, 2, 6)$, $(2, 3, 3)$, $(-1, 6, 3)$, $(5, -1, 4)$ respectively, show that the lines AC and BD intersect.

(8) The position vectors of the points A, B, C, D are $\mathbf{a}$, $\mathbf{b}$, $\mathbf{c}$, $\mathbf{d}$ respectively. Prove that the vector

$$\mathbf{b}\times\mathbf{c}+\mathbf{c}\times\mathbf{a}+\mathbf{a}\times\mathbf{b}$$

has magnitude equal to twice the area of the triangle ABC and direction perpendicular to the plane of A, B, C.

Also prove that the volume of the tetrahedron $ABCD$ is

$$\pm\tfrac{1}{6}\{[\mathbf{a}, \mathbf{b}, \mathbf{c}]-[\mathbf{b}, \mathbf{c}, \mathbf{d}]-[\mathbf{c}, \mathbf{a}, \mathbf{d}]-[\mathbf{a}, \mathbf{b}, \mathbf{d}]\}.$$

(9) Find the equation of the plane through the line

$$\frac{x-1}{3} = \frac{y+4}{4} = \frac{z-1}{1}$$

and which is parallel to the line

$$\frac{x}{4} = \frac{y}{3} = \frac{z}{12}.$$

(10) Find the equation of the plane through the origin and through the line

$$\frac{x+1}{2} = \frac{y-3}{-1} = \frac{z-2}{3}.$$

(11) A force of 21 units acts along the line joining $A\,(2, 1, 1)$ and $B\,(10, -3, 2)$ and in the direction of $\mathbf{AB}$. Find the moment of the force about the axes and about the line

$$\frac{x}{6} = \frac{y}{3} = \frac{z}{2}.$$

(12) P, P' are two points with position vectors $\mathbf{r}$, $\mathbf{r}'$ and on the lines l, l' respectively. The unit vectors along the two lines are $\hat{\mathbf{l}}, \hat{\mathbf{l}}'$. Show that the scalar moment about the line l of a force of magnitude F along l' is given by $\hat{\mathbf{l}}.(\mathbf{r}'-\mathbf{r})\times F\hat{\mathbf{l}}'$.

OA, OB, OC are adjacent edges of a cube of edge a. AA', BB', CC' are diagonals of the cube. A force of magnitude F acts along $\overrightarrow{AA'}$. Find its moment about BB' and BC.

The vector triple product

In terms of components we have for the vector triple product

$$\mathbf{a}\times(\mathbf{b}\times\mathbf{c}) = (a_1\mathbf{i}+a_2\mathbf{j}+a_3\mathbf{k})\times\{(b_2c_3-b_3c_2)\mathbf{i}+(b_3c_1-b_1c_3)\mathbf{j} + (b_1c_2-b_2c_1)\mathbf{k}\}.$$

This on expansion gives

$$\begin{aligned}\mathbf{a}\times(\mathbf{b}\times\mathbf{c}) = {} & (a_2b_1c_2-a_2b_2c_1-a_3b_3c_1+a_3b_1c_3)\mathbf{i}\\ & +(a_3b_2c_3-a_3b_3c_2-a_1b_1c_2+a_1b_2c_1)\mathbf{j}\\ & +(a_1b_3c_1-a_1b_1c_3-a_2b_2c_3+a_2b_3c_2)\mathbf{k}.\end{aligned}$$

We shall now show that the vector triple product is frame-invariant. Rearranging we have

$$\begin{aligned}\mathbf{a}\times(\mathbf{b}\times\mathbf{c}) &= (a_2c_2+a_3c_3)b_1\mathbf{i}+(a_1c_1+a_3c_3)b_2\mathbf{j}+(a_1c_1+a_2c_2)b_3\mathbf{k}\\ &\qquad -(a_2b_2+a_3b_3)c_1\mathbf{i}-(a_1b_1+a_3b_3)c_2\mathbf{j}-(a_1b_1+a_2b_2)c_3\mathbf{k}\\ &= (a_1c_1+a_2c_2+a_3c_3)\,(b_1\mathbf{i}+b_2\mathbf{j}+b_3\mathbf{k})\\ &\qquad -(a_1b_1+a_2b_2+a_3b_3)\,(c_1\mathbf{i}+c_2\mathbf{j}+c_3\mathbf{k})\\ &= (\mathbf{a}.\mathbf{c})\mathbf{b}-(\mathbf{a}.\mathbf{b})\mathbf{c},\end{aligned}$$

which is frame-invariant since the scalar products $\mathbf{a}.\mathbf{c}$, $\mathbf{a}.\mathbf{b}$ are frame-invariant.

The expansion for $\mathbf{a}\times(\mathbf{b}\times\mathbf{c})$, namely

$$\mathbf{a}\times(\mathbf{b}\times\mathbf{c}) = (\mathbf{a}.\mathbf{c})\mathbf{b}-(\mathbf{a}.\mathbf{b})\mathbf{c},$$

should be remembered.

Alternatively we may proceed as follows. Since $\mathbf{b}\times\mathbf{c}$ is perpendicular to the plane of $\mathbf{b}$ and $\mathbf{c}$, and $\mathbf{a}\times(\mathbf{b}\times\mathbf{c})$ is perpendicular to $\mathbf{b}\times\mathbf{c}$, the vector triple product $\mathbf{a}\times(\mathbf{b}\times\mathbf{c})$ must be a vector coplanar with $\mathbf{b}$ and $\mathbf{c}$ and we may therefore express $\mathbf{a}\times(\mathbf{b}\times\mathbf{c})$ as a linear sum of $\mathbf{b}$ and $\mathbf{c}$. Let

$$\mathbf{a}\times(\mathbf{b}\times\mathbf{c}) = \lambda\mathbf{b}+\mu\mathbf{c},$$

where λ, μ are numbers to be determined. Since we require the vector triple product to be frame-invariant the values of λ, μ must be independent of the frame of reference. In order to determine their values we may therefore choose any convenient frame. Take the usual right-handed axes Ox, Oy, Oz such that Ox is along the direction of $\mathbf{b}$ and Oy is in the plane of $\mathbf{b}$ and $\mathbf{c}$. Then

$$\mathbf{b} = b_1\mathbf{i}, \quad \mathbf{c} = c_1\mathbf{i}+c_2\mathbf{j}, \quad \mathbf{a} = a_1\mathbf{i}+a_2\mathbf{j}+a_3\mathbf{k}.$$

Therefore $\qquad \mathbf{b}\times\mathbf{c} = b_1c_2\mathbf{k}.$

Hence $\qquad \mathbf{a}\times(\mathbf{b}\times\mathbf{c}) = a_2b_1c_2\mathbf{i}-a_1b_1c_2\mathbf{j}.$

But
$$\begin{aligned}\mathbf{a}\times(\mathbf{b}\times\mathbf{c}) &= \lambda\mathbf{b}+\mu\mathbf{c}\\ &= \lambda b_1\mathbf{i}+\mu c_1\mathbf{i}+\mu c_2\mathbf{j}\\ &= (\lambda b_1+\mu c_1)\mathbf{i}+\mu c_2\mathbf{j}.\end{aligned}$$

Hence $\qquad \lambda b_1+\mu c_1 = a_2b_1c_2 \quad \text{and} \quad \mu c_2 = -a_1b_1c_2.$

Solving for λ, μ we obtain

$$\lambda = a_1c_1+a_2c_2, \quad \mu = -a_1b_1 \qquad (b_1 \neq 0,\ c_2 \neq 0).$$

Now $\qquad a_1c_1+a_2c_2 = \mathbf{a}.\mathbf{c} \quad \text{and} \quad a_1b_1 = \mathbf{a}.\mathbf{b}.$

Therefore $\qquad \lambda = \mathbf{a}.\mathbf{c} \quad \text{and} \quad \mu = -\mathbf{a}.\mathbf{b}.$

Since the scalar products $\mathbf{a}.\mathbf{c}$, $\mathbf{a}.\mathbf{b}$ are frame-invariant these values of λ, μ hold for all frames of reference, i.e. they are independent of the choice of axes. Hence

$$\mathbf{a}\times(\mathbf{b}\times\mathbf{c}) = (\mathbf{a}.\mathbf{c})\mathbf{b}-(\mathbf{a}.\mathbf{b})\mathbf{c}.$$

The corresponding expression for $(\mathbf{a}\times\mathbf{b})\times\mathbf{c}$ is obtained thus:

$$\begin{aligned}(\mathbf{a}\times\mathbf{b})\times\mathbf{c} &= -\mathbf{c}\times(\mathbf{a}\times\mathbf{b})\\ &= -\{(\mathbf{c}.\mathbf{b})\mathbf{a}-(\mathbf{c}.\mathbf{a})\mathbf{b}\}\\ &= (\mathbf{c}.\mathbf{a})\mathbf{b}-(\mathbf{c}.\mathbf{b})\mathbf{a}.\end{aligned}$$

From this we again see that $\mathbf{a}\times(\mathbf{b}\times\mathbf{c}) \neq (\mathbf{a}\times\mathbf{b})\times\mathbf{c}$. Hence the position of the brackets in the vector triple product is important. As an aid to remembering the expansion for the vector triple product note that each scalar product contains the factor outside the bracket, the first being formed by the extreme factors.

Example

If $\mathbf{a} = (2,\ 3,\ -1)$, $\mathbf{b} = (1,\ -1,\ 1)$, $\mathbf{c} = (3,\ -2,\ -2)$ *obtain* $\mathbf{a}\times(\mathbf{b}\times\mathbf{c})$.

The vector triple product may be obtained either by evaluating $(\mathbf{b}\times\mathbf{c}) = \mathbf{d}$ and then $\mathbf{a}\times\mathbf{d}$ or by evaluating $(\mathbf{a}.\mathbf{c})\mathbf{b}-(\mathbf{a}.\mathbf{b})\mathbf{c}$.
We have

$$\mathbf{b}\times\mathbf{c} = (1,\ -1,\ 1)\times(3,\ -2,\ -2) = (4,\ 5,\ 1).$$

Therefore

$$\mathbf{a}\times(\mathbf{b}\times\mathbf{c}) = (2,\ 3,\ -1)\times(4,\ 5,\ 1) = (8,\ -6,\ -2).$$

Alternatively

$$\mathbf{a}.\mathbf{c} = (2,\ 3,\ -1).(3,\ -2,\ -2) = 6+(-6)+2 = 2,$$

$$\mathbf{a}.\mathbf{b} = (2,\ 3,\ -1).(1,\ -1,\ 1) = 2+(-3)+(-1) = -2.$$

Therefore

$$\mathbf{a}\times(\mathbf{b}\times\mathbf{c}) = 2(1,\ -1,\ 1)-(-2)\,(3,\ -2,\ -2) = (8,\ -6,\ -2).$$

Properties of the vector triple product

The following properties

(1) $k\{\mathbf{a}\times(\mathbf{b}\times\mathbf{c})\} = (k\mathbf{a})\times(\mathbf{b}\times\mathbf{c}) = \mathbf{a}\times\{(k\mathbf{b})\times\mathbf{c}\} = \mathbf{a}\times\{\mathbf{b}\times(k\mathbf{c})\}$,

(2) $(\mathbf{a}+\mathbf{b})\times(\mathbf{c}\times\mathbf{d}) = \mathbf{a}\times(\mathbf{c}\times\mathbf{d})+\mathbf{b}\times(\mathbf{c}\times\mathbf{d})$,

are easily proved by using the properties of the scalar and vector products.

Let $$\mathbf{b}\times\mathbf{c} = \mathbf{p}.$$

Then $$k\{\mathbf{a}\times(\mathbf{b}\times\mathbf{c})\} = k(\mathbf{a}\times\mathbf{p}).$$

Now $$k(\mathbf{a}\times\mathbf{p}) = (k\mathbf{a})\times\mathbf{p} = (k\mathbf{a})\times(\mathbf{b}\times\mathbf{c}).$$

Also $$k(\mathbf{a}\times\mathbf{p}) = \mathbf{a}\times(k\mathbf{p}) = \mathbf{a}\times\{k(\mathbf{b}\times\mathbf{c})\}$$
$$= \mathbf{a}\times\{(k\mathbf{b})\times\mathbf{c}\} = \mathbf{a}\times\{\mathbf{b}\times(k\mathbf{c})\}.$$

Therefore

$$k\{\mathbf{a}\times(\mathbf{b}\times\mathbf{c})\} = (k\mathbf{a})\times(\mathbf{b}\times\mathbf{c}) = \mathbf{a}\times\{(k\mathbf{b})\times\mathbf{c}\} = \mathbf{a}\times\{\mathbf{b}\times(k\mathbf{c})\}.$$

When $k = 0$, the result is obviously true.

Let $$\mathbf{c}\times\mathbf{d} = \mathbf{q}.$$

Then $$(\mathbf{a}+\mathbf{b})\times(\mathbf{c}\times\mathbf{d}) = (\mathbf{a}+\mathbf{b})\times\mathbf{q}$$
$$= \mathbf{a}\times\mathbf{q}+\mathbf{b}\times\mathbf{q}$$
$$= \mathbf{a}\times(\mathbf{c}\times\mathbf{d})+\mathbf{b}\times(\mathbf{c}\times\mathbf{d}).$$

These may also be proved by using the result

$$\mathbf{a}\times(\mathbf{b}\times\mathbf{c}) = (\mathbf{a}.\mathbf{c})\mathbf{b}-(\mathbf{a}.\mathbf{b})\mathbf{c}.$$

Thus $$k\{\mathbf{a}\times(\mathbf{b}\times\mathbf{c})\} = k\{(\mathbf{a}.\mathbf{c})\mathbf{b}-(\mathbf{a}.\mathbf{b})\mathbf{c}\}$$
$$= \{(k\mathbf{a}).\mathbf{c}\}\mathbf{b}-\{(k\mathbf{a}).\mathbf{b}\}\mathbf{c}$$
$$= (k\mathbf{a})\times(\mathbf{b}\times\mathbf{c}).$$

Similarly, $$k\{\mathbf{a}\times(\mathbf{b}\times\mathbf{c})\} = \mathbf{a}\times\{(k\mathbf{b})\times\mathbf{c}\} = \mathbf{a}\times\{\mathbf{b}\times(k\mathbf{c})\}.$$

Again $$(\mathbf{a}+\mathbf{b})\times(\mathbf{c}\times\mathbf{d}) = \{(\mathbf{a}+\mathbf{b}).\mathbf{d}\}\mathbf{c}-\{(\mathbf{a}+\mathbf{b}).\mathbf{c}\}\mathbf{d}$$
$$= (\mathbf{a}.\mathbf{d}+\mathbf{b}.\mathbf{d})\mathbf{c}-(\mathbf{a}.\mathbf{c}+\mathbf{b}.\mathbf{c})\mathbf{d}$$
$$= (\mathbf{a}.\mathbf{d})\mathbf{c}-(\mathbf{a}.\mathbf{c})\mathbf{d}+(\mathbf{b}.\mathbf{d})\mathbf{c}-(\mathbf{b}.\mathbf{c})\mathbf{d}$$
$$= \mathbf{a}\times(\mathbf{c}\times\mathbf{d})+\mathbf{b}\times(\mathbf{c}\times\mathbf{d}).$$

An application of the vector triple product

Consider a particle of mass m rigidly connected to a fixed rotating axis. Let O be any point on the axis and P the position of the particle at any time (Fig. 12.4). If $\boldsymbol{\omega}$ is the angular velocity vector about the axis of the particle we have

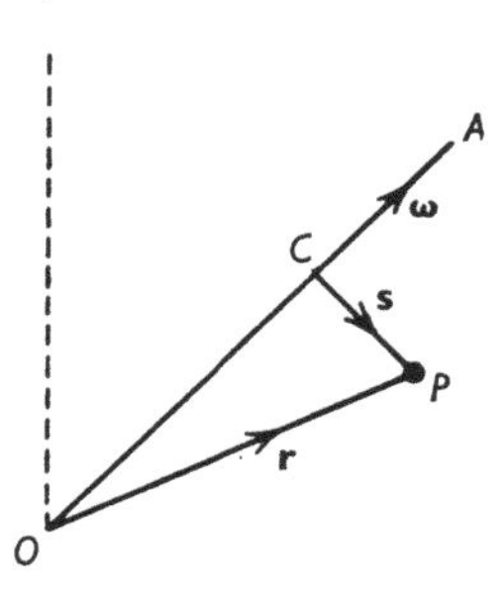

Fig. 12.4

$$\dot{\mathbf{r}} = \boldsymbol{\omega}\times\mathbf{r}.$$

The momentum of the particle about O is

$$m\dot{\mathbf{r}} = m(\boldsymbol{\omega}\times\mathbf{r}) = (m\boldsymbol{\omega}\times\mathbf{r}).$$

The moment of momentum of the particle about O is defined by

$$\mathbf{r}\times m\dot{\mathbf{r}} = \mathbf{r}\times(m\boldsymbol{\omega}\times\mathbf{r}).$$

Let PC be the perpendicular from P to the axis. Let $\mathbf{OC} = \mathbf{c}$ and $\mathbf{CP} = \mathbf{s}$. Since $\mathbf{c}+\mathbf{s} = \mathbf{r}$ and $\mathbf{c}\times\boldsymbol{\omega} = \mathbf{0}$, the moment of momentum of the particle about O can be restated as

$$\begin{aligned}\mathbf{r}\times(m\boldsymbol{\omega}\times\mathbf{r}) &= m\mathbf{r}\times\{\boldsymbol{\omega}\times(\mathbf{c}+\mathbf{s})\}\\ &= m\mathbf{r}\times(\boldsymbol{\omega}\times\mathbf{c}+\boldsymbol{\omega}\times\mathbf{s})\\ &= m\mathbf{r}\times(\boldsymbol{\omega}\times\mathbf{s}).\end{aligned}$$

Expanding the vector triple product we have,

$$\begin{aligned}\text{moment of momentum about } O &= m(\mathbf{r}\,.\,\mathbf{s})\boldsymbol{\omega}-m(\mathbf{r}\,.\,\boldsymbol{\omega})\mathbf{s}\\ &= ms^2\boldsymbol{\omega}+\omega OC(-m\mathbf{s}),\end{aligned}$$

since $\qquad \mathbf{r}\,.\,\mathbf{s} = rs\cos C\hat{P}O = (r\cos C\hat{P}O)s = s^2$

and $\qquad \mathbf{r}\,.\,\boldsymbol{\omega} = r\omega\cos C\hat{O}P = (r\cos C\hat{O}P)\omega = \omega OC.$

Suppose now we have a plane of masses rigidly connected to the fixed rotating axis, with C as the centre of mass. Summing for all the masses we have

$$\Sigma(m\mathbf{s}) = \mathbf{0},$$

and the moment of momentum $\mathbf{H}$ of the masses about O is given by

$$\begin{aligned}\mathbf{H} &= \Sigma(ms^2)\,\boldsymbol{\omega}\\ &= I\boldsymbol{\omega}\end{aligned}$$

where I is the moment of inertia of the plane of masses about the axis.

Differentiation of triple products

Either from first principles or from the results of differentiating scalar and vector products we have if $\mathbf{a}$, $\mathbf{b}$, $\mathbf{c}$ are differentiable functions of the scalar variable t

(1) $$\frac{d}{dt}\{\mathbf{a}\,.\,(\mathbf{b}\times\mathbf{c})\} = \frac{d\mathbf{a}}{dt}\,.\,(\mathbf{b}\times\mathbf{c})+\mathbf{a}\,.\left(\frac{d\mathbf{b}}{dt}\times\mathbf{c}\right)+\mathbf{a}\,.\left(\mathbf{b}\times\frac{d\mathbf{c}}{dt}\right),$$

i.e. $$\frac{d}{dt}[\mathbf{a},\,\mathbf{b},\,\mathbf{c}] = \left[\frac{d\mathbf{a}}{dt},\,\mathbf{b},\,\mathbf{c}\right]+\left[\mathbf{a},\,\frac{d\mathbf{b}}{dt},\,\mathbf{c}\right]+\left[\mathbf{a},\,\mathbf{b},\,\frac{d\mathbf{c}}{dt}\right].$$

(2) $$\frac{d}{dt}\{\mathbf{a}\times(\mathbf{b}\times\mathbf{c})\} = \frac{d\mathbf{a}}{dt}\times(\mathbf{b}\times\mathbf{c})+\mathbf{a}\times\left(\frac{d\mathbf{b}}{dt}\times\mathbf{c}\right)+\mathbf{a}\times\left(\mathbf{b}\times\frac{d\mathbf{c}}{dt}\right).$$

It is essential to maintain the cyclic order of the vectors in each term of (1) and to maintain the order of the factors in each term of (2).

We now prove (1) by using the derivatives of the scalar and vector products.

$$\begin{aligned}\frac{d}{dt}\{\mathbf{a}.(\mathbf{b}\times\mathbf{c})\} &= \frac{d\mathbf{a}}{dt}.(\mathbf{b}\times\mathbf{c})+\mathbf{a}.\frac{d}{dt}(\mathbf{b}\times\mathbf{c})\\ &= \frac{d\mathbf{a}}{dt}.(\mathbf{b}\times\mathbf{c})+\mathbf{a}.\left(\frac{d\mathbf{b}}{dt}\times\mathbf{c}+\mathbf{b}\times\frac{d\mathbf{c}}{dt}\right)\\ &= \frac{d\mathbf{a}}{dt}.(\mathbf{b}\times\mathbf{c})+\mathbf{a}.\left(\frac{d\mathbf{b}}{dt}\times\mathbf{c}\right)+\mathbf{a}.\left(\mathbf{b}\times\frac{d\mathbf{c}}{dt}\right).\end{aligned}$$

In the same way (2) may be proved.

Example

If $\mathbf{a} = 2t^2\mathbf{i}+3t\mathbf{j}-\mathbf{k}$, $\mathbf{b} = t^2\mathbf{i}-t\mathbf{j}+\mathbf{k}$, $\mathbf{c} = 3t^2\mathbf{i}-2t\mathbf{j}-2\mathbf{k}$ *obtain* $d\{\mathbf{a}.(\mathbf{b}\times\mathbf{c})\}/dt$ and $d\{\mathbf{a}\times(\mathbf{b}\times\mathbf{c})\}/dt$.

In practice it is quicker to evaluate the triple products before differentiating

$$\begin{aligned}\mathbf{a}.(\mathbf{b}\times\mathbf{c}) &= \begin{vmatrix} 2t^2 & 3t & -1\\ t^2 & -t & 1\\ 3t^2 & -2t & -2\end{vmatrix}\\ &= 2t^2(2t+2t)-3t(-2t^2-3t^2)-(-2t^3+3t^3)\\ &= 8t^3+15t^3-t^3\\ &= 22t^3.\end{aligned}$$

Therefore

$$\frac{d}{dt}\{\mathbf{a}.(\mathbf{b}\times\mathbf{c})\} = 66t^2.$$

$$\begin{aligned}\mathbf{a}\times(\mathbf{b}\times\mathbf{c}) &= (\mathbf{a}.\mathbf{c})\mathbf{b}-(\mathbf{a}.\mathbf{b})\mathbf{c}\\ &= (6t^4-6t^2+2)\,(t^2\mathbf{i}-t\mathbf{j}+\mathbf{k})\\ &\qquad -(2t^4-3t^2-1)\,(3t^2\mathbf{i}-2t\mathbf{j}-2\mathbf{k})\\ &= (3t^4+5t^2)\mathbf{i}+(-2t^5-4t)\mathbf{j}+(10t^4-12t^2)\mathbf{k}.\end{aligned}$$

Therefore

$$\begin{aligned}\frac{d}{dt}\{\mathbf{a}\times(\mathbf{b}\times\mathbf{c})\} &= (12t^3+10t)\mathbf{i}-(10t^4+4)\mathbf{j}+(40t^3-24t)\mathbf{k}\\ &= 2t(6t^2+5)\mathbf{i}-2(5t^4+2)\mathbf{j}+8t(5t^2-3)\mathbf{k}.\end{aligned}$$

Exercise 12(*b*)

(1) If $\mathbf{a} = (1, 1, 2)$, $\mathbf{b} = (2, -1, 1)$, $\mathbf{c} = (3, 2, -4)$ verify that

$$\mathbf{a}\times(\mathbf{b}\times\mathbf{c}) = (\mathbf{a}.\mathbf{c})\mathbf{b}-(\mathbf{a}.\mathbf{b})\mathbf{c}$$

and
$$(\mathbf{a}\times\mathbf{c})\times\mathbf{b} = (\mathbf{a}.\mathbf{b})\mathbf{c}-(\mathbf{c}.\mathbf{b})\mathbf{a}.$$

(2) Prove that a necessary and sufficient condition for $\mathbf{a}\times(\mathbf{b}\times\mathbf{c})$ to be equal to $(\mathbf{a}\times\mathbf{b})\times\mathbf{c}$ is that $(\mathbf{a}\times\mathbf{c})\times\mathbf{b} = \mathbf{0}$.

Verify this is so when $\mathbf{c} = k\mathbf{a}$, where k is a constant.

Discuss the cases where $\mathbf{a}.\mathbf{b} = 0$ or $\mathbf{b}.\mathbf{c} = 0$.

(3) Prove that

$$\mathbf{a}\times(\mathbf{b}\times\mathbf{c})+\mathbf{b}\times(\mathbf{c}\times\mathbf{a})+\mathbf{c}\times(\mathbf{a}\times\mathbf{b}) = \mathbf{0}.$$

(4) Prove that $(\mathbf{a}+\mathbf{b})\times(\mathbf{a}\times\mathbf{c}) = [\mathbf{a}, \mathbf{b}, \mathbf{c}]\mathbf{a}$.

(5) If a vector $\mathbf{a}$ is resolved into components parallel and perpendicular to another vector $\mathbf{b}$, show that the component perpendicular to $\mathbf{b}$ is $\mathbf{b}\times(\mathbf{a}\times\mathbf{b})/\mathbf{b}^2$.

(6) By multiplying vectorially by $\mathbf{c}$ the equation $\mathbf{x}\times\mathbf{a} = \mathbf{b}$ and expanding the vector triple product, show that the vector $\mathbf{x}$ satisfying the equations

$$\mathbf{x}\times\mathbf{a} = \mathbf{b}, \quad \mathbf{x}.\mathbf{c} = k,$$

where k is a given scalar, and $\mathbf{a}.\mathbf{c} \neq 0$, is given by

$$\mathbf{x} = (k\mathbf{a}-\mathbf{b}\times\mathbf{c})/\mathbf{a}.\mathbf{c}.$$

(7) (i) Prove that $\mathbf{a}\times(\mathbf{b}\times\mathbf{c}) = (\mathbf{a}.\mathbf{c})\mathbf{b}-(\mathbf{a}.\mathbf{b})\mathbf{c}$, and that $\mathbf{a}.\mathbf{b}\times\mathbf{c} = \mathbf{a}\times\mathbf{b}.\mathbf{c}$. Hence prove that

$$(\mathbf{a}\times\mathbf{b}).(\mathbf{c}\times\mathbf{d}) = (\mathbf{a}.\mathbf{c})(\mathbf{b}.\mathbf{d})-(\mathbf{b}.\mathbf{c})(\mathbf{a}.\mathbf{d}).$$

(ii) The vectors $\mathbf{a}'$, $\mathbf{b}'$, $\mathbf{c}'$ are defined in terms of three non-coplanar vectors $\mathbf{a}$, $\mathbf{b}$, $\mathbf{c}$ by

$$\mathbf{a}' = \frac{\mathbf{b}\times\mathbf{c}}{\{\mathbf{a}, \mathbf{b}, \mathbf{c}\}}, \quad \mathbf{b}' = \frac{\mathbf{c}\times\mathbf{a}}{\{\mathbf{a}, \mathbf{b}, \mathbf{c}\}}, \quad \mathbf{c}' = \frac{\mathbf{a}\times\mathbf{b}}{\{\mathbf{a}, \mathbf{b}, \mathbf{c}\}}.$$

Prove that $\mathbf{a} = k\mathbf{b}'\times\mathbf{c}'$ for some constant k.

Evaluate $\mathbf{a}'.\mathbf{a}$ and $\{\mathbf{a}', \mathbf{b}', \mathbf{c}'\}$ and hence find $\mathbf{a}$, $\mathbf{b}$, $\mathbf{c}$ in terms of $\mathbf{a}'$, $\mathbf{b}'$, $\mathbf{c}'$.

[$\{\mathbf{a}, \mathbf{b}, \mathbf{c}\}$ stands for $\mathbf{a}\times\mathbf{b}.\mathbf{c}$.] (M.A.)

(8) $\mathbf{F}$ and $\mathbf{G}$ are two given vectors. Find a vector $\mathbf{r}$ such that $\mathbf{r}$ is

perpendicular to $\mathbf{F}$ and that $\mathbf{G}-(\mathbf{r}\times\mathbf{F})$ is parallel to $\mathbf{F}$. Show that $\mathbf{G}-(\mathbf{r}\times\mathbf{F}) = \lambda\mathbf{F}$, where

$$\lambda = \frac{\mathbf{F}.\mathbf{G}}{\mathbf{F}^2}. \qquad \text{(M.A.)}$$

(9) Show that the condition for the relations

$$\mathbf{x} = \lambda(\mathbf{b}\times\mathbf{c})+\mu(\mathbf{c}\times\mathbf{a})+\nu(\mathbf{a}\times\mathbf{b}),$$

$$\mathbf{x}\times\mathbf{b} = \mathbf{0}, \quad \mathbf{x}.\mathbf{a} = k \qquad (k \neq 0),$$

to be consistent is $\mathbf{b}.\mathbf{c} = 0$. Hence show that

$$\lambda = \frac{k}{[\mathbf{a}, \mathbf{b}, \mathbf{c}]}, \quad \mu = -\frac{[\mathbf{a}, \mathbf{b}, \mathbf{c}]-k\mathbf{b}^2}{[\mathbf{a}, \mathbf{b}, \mathbf{c}]\,\mathbf{a}.\mathbf{b}}, \quad \nu = 0.$$

Further vector equations

Equations containing vectors are of two types:

(*a*) those in which the unknowns are vectors and

(*b*) those in which the unknowns are numbers, that is, scalars.

There are no general rules for the solutions of vector equations. The following examples show that the forming of triple products often gives a useful method of solution.

Vector equations containing unknown vectors

Examples

(1) *Solve for* $\mathbf{x}$ *the equation* $\mathbf{a}\times\mathbf{x} = \mathbf{a}\times\mathbf{b}$, $\mathbf{a} \neq \mathbf{0}$.

In addition to the two methods given on page 218 for the solution of this equation we may proceed as follows.

Multiplying

$$\mathbf{a}\times\mathbf{x} = \mathbf{a}\times\mathbf{b}$$

vectorially by $\mathbf{a}$ we have

$$(\mathbf{a}\times\mathbf{x})\times\mathbf{a} = (\mathbf{a}\times\mathbf{b})\times\mathbf{a}.$$

Expanding,
$$\mathbf{a}^2\mathbf{x}-(\mathbf{a}.\mathbf{x})\mathbf{a} = \mathbf{a}^2\mathbf{b}-(\mathbf{a}.\mathbf{b})\mathbf{a}.$$

Therefore
$$\mathbf{x} = \mathbf{b}+\frac{\mathbf{a}.\mathbf{x}-\mathbf{a}.\mathbf{b}}{\mathbf{a}^2}\,\mathbf{a}.$$

The form of this suggests that $\mathbf{x}$ is expressible as the sum of vectors in the directions of $\mathbf{b}$ and $\mathbf{a}$, that is

$$\mathbf{x} = \mathbf{b}+\lambda\mathbf{a}$$

where λ is a parameter.

(2) *Find the vector* $\mathbf{x}$ *satisfying the equations* $\mathbf{a}\times\mathbf{x} = \mathbf{a}\times\mathbf{b}$, $\mathbf{a}.\mathbf{x} = k$.

The following is an alternative solution to that given on page 218. Proceeding as in the previous example we obtain

$$\mathbf{x} = \mathbf{b}+\frac{\mathbf{a}.\mathbf{x}-\mathbf{a}.\mathbf{b}}{\mathbf{a}^2}\mathbf{a}.$$

But $$\mathbf{a}.\mathbf{x} = k.$$

Therefore $$\mathbf{x} = \mathbf{b}+\frac{k-\mathbf{a}.\mathbf{b}}{\mathbf{a}^2}\mathbf{a}.$$

(3) *The equation* $\mathbf{x}\times\mathbf{a} = \mathbf{b}$.

In this important vector equation, $\mathbf{a}$, $\mathbf{b}$ are given and $\mathbf{x}$ has to be found. There are two cases.

(i) $\mathbf{a}.\mathbf{b} \neq 0$. In this case there is no solution, that is, there is no value of $\mathbf{x}$ satisfying the equation.

(ii) $\mathbf{a}.\mathbf{b} = 0$. In this case there is an infinite number of solutions for $\mathbf{x}$, as may be seen on consideration of the definition of the vector product. That this is so may also be shown as follows.

Suppose $\mathbf{x}\times\mathbf{a} = \mathbf{b}$ where $\mathbf{a} = (3,\ -5,\ 4)$, $\mathbf{b} = (3,\ 1,\ -1)$. Let

$$\mathbf{x} = (x_1,\ x_2,\ x_3).$$

Then $$(x_1,\ x_2,\ x_3)\times(3,\ -5,\ 4) = (3,\ 1,\ -1).$$

Therefore $(5x_3+4x_2,\ 3x_3-4x_1, -3x_2-5x_1) = (3,\ 1,\ -1)$.

Hence $5x_3+4x_2 = 3,\quad 3x_3-4x_1 = 1,\quad -3x_2-5x_1 = -1.$

On elimination of x_3 from the first two equations we get

$$3x_2+5x_1 = 1$$

which is the third equation. Hence x_1, x_2, x_3 are indeterminate, there being an infinite number of solutions.

Expressing x_1, x_2 in terms of x_3 we have

$$x_1 = \tfrac{1}{4}(-1+3x_3) = -\tfrac{1}{4}+\tfrac{3}{4}x_3,$$

$$x_2 = \tfrac{1}{4}(3-5x_3) = \tfrac{3}{4}-\tfrac{5}{4}x_3.$$

Writing $x_3 = 4p$ where p is a parameter we have

$$x_1 = -\tfrac{1}{4}+3p,\ x_2 = \tfrac{3}{4}-5p,\ x_3 = 4p.$$

A unique solution is obtainable if in addition we are given, say, the equation $\mathbf{a}.\mathbf{x} = -17$. We then have

$$(3,\ -5,\ 4).(-\tfrac{1}{4}+3p,\ \tfrac{3}{4}-5p,\ 4p) = -17.$$

Hence $\quad -\frac{3}{4}+9p-\frac{15}{4}+25p+16p = -17.$

Solving $\quad p = -\frac{1}{4}.$

Solution is therefore $\mathbf{x} = (-\frac{1}{4}-\frac{3}{4}, \frac{3}{4}+\frac{5}{4}, -1)$

$$= (-1, 2, -1).$$

The general solution of $\mathbf{x}\times\mathbf{a} = \mathbf{b}$, $(\mathbf{a}.\mathbf{b} = 0)$, may be obtained by multiplying the equation vectorially by $\mathbf{a}$.

$$\mathbf{a}\times(\mathbf{x}\times\mathbf{a}) = \mathbf{a}\times\mathbf{b}.$$

Expanding, $\quad (\mathbf{a}.\mathbf{a})\mathbf{x}-(\mathbf{a}.\mathbf{x})\mathbf{a} = \mathbf{a}\times\mathbf{b}.$

As before this suggests writing

$$\mathbf{x} = \lambda\mathbf{a}+\mu(\mathbf{a}\times\mathbf{b}).$$

Substituting in $\mathbf{x}\times\mathbf{a} = \mathbf{b}$ we get

$$\mu(\mathbf{a}\times\mathbf{b})\times\mathbf{a} = \mathbf{b} \quad (\lambda\mathbf{a}\times\mathbf{a} = \mathbf{0}).$$

Expanding $\quad \mu\{(\mathbf{a}.\mathbf{a})\mathbf{b}-(\mathbf{a}.\mathbf{b})\mathbf{a}\} = \mathbf{b}.$

Therefore $\quad \mu\mathbf{a}^2\mathbf{b} = \mathbf{b} \quad (\mathbf{a}.\mathbf{b} = 0).$

Hence $\quad \mu = \dfrac{1}{\mathbf{a}^2}.$

Thus the general solution of $\mathbf{x}\times\mathbf{a} = \mathbf{b}$ $(\mathbf{a}.\mathbf{b} = 0)$ is

$$\mathbf{x} = \lambda\mathbf{a}+\frac{\mathbf{a}\times\mathbf{b}}{\mathbf{a}^2}$$

where λ is a parameter.

When $\mathbf{a} = (3, -5, 4)$, $\mathbf{b} = (3, 1, -1)$ we have

$$\mathbf{x} = \lambda(3, -5, 4)+\frac{(1, 15, 18)}{50},$$

that is, $\quad \mathbf{x} = \dfrac{1}{50}(150\lambda+1, -250\lambda+15, 200\lambda+18).$

Again if $\mathbf{a}.\mathbf{x} = -17$ we have a unique solution. In this case

$$(3, -5, 4).\frac{1}{50}(150\lambda+1, -250\lambda+15, 200\lambda+18) = -17.$$

Evaluating the scalar product,

$$\frac{1}{50}(450\lambda+3+1250\lambda-75+800\lambda+72) = -17.$$

Solving, $$\lambda = -\frac{17}{50}.$$

Therefore $$\mathbf{x} = \frac{1}{50}(-50,\ 100,\ -50)$$
$$= (-1,\ 2,\ -1).$$

We give another solution which is of interest of the equation $\mathbf{x}\times\mathbf{a} = \mathbf{b}$ $(\mathbf{a}.\mathbf{b} = 0)$. Since $\mathbf{a}\times\mathbf{b}$ is perpendicular to both $\mathbf{a}$ and $\mathbf{b}$, we can express $\mathbf{x}$ in terms of the three mutually perpendicular vectors $\mathbf{a}$, $\mathbf{b}$, $\mathbf{a}\times\mathbf{b}$, that is,
$$\mathbf{x} = \lambda\mathbf{a}+\mu\mathbf{b}\times\nu(\mathbf{a}\times\mathbf{b}).$$

Substituting for $\mathbf{x}$ in the given equation $\mathbf{x}\times\mathbf{a} = \mathbf{b}$, we obtain
$$\{\lambda\mathbf{a}+\mu\mathbf{b}+\nu(\mathbf{a}\times\mathbf{b})\}\times\mathbf{a} = \mathbf{b},$$
which on expansion gives
$$\mu(\mathbf{b}\times\mathbf{a})+\nu(\mathbf{a}\times\mathbf{b})\times\mathbf{a} = \mathbf{b}.$$
Expanding the vector triple product, we have
$$\mu(\mathbf{b}\times\mathbf{a})+\nu\{(\mathbf{a}.\mathbf{a})\mathbf{b}-(\mathbf{a}.\mathbf{b})\mathbf{a}\} = \mathbf{b}.$$
Therefore $$\mu(\mathbf{b}\times\mathbf{a})+\nu\mathbf{a}^2\mathbf{b} = \mathbf{b}, \quad (\mathbf{a}.\mathbf{b} = 0),$$
from which we deduce
$$\mu = 0 \quad \text{and} \quad \nu\mathbf{a}^2 = 1.$$

Hence $$\mathbf{x} = \lambda\mathbf{a}+\frac{\mathbf{a}\times\mathbf{b}}{\mathbf{a}^2}$$
where λ is a parameter.

(4) *The equations* $\mathbf{x}\times\mathbf{a} = \mathbf{b}$, $\mathbf{x}.\mathbf{c} = p$, $(\mathbf{a}.\mathbf{b} = 0)$.

Forming the vector triple product with $\mathbf{c}$ we have
$$(\mathbf{x}\times\mathbf{a})\times\mathbf{c} = \mathbf{b}\times\mathbf{c}.$$

Expanding, $$(\mathbf{x}.\mathbf{c})\mathbf{a}-(\mathbf{c}.\mathbf{a})\mathbf{x} = \mathbf{b}\times\mathbf{c}.$$

Substituting $\mathbf{x}.\mathbf{c} = p$, this becomes
$$p\mathbf{a}-(\mathbf{a}.\mathbf{c})\mathbf{x} = \mathbf{b}\times\mathbf{c}.$$

Therefore $$\mathbf{x} = \frac{p\mathbf{a}+(\mathbf{c}\times\mathbf{b})}{\mathbf{a}.\mathbf{c}}$$

provided $\mathbf{a}.\mathbf{c} \neq 0$. Thus the solution is unique.

If however $\mathbf{a}.\mathbf{c} = 0$ we proceed as follows. The general solution of $\mathbf{x}\times\mathbf{a} = \mathbf{b}$, $(\mathbf{a}.\mathbf{b} = 0)$ is

$$\mathbf{x} = \lambda\mathbf{a}+\frac{\mathbf{a}\times\mathbf{b}}{\mathbf{a}^2}. \tag{12.1}$$

Forming the scalar product with $\mathbf{c}$ we have

$$\mathbf{x}.\mathbf{c} = \lambda\mathbf{a}.\mathbf{c}+\frac{(\mathbf{a}\times\mathbf{b}).\mathbf{c}}{\mathbf{a}^2}.$$

When $\mathbf{a}.\mathbf{c} = 0$ we have the condition

$$p = \frac{(\mathbf{a}\times\mathbf{b}).\mathbf{c}}{\mathbf{a}^2}$$

for (12.1) to satisfy the equation $\mathbf{x}.\mathbf{c} = p$. Thus

$$\mathbf{x} = \lambda\mathbf{a}+\frac{\mathbf{a}\times\mathbf{b}}{\mathbf{a}^2}$$

will satisfy the second equation $\mathbf{x}.\mathbf{c} = p$ for any value of λ and in this case the solution is not unique.

In particular the solution of

$$\mathbf{x}\times\mathbf{a} = \mathbf{b},\ \mathbf{x}.\mathbf{a} = -17$$

where $\mathbf{a} = (3,\ -5,\ 4)$, $\mathbf{b} = (3,\ 1,\ -1)$ is given by

$$\begin{aligned}\mathbf{x} &= \frac{p\mathbf{a}+(\mathbf{a}\times\mathbf{b})}{\mathbf{a}.\mathbf{a}}\\ &= \frac{-17(3,\ -5,\ 4)+(1,\ 15,\ 18)}{50}\\ &= \frac{(-50,\ 100,\ -50)}{50}\\ &= (-1,\ 2,\ -1).\end{aligned}$$

(5) *Solve the simultaneous equations*

$$p\mathbf{x}+q\mathbf{y} = \mathbf{a},\quad \mathbf{x}\times\mathbf{y} = \mathbf{b},\quad (\mathbf{a}.\mathbf{b} = 0).$$

Forming the vector product with $\mathbf{x}$ the first equation becomes

$$(p\mathbf{x}+q\mathbf{y})\times\mathbf{x} = \mathbf{a}\times\mathbf{x}.$$

Expanding, $$q(\mathbf{y}\times\mathbf{x}) = \mathbf{a}\times\mathbf{x}.$$

Using the second equation we have

$$q\mathbf{b} = \mathbf{x}\times\mathbf{a}.$$

From Example (3), since $\mathbf{a}.\mathbf{b} = 0$, the general solution of this equation is given by

$$\mathbf{x} = \lambda\mathbf{a}+\frac{\mathbf{a}\times q\mathbf{b}}{\mathbf{a}^2}$$

$$= \lambda\mathbf{a}+q\,\frac{\mathbf{a}\times\mathbf{b}}{\mathbf{a}^2},$$

where λ is a variable scalar. Substituting in $p\mathbf{x}+q\mathbf{y} = \mathbf{a}$ we obtain

$$\mathbf{y} = \frac{1-p\lambda}{q}\,\mathbf{a}-p\,\frac{\mathbf{a}\times\mathbf{b}}{\mathbf{a}^2}.$$

Thus the solution is not unique.

Vector equations containing unknown scalars

Examples

(1) *Find the scalars x, y, z satisfying the equation*

$$\mathbf{a} = x\mathbf{b}+y\mathbf{c}+z\mathbf{d}$$

where $\mathbf{a} = (-3, 2, 6)$, $\mathbf{b} = (-1, 1, 2)$, $\mathbf{c} = (3, 1, -1)$, $\mathbf{d} = (2, 1, 1)$.

Form the scalar triple product by multiplying scalarly by $\mathbf{c}\times\mathbf{d}$ each term of the given equation. Doing this we obtain

$$\mathbf{a}.(\mathbf{c}\times\mathbf{d}) = x\mathbf{b}.(\mathbf{c}\times\mathbf{d})+y\mathbf{c}.(\mathbf{c}\times\mathbf{d})+z\mathbf{d}.(\mathbf{c}\times\mathbf{d}).$$

Since $$\mathbf{c}.(\mathbf{c}\times\mathbf{d}) = \mathbf{d}.(\mathbf{c}\times\mathbf{d}) = 0$$

we have $$x = \frac{\mathbf{a}.(\mathbf{c}\times\mathbf{d})}{\mathbf{b}.(\mathbf{c}\times\mathbf{d})},$$

provided $\mathbf{b}.(\mathbf{c}\times\mathbf{d}) \neq 0$.

Similarly $$y = \frac{\mathbf{a}.(\mathbf{d}\times\mathbf{b})}{\mathbf{c}.(\mathbf{d}\times\mathbf{b})},\quad z = \frac{\mathbf{a}.(\mathbf{b}\times\mathbf{c})}{\mathbf{d}.(\mathbf{b}\times\mathbf{c})}.$$

Thus the general solution of the equation

$$\mathbf{a} = x\mathbf{b}+y\mathbf{c}+z\mathbf{d}$$

is $$x = \frac{[\mathbf{a}, \mathbf{c}, \mathbf{d}]}{[\mathbf{b}, \mathbf{c}, \mathbf{d}]},\quad y = \frac{[\mathbf{a}, \mathbf{d}, \mathbf{b}]}{[\mathbf{b}, \mathbf{c}, \mathbf{d}]},\quad z = \frac{[\mathbf{a}, \mathbf{b}, \mathbf{c}]}{[\mathbf{b}, \mathbf{c}, \mathbf{d}]},$$

provided that $[\mathbf{b}, \mathbf{c}, \mathbf{d}] \neq 0$.

Evaluating,

$$\mathbf{b}.(\mathbf{c}\times\mathbf{d}) = \begin{vmatrix} -1 & 1 & 2 \\ 3 & 1 & -1 \\ 2 & 1 & 1 \end{vmatrix} = -(1+1)-(3+2)+2(3-2) = -5.$$

$$\mathbf{a}.(\mathbf{c}\times\mathbf{d}) = \begin{vmatrix} -3 & 2 & 6 \\ 3 & 1 & -1 \\ 2 & 1 & 1 \end{vmatrix} = -3(1+1)-2(3+2)+6(3-2) = -10.$$

$$\mathbf{a}.(\mathbf{d}\times\mathbf{b}) = \begin{vmatrix} -3 & 2 & 6 \\ 2 & 1 & 1 \\ -1 & 1 & 2 \end{vmatrix} = -3(2-1)-2(4+1)+6(2+1) = 5.$$

$$\mathbf{a}.(\mathbf{b}\times\mathbf{c}) = \begin{vmatrix} -3 & 2 & 6 \\ -1 & 1 & 2 \\ 3 & 1 & -1 \end{vmatrix} = -3(-1-2)-2(1-6)+6(-1-3) = -5.$$

Hence $\qquad x = 2, \quad y = -1, \quad z = 1.$

(2) *Find the scalars x, y satisfying the equation*

$$\mathbf{a}\times\mathbf{b} = x\mathbf{b}+y\mathbf{c}$$

where $\mathbf{a} = (3, 1, 1)$, $\mathbf{b} = (2, 0, -2)$, $\mathbf{c} = (1, -2, 0)$.

Multiplying scalarly by $\mathbf{a}$ we obtain

$$x\mathbf{a}.\mathbf{b}+y\mathbf{a}.\mathbf{c} = 0.$$

Similarly multiplying scalarly by $\mathbf{b}$ gives

$$x\mathbf{b}^2+y\mathbf{b}.\mathbf{c} = 0.$$

Hence there is no solution unless $\mathbf{a}.\mathbf{b}:\mathbf{a}.\mathbf{c} = \mathbf{b}^2:\mathbf{b}.\mathbf{c}$.

Since $\mathbf{a}.\mathbf{b}:\mathbf{a}.\mathbf{c} = 4:1$ and $\mathbf{b}^2:\mathbf{b}.\mathbf{c} = 4:1$, there are scalars satisfying the given equation.

Multiplying scalarly by $\mathbf{c}\times\mathbf{a}$, we have

$$\begin{aligned}(\mathbf{a}\times\mathbf{b}).(\mathbf{c}\times\mathbf{a}) &= x\mathbf{b}.(\mathbf{c}\times\mathbf{a})+y\mathbf{c}.(\mathbf{c}\times\mathbf{a}) \\ &= x\mathbf{b}.(\mathbf{c}\times\mathbf{a}).\end{aligned}$$

Therefore, provided $\mathbf{a}.(\mathbf{b}\times\mathbf{c}) \neq 0$,

$$x = \frac{(\mathbf{a}\times\mathbf{b}).(\mathbf{c}\times\mathbf{a})}{\mathbf{a}.(\mathbf{b}\times\mathbf{c})}.$$

Similarly multiplying scalarly by $\mathbf{a}\times\mathbf{b}$, we obtain

$$y = \frac{(\mathbf{a}\times\mathbf{b}).(\mathbf{a}\times\mathbf{b})}{\mathbf{a}.(\mathbf{b}\times\mathbf{c})}.$$

Evaluating,

$$\mathbf{a}.(\mathbf{b}\times\mathbf{c}) = \begin{vmatrix} 3 & 1 & 1 \\ 2 & 0 & -2 \\ 1 & -2 & 0 \end{vmatrix} = 3(0-4)-(0+2)+(-4-0) = -18.$$

$$(\mathbf{a}\times\mathbf{b}).(\mathbf{c}\times\mathbf{a}) = (-2,\ 8,\ -2).(-2,\ -1,\ 7) = -18.$$

$$(\mathbf{a}\times\mathbf{b}).(\mathbf{a}\times\mathbf{b}) = (-2,\ 8,\ -2).(-2,\ 8,\ -2) = 72.$$

Hence $\qquad x = 1, \quad y = -4.$

Exercise 12(*c*)

(1) Find p, q such that $\mathbf{a}\times\mathbf{b} = p\mathbf{c}+q\mathbf{d}$, where $\mathbf{a} = 2\mathbf{i}+3\mathbf{j}-\mathbf{k}$, $\mathbf{b} = 5\mathbf{i}+\mathbf{j}+2\mathbf{k}$, $\mathbf{c} = \mathbf{i}-2\mathbf{j}-4\mathbf{k}$, $\mathbf{d} = \mathbf{i}-\mathbf{j}-\mathbf{k}$.

(2) Find $\mathbf{x}$ satisfying the equations

$$\mathbf{x}\times\mathbf{a} = \mathbf{b}, \quad \mathbf{x}.\mathbf{a} = 5,$$

where $\mathbf{a} = (2,\ 2,\ -3)$, $\mathbf{b} = (1,\ 17,\ 12)$.

(3) Given that $\mathbf{a}$, $\mathbf{b}$ are perpendicular vectors, show that any vector $\mathbf{x}$ can be expressed in the form

$$\mathbf{x} = \lambda\mathbf{a}+\mu\mathbf{b}+\nu(\mathbf{a}\times\mathbf{b}),$$

where λ, μ, ν are scalars.

Hence by writing $\mathbf{a} = a\mathbf{i}$, $\mathbf{b} = b\mathbf{j}$, show that the solution of the equation $\mathbf{x}\times\mathbf{a} = \mathbf{b}$ is given by

$$\mathbf{x} = \lambda\mathbf{a}+\frac{1}{a^2}\mathbf{a}\times\mathbf{b}.$$

(4) A vector $\mathbf{x}$ satisfies the equations

$$\mathbf{x}\times\mathbf{b} = \mathbf{c}\times\mathbf{b} \quad \text{and} \quad \mathbf{x}.\mathbf{a} = 0.$$

Prove that $\mathbf{x} = \mathbf{c}-(\mathbf{a}.\mathbf{c}/\mathbf{a}.\mathbf{b})\mathbf{b}$ provided that $\mathbf{a}.\mathbf{b} \neq 0$. (L.U.)

(5) If $\mathbf{a}$, $\mathbf{b}$ are given vectors, $\mathbf{b}$ being perpendicular to $\mathbf{a}$, and if k is a given scalar, show that the solution of the equations

$$\mathbf{a}.\mathbf{x} = k \text{ and } \mathbf{a}\times\mathbf{x} = \mathbf{b}$$

for an unknown vector $\mathbf{x}$ is unique, and find it.

(6) Given that $\mathbf{x}$ satisfies the equation

$$\mathbf{x} = p\mathbf{a}+\mathbf{x}\times\mathbf{a},$$

where p is a non-zero scalar, find $\mathbf{x}$.

(7) Show that
$$\mathbf{a}\times(\mathbf{b}\times\mathbf{c}) = (\mathbf{a}.\mathbf{c})\mathbf{b}-(\mathbf{a}.\mathbf{b})\mathbf{c}.$$

Find the vector $\mathbf{x}$ and the scalar λ which satisfy the equations
$$\mathbf{a}\times\mathbf{x} = \mathbf{b}+\lambda\mathbf{a}, \quad \mathbf{a}.\mathbf{x} = 2,$$
where
$$\mathbf{a} = \mathbf{i}+2\mathbf{j}-\mathbf{k}, \quad \mathbf{b} = 2\mathbf{i}-\mathbf{j}+\mathbf{k}. \qquad \text{(L.U.)}$$

(8) Find all the vectors $\mathbf{x}$ such that
$$\mathbf{a}\times\mathbf{x} = \mathbf{a}\times\mathbf{b}.$$

(9) If $\mathbf{a}$ is a given vector and k is a given scalar show that the complete solution of the equation $\mathbf{x}.\mathbf{a} = k$ is
$$\mathbf{x} = \frac{k}{\mathbf{a}^2}\mathbf{a}+(\mathbf{a}\times\mathbf{y}),$$
where $\mathbf{y}$ is an arbitrary vector.

(10) Show that the solution of the equation
$$\mathbf{x}^2+\mathbf{a}.\mathbf{x}+k = 0$$
is given by
$$\mathbf{x} = \tfrac{1}{2}\{\mathbf{u}(\mathbf{a}^2-4k)^{\frac{1}{2}}-\mathbf{a}\},$$
where $\mathbf{u}$ is an arbitrary unit vector.

(11) Solve the following equations for $\mathbf{x}$.

(i) $\mathbf{x}\times\mathbf{a} = \mathbf{b}$, where $\mathbf{a}.\mathbf{b} = 0$.

(ii) $\lambda\mathbf{x}+\mathbf{x}\times\mathbf{c} = \mathbf{a}$.

(iii) $a\mathbf{x}+(\mathbf{x}.\mathbf{u})\mathbf{v} = \mathbf{w}$,

considering the cases in which $a+\mathbf{u}.\mathbf{v}$ is not zero and is zero separately. (L.U.)

(12) Find the vectors $\mathbf{x}$ and $\mathbf{y}$ satisfying the simultaneous equations
$$\mathbf{x}+(\mathbf{a}\times\mathbf{y}) = \mathbf{b},$$
$$\mathbf{y}+(\mathbf{a}\times\mathbf{x}) = \mathbf{c},$$
where $\mathbf{a}$, $\mathbf{b}$ and $\mathbf{c}$ are given vectors.

13

FURTHER APPLICATIONS OF THE VECTOR PRODUCT

Statics

Equivalent systems of forces

Two free vectors are said to be equal if they have the same magnitude, direction and sense. Two localized vectors are also said to be equal if they have the same magnitude, direction and sense. Furthermore, two localized vectors are said to be equivalent if they are equal and also act along the same line. It follows that if two localized vectors are equivalent their moments about all points are equal. In particular, since force is a localized vector, consider a force $\mathbf{F}$ acting at the point $P(\mathbf{r})$ and a force $\mathbf{F}$ acting at the point $P'(\mathbf{r}')$, where P' is on the line of action of the force $\mathbf{F}$ acting at P (Fig. 13.1).

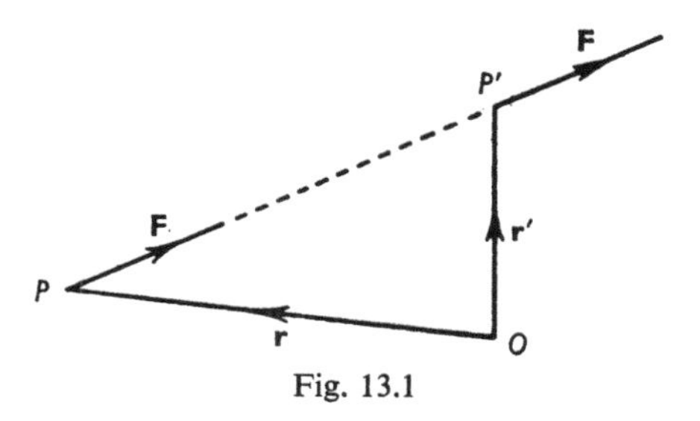

Fig. 13.1

The moment about the origin O of the force $\mathbf{F}$ acting at P is $\mathbf{r}\times\mathbf{F}$, and the moment of the force $\mathbf{F}$ acting at P' is $\mathbf{r}'\times\mathbf{F}$. Now

$$\begin{aligned}\mathbf{r}\times\mathbf{F} &= (\mathbf{r}'-\mathbf{PP}')\times\mathbf{F}\\ &= \mathbf{r}'\times\mathbf{F}-\mathbf{PP}'\times\mathbf{F}\\ &= \mathbf{r}'\times\mathbf{F},\end{aligned}$$

since $\mathbf{PP}'$, $\mathbf{F}$ are in the same direction. This means that we may slide a force along its line of action without changing its effect. This is known as the principle of transmissibility, and it enables us to regard force as a sliding vector.

Extending the concept of equivalence to two systems of forces it can be shown that:

The necessary and sufficient conditions for two systems of forces to be equivalent are that (i) *their vector sums should be equal and* (ii) *the vector sums of their moments about any one point should be equal.*

Equilibrium of two forces

Suppose two forces $\mathbf{F}_1$, $\mathbf{F}_2$ act on a particle. The necessary and sufficient condition for equilibrium is that the forces must have the same magnitude and direction but be opposite in sense, that is

$$\mathbf{F}_1+\mathbf{F}_2 = \mathbf{0}.$$

However, in the case of two forces $\mathbf{F}_1$, $\mathbf{F}_2$ acting on a rigid body, the necessary and sufficient conditions for equilibrium are that (i) the forces must have the same magnitude and direction but be opposite in sense, and (ii) the forces must be localized in the same line.

From (i) we have

$$\mathbf{F}_1+\mathbf{F}_2 = \mathbf{0}.$$

If the forces $\mathbf{F}_1$, $\mathbf{F}_2$ act at the points $A(\mathbf{r}_1)$, $B(\mathbf{r}_2)$ respectively, we have from (ii), since $\mathbf{AB}$ is parallel to $\mathbf{F}_1$,

$$\mathbf{AB}\times\mathbf{F}_1 = \mathbf{0},$$

that is,

$$(\mathbf{r}_1-\mathbf{r}_2)\times\mathbf{F}_1 = \mathbf{0}.$$

This result has a statical significance; for if we take moments about the origin O, we have

$$\mathbf{G} = \mathbf{r}_1\times\mathbf{F}_1+\mathbf{r}_2\times\mathbf{F}_2,$$

where $\mathbf{G}$ is the vector sum of the moments. Since $\mathbf{F}_2 = -\mathbf{F}_1$ we have

$$\mathbf{G} = (\mathbf{r}_1-\mathbf{r}_2)\times\mathbf{F}_1.$$

Furthermore, since the forces are localized in the same line, $(\mathbf{r}_1-\mathbf{r}_2)$ is parallel to $\mathbf{F}_1$. Hence

$$\mathbf{G} = \mathbf{0},$$

showing that the forces together have no resultant turning effect.

Now consider two forces $\mathbf{F}_1$, $\mathbf{F}_2$ acting at $A(\mathbf{r}_1)$, $B(\mathbf{r}_2)$ respectively, and suppose

$$\text{(i)}\quad \mathbf{F}_1+\mathbf{F}_2 = \mathbf{0},$$

implying that $\mathbf{F}_1$, $\mathbf{F}_2$ are parallel, equal in magnitude and opposite in sense, and

$$\text{(ii)}\quad \mathbf{r}_1 \times \mathbf{F}_1 + \mathbf{r}_2 \times \mathbf{F}_2 = \mathbf{0}.$$

Hence
$$(\mathbf{r}_1 - \mathbf{r}_2) \times \mathbf{F}_1 = \mathbf{0}.$$

Since $\mathbf{F}_1 \neq \mathbf{0}$, either $\mathbf{r}_1 - \mathbf{r}_2 = \mathbf{0}$ or $(\mathbf{r}_1 - \mathbf{r}_2)$ is parallel to $\mathbf{F}_1$, and therefore to $\mathbf{F}_2$. Hence $\mathbf{F}_1$, $\mathbf{F}_2$ are localized in the same line. Therefore $\mathbf{F}_1$, $\mathbf{F}_2$ are in equilibrium.

Thus we see that the necessary and sufficient conditions for two forces to be in equilibrium are that (i) their vector sum is zero and (ii) the vector sum of their moments is zero.

It can be shown that these conditions apply in the case of the equilibrium of a system of forces acting on a rigid body. Also the effect of a system of forces acting on a rigid body is unaltered if we introduce a pair of equal and opposite forces localized in the same line.

Resultant of two forces

If a system of forces is equivalent to a single force then this force is called the resultant of the system. Two forces acting on a particle always have a resultant given by the vector sum of the forces. However, two forces acting on a rigid body do not necessarily have a resultant.

Consider forces $\mathbf{F}_1$, $\mathbf{F}_2$ acting at $A(\mathbf{r}_1)$, $B(\mathbf{r}_2)$ respectively, $\mathbf{F}_1 + \mathbf{F}_2 \neq \mathbf{0}$. The vector sum of the forces is given by

$$\mathbf{R} = \mathbf{F}_1 + \mathbf{F}_2,$$

and the vector sum of the moments of the forces about the origin O is given by

$$\mathbf{G} = \mathbf{r}_1 \times \mathbf{F}_1 + \mathbf{r}_2 \times \mathbf{F}_2.$$

The two forces will have a resultant if we can find a position vector $\mathbf{r}$ on the line of action of $\mathbf{R}$ such that

$$\mathbf{r} \times \mathbf{R} = \mathbf{G}.$$

For this equation to be satisfied we must have $\mathbf{R}$ perpendicular to $\mathbf{G}$, that is,

$$\mathbf{R} . \mathbf{G} = 0.$$

If $\mathbf{G} = \mathbf{0}$, then the equation $\mathbf{r} \times \mathbf{R} = \mathbf{G}$ is clearly satisfied by $\mathbf{r} = \mathbf{0}$. Hence we have a resultant force $\mathbf{F}_1 + \mathbf{F}_2$ acting through the origin O when $\mathbf{G} = \mathbf{0}$.

Thus two forces $\mathbf{F}_1$, $\mathbf{F}_2$ ($\mathbf{F}_1+\mathbf{F}_2 \neq \mathbf{0}$) have a resultant if either (i) $\mathbf{r}_1\times\mathbf{F}_1+\mathbf{r}_2\times\mathbf{F}_2 = \mathbf{0}$, or (ii) $\mathbf{r}_1\times\mathbf{F}_1+\mathbf{r}_2\times\mathbf{F}_2$ is perpendicular to $\mathbf{F}_1+\mathbf{F}_2$.

Resultant of two parallel forces

Two forces $\mathbf{F}_1$, $\mathbf{F}_2$ acting at $A(\mathbf{r}_1)$, $B(\mathbf{r}_2)$ are collinear if $\mathbf{F}_1$ is parallel to $\mathbf{F}_2$ and $\mathbf{AB}$ is parallel to either $\mathbf{F}_1$ or $\mathbf{F}_2$, that is, if $\mathbf{F}_1 = k\mathbf{F}_2$ and $(\mathbf{r}_1-\mathbf{r}_2) = k'\mathbf{F}_1$, where k, k' are numbers. Alternatively, in terms of vector products, these conditions are $\mathbf{F}_1\times\mathbf{F}_2 = \mathbf{0}$ and $(\mathbf{r}_1-\mathbf{r}_2)\times\mathbf{F}_1 = \mathbf{0}$.

In general, consider the parallel forces $\mathbf{F}_1 = k_1\mathbf{F}$, $\mathbf{F}_2 = k_2\mathbf{F}$ acting at the points $A(\mathbf{r}_1)$, $B(\mathbf{r}_2)$ respectively, $\mathbf{F}_1+\mathbf{F}_2 \neq \mathbf{0}$, that is, $k_1+k_2 \neq 0$. Then

$$\mathbf{R} = \mathbf{F}_1+\mathbf{F}_2 = (k_1+k_2)\mathbf{F},$$

and

$$\mathbf{G} = \mathbf{r}_1\times\mathbf{F}_1+\mathbf{r}_2\times\mathbf{F}_2$$

$$= k_1(\mathbf{r}_1\times\mathbf{F})+k_2(\mathbf{r}_2\times\mathbf{F}).$$

Since $\mathbf{R}.\mathbf{G} = 0$, we can find a point $\mathbf{r}$ such that $\mathbf{r}\times\mathbf{R} = \mathbf{G}$, that is,

$$\mathbf{r}\times(k_1+k_2)\mathbf{F} = k_1(\mathbf{r}_1\times\mathbf{F})+k_2(\mathbf{r}_2\times\mathbf{F}).$$

Since $k_1+k_2 \neq 0$, we have

$$\mathbf{r}\times\mathbf{F} = \frac{k_1\mathbf{r}_1+k_2\mathbf{r}_2}{k_1+k_2}\times\mathbf{F}.$$

The solution of this equation is

$$\mathbf{r} = \frac{k_1\mathbf{r}_1+k_2\mathbf{r}_2}{k_1+k_2}+t\mathbf{F},$$

where t is a parameter. This is the equation of the line of action of the resultant with $(k_1\mathbf{r}_1+k_2\mathbf{r}_2)/(k_1+k_2)$ a point on this line. Hence two parallel forces $k_1\mathbf{F}$, $k_2\mathbf{F}$ ($k_1 \neq -k_2$) have a resultant $(k_1+k_2)\mathbf{F}$ acting through the point $(k_1\mathbf{r}_1+k_2\mathbf{r}_2)/(k_1+k_2)$.

Couple

A couple is a pair of parallel forces, equal in magnitude but opposite in sense, and localized in different lines.

Consider the couple formed by the forces $\mathbf{F}$ acting at the point $A(\mathbf{r}_1)$ and $-\mathbf{F}$ acting at the point $B(\mathbf{r}_2)$ (Fig. 13.2). We have

$$\mathbf{R} = \mathbf{F}+(-\mathbf{F}) = \mathbf{0},$$

and $$\mathbf{G} = \mathbf{r}_1 \times \mathbf{F} + \mathbf{r}_2 \times (-\mathbf{F})$$
$$= (\mathbf{r}_1 - \mathbf{r}_2) \times \mathbf{F}.$$

Since $\mathbf{r}_1 \neq \mathbf{r}_2$ and $\mathbf{F} \neq \mathbf{0}$ we see that $\mathbf{G} \neq \mathbf{0}$, showing that a body acted on by a couple is not in equilibrium. Also since $\mathbf{R} = \mathbf{0}$ and $\mathbf{G} \neq \mathbf{0}$, a couple cannot be reduced to a single resultant force. It has, however, a turning effect given by $\mathbf{G}$, the moment or torque of the couple.

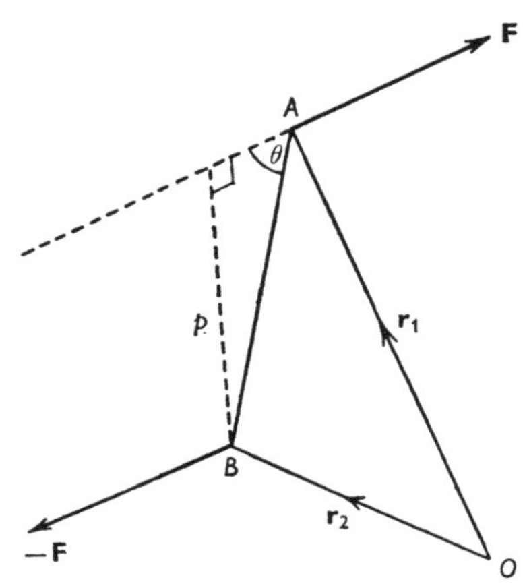

Fig. 13.2

Also $$\mathbf{r}_1 - \mathbf{r}_2 = \mathbf{BA}.$$

Therefore $$\mathbf{G} = \mathbf{BA} \times \mathbf{F},$$

showing that the moment of a couple is independent of the origin O. We may write
$$\mathbf{G} = |\mathbf{BA}|\ |\mathbf{F}| \sin\theta \hat{\mathbf{n}}$$

where $\hat{\mathbf{n}}$ is the unit vector perpendicular to the plane of the couple, that is, to the plane containing the forces. Since $|\mathbf{BA}| \sin\theta$ is the perpendicular distance p between the lines of action of the forces, the magnitude of the moment of the couple is $p|\mathbf{F}|$.

Exercise 13(*a*)

(1) Which of the following pairs of forces are in equilibrium?

(i) $\mathbf{P} = \mathbf{i} - 2\mathbf{j} + \mathbf{k}$ acting at the point (6, 3, 2), $\mathbf{Q} = -\mathbf{i} + 2\mathbf{j} - \mathbf{k}$ acting at the point (5, 5, 1).

(ii) $\mathbf{X} = 3\mathbf{i}-\mathbf{j}-2\mathbf{k}$ acting at the point $(2, 3, -1)$,
$\mathbf{Y} = -3\mathbf{i}+\mathbf{j}+2\mathbf{k}$ acting at the point $(-1, 4, -3)$.

(2) Which of the following pairs of forces are collinear?

(i) $\mathbf{P} = 6\mathbf{i}-2\mathbf{j}+4\mathbf{k}$ acting at the point $(1, 2, 3)$,
$\mathbf{Q} = -9\mathbf{i}+3\mathbf{j}-6\mathbf{k}$ acting at the point $(4, 3, 3)$.

(ii) $\mathbf{X} = \mathbf{i}-\mathbf{j}-\mathbf{k}$ acting at the point $(3, 0, 6)$, $\mathbf{Y} = 2\mathbf{i}-2\mathbf{j}-2\mathbf{k}$ acting at the point $(4, -1, 5)$.

In each case find their resultant.

(3) A particle is in equilibrium under the action of three coplanar forces $F_1\hat{\mathbf{u}}_1$, $F_2\hat{\mathbf{u}}_2$, $F_3\hat{\mathbf{u}}_3$ where $\hat{\mathbf{u}}_1$, $\hat{\mathbf{u}}_2$, $\hat{\mathbf{u}}_3$ are unit vectors. Express the condition of equilibrium by a vector equation. By multiplying this equation vectorially show that $(\hat{\mathbf{u}}_2 \times \hat{\mathbf{u}}_3)/F_1 = (\hat{\mathbf{u}}_3 \times \hat{\mathbf{u}}_1)/F_2 = (\hat{\mathbf{u}}_1 \times \hat{\mathbf{u}}_2)/F_3$. Hence deduce Lami's theorem, namely: If three concurrent forces are in equilibrium they are coplanar, and each force is proportional to the sine of the angle between the other two.

(4) Prove that the system of forces $2\mathbf{i}+5\mathbf{j}+3\mathbf{k}$ at the point $(-1, -9, -5)$ and $\mathbf{i}-2\mathbf{j}+4\mathbf{k}$ at the point $(-1, 0, -10)$ can be reduced to a single resultant. Find this resultant.

(5) Find the moment of the couple formed by the forces $(2\mathbf{i}+\mathbf{j}-4\mathbf{k})$ at the point $(1, 2, -4)$ and $(-2\mathbf{i}-\mathbf{j}+4\mathbf{k})$ at the point $(0, -1, 3)$.

(6) Prove that a force $\mathbf{F}$ acting at the point whose position vector is $\mathbf{r}$ is equivalent to a force $\mathbf{F}$ at the origin together with a couple $\mathbf{r}\times\mathbf{F}$.

(7) A system consists of forces $(3, 0, -1)$ acting at $(1, 2, 0)$ and $(2, 3, 0)$ acting at $(0, -1, 3)$. Reduce this system to

(i) a force at the origin O together with a couple,

(ii) a force at $(1, 2, 3)$ together with a couple.

Prove that the system cannot be reduced to a single resultant.

(8) Forces $\alpha\mathbf{BC}$, $\beta\mathbf{CA}$, $\gamma\mathbf{AB}$, $\lambda\mathbf{OA}$, $\mu\mathbf{OB}$, $\nu\mathbf{OC}$ act along the edges of a tetrahedron $OABC$. If they reduce to a couple show that

$$\lambda+\beta-\gamma = \mu+\gamma-\alpha = \nu+\alpha-\beta = 0,$$

and if they reduce to a single force show that

$$\lambda\alpha+\mu\beta+\nu\gamma = 0.$$

Kinematics and dynamics

Motion of a particle in a circle

Consider a particle moving in a circle, centre O and radius a. Let $\mathbf{r}$ be its position vector, at any instant, relative to O. Since

$$|\mathbf{r}| = a$$

we have
$$\mathbf{r}^2 = a^2.$$

Differentiating with respect to time, we obtain

$$\mathbf{r}.\dot{\mathbf{r}} = 0. \tag{13·1}$$

If $\hat{\mathbf{n}}$ is the unit vector perpendicular to the plane of motion, we have since $\hat{\mathbf{n}}$ is perpendicular to $\mathbf{r}$,

$$\hat{\mathbf{n}}.\mathbf{r} = 0.$$

Differentiating with respect to time, we obtain

$$\hat{\mathbf{n}}.\dot{\mathbf{r}} = 0. \tag{13·2}$$

Hence from (13·1) and (13·2), $\hat{\mathbf{n}}$, $\mathbf{r}$, $\dot{\mathbf{r}}$ are mutually orthogonal. Thus

$$\dot{\mathbf{r}} = \omega\hat{\mathbf{n}}\times\mathbf{r}$$

where ω is a scalar, not necessarily a constant. Since $\dot{\mathbf{r}}$ is the velocity of the particle and $\mathbf{r}$ is its position vector, $\omega\hat{\mathbf{n}}$ is by definition the angular velocity of the motion, and $\omega = \dot{\theta}$ where θ is the angle which the radius vector makes with some fixed line in the plane of motion. The particle has therefore a velocity of $a\omega$ in the direction of $\hat{\mathbf{n}}\times\mathbf{r}$, that is, in the direction of the positive tangent.

On further differentiation we obtain

$$\begin{aligned}\ddot{\mathbf{r}} &= \dot{\omega}\hat{\mathbf{n}}\times\mathbf{r}+\omega\hat{\mathbf{n}}\times\dot{\mathbf{r}}\\ &= \dot{\omega}\hat{\mathbf{n}}\times\mathbf{r}+\omega\hat{\mathbf{n}}\times(\omega\hat{\mathbf{n}}\times\mathbf{r})\\ &= \dot{\omega}\hat{\mathbf{n}}\times\mathbf{r}+\omega\hat{\mathbf{n}}\times\omega|\mathbf{r}|\hat{\mathbf{n}}\times\hat{\mathbf{r}}\\ &= \dot{\omega}\hat{\mathbf{n}}\times\mathbf{r}+\omega^2|\mathbf{r}|\hat{\mathbf{n}}\times\hat{\mathbf{p}},\end{aligned}$$

where $\hat{\mathbf{r}}$ is the unit vector having the same direction and sense as $\mathbf{r}$, and $\hat{\mathbf{p}} = \hat{\mathbf{n}}\times\hat{\mathbf{r}}$. Since $\hat{\mathbf{n}}\times\hat{\mathbf{p}} = |\hat{\mathbf{n}}|\ |\hat{\mathbf{p}}|\sin 90°(-\hat{\mathbf{r}}) = -\hat{\mathbf{r}}$, we have

$$\begin{aligned}\ddot{\mathbf{r}} &= \dot{\omega}\hat{\mathbf{n}}\times\mathbf{r}-\omega^2|\mathbf{r}|\hat{\mathbf{r}}\\ &= \dot{\omega}\hat{\mathbf{n}}\times\mathbf{r}-\omega^2\mathbf{r}\\ &= a\dot{\omega}\hat{\mathbf{n}}\times\hat{\mathbf{r}}-a\omega^2\hat{\mathbf{r}}, \qquad (\mathbf{r} = a\hat{\mathbf{r}}),\end{aligned}$$

showing that the acceleration consists of a transverse component $a\dot{\omega}$ along the positive tangent and a radial component $a\omega^2$ towards the centre of the circle.

N.B. The term $\omega\hat{\mathbf{n}} \times (\omega\hat{\mathbf{n}} \times \mathbf{r})$ may be expanded by use of the vector triple product.

Moment of momentum of a particle

Suppose a particle at any time t has mass m, position vector $\mathbf{r}$ relative to a fixed origin O, and velocity $\mathbf{v} = \dot{\mathbf{r}}$. Then from Newton's second law we may write

$$\frac{d}{dt}(m\mathbf{v}) = \mathbf{F}$$

where $\mathbf{F}$ is the force on the particle. Multiplying vectorially by $\mathbf{r}$ we obtain

$$\mathbf{r} \times \frac{d}{dt}(m\mathbf{v}) = \mathbf{r} \times \mathbf{F}.$$

We define the moment of momentum $\mathbf{H}$ of the particle about O as the vector quantity given by the moment about O of the linear momentum $m\mathbf{v}$ of the particle, that is,

$$\mathbf{H} = \mathbf{r} \times m\mathbf{v} = \mathbf{r} \times m\dot{\mathbf{r}}.$$

Differentiating,
$$\frac{d\mathbf{H}}{dt} = \dot{\mathbf{r}} \times m\dot{\mathbf{r}} + \mathbf{r} \times \frac{d}{dt}(m\dot{\mathbf{r}})$$

$$= \mathbf{r} \times \mathbf{F},$$

since $\dot{\mathbf{r}} \times m\dot{\mathbf{r}} = \mathbf{0}$. But the moment $\mathbf{G}$ of the force $\mathbf{F}$ about the origin O is given by

$$\mathbf{G} = \mathbf{r} \times \mathbf{F}.$$

Therefore
$$\mathbf{G} = \frac{d\mathbf{H}}{dt}.$$

Thus *the moment about O of the force acting on a particle is equal to the rate of change of the moment of momentum of the particle about O.*

If $\mathbf{G} = \mathbf{0}$ then $\mathbf{H}$ is constant and we have the principle of the conservation of moment of momentum namely:

The moment of momentum of a particle about O is constant, if the moment of the force acting on the particle about O is zero.

Also if the component of the moment of the force in any direction is zero then the component of the moment of momentum in this direction is conserved.

The moment of momentum is also known as the angular momentum.

Central orbits

Consider a particle acted upon by a force $\mathbf{F}$ directed toward or away from a fixed point O. The force $\mathbf{F}$ is called a central force and the point O is called the centre of force. The path described by the particle is called the central orbit. Examples of this type of motion occur in the study of the motion of planets and comets under the gravitational attraction of the sun.

We shall now show that a particle acted upon by a central force moves in the plane containing the centre of force. Let the position vector of the particle relative to O be $\mathbf{r}$. Let $\mathbf{H}$ be the moment of momentum of the particle and $\mathbf{G}$ be the moment of $\mathbf{F}$ about O. We have

$$\frac{d\mathbf{H}}{dt} = \mathbf{G}$$

$$= \mathbf{r}\times\mathbf{F}$$

$$= \mathbf{0},$$

since $\mathbf{r}$, $\mathbf{F}$ are parallel. Thus the moment of momentum is constant and we can write

$$\mathbf{H} = \mathbf{r}\times(m\dot{\mathbf{r}}) = \text{constant},$$

implying $$\mathbf{H}.\mathbf{r} = 0, \quad \mathbf{H}.\dot{\mathbf{r}} = 0.$$

Since throughout the motion, the position vector $\mathbf{r}$ is perpendicular to the constant vector $\mathbf{H}$, we see that the particle moves in the plane through O perpendicular to the constant vector $\mathbf{H}$. This result is immediately obtained by realizing that $\mathbf{H}.\mathbf{r} = 0$ is the equation of the plane through O with normal-vector $\mathbf{H}$.

Alternatively the central force may be written

$$\mathbf{F} = \lambda\mathbf{r},$$

where λ need not be a constant and may be positive if $\mathbf{F}$ is directed away from O and negative if $\mathbf{F}$ is directed toward O. Therefore

$$m\ddot{\mathbf{r}} = \lambda\mathbf{r}.$$

Multiplying this vectorially by $\mathbf{r}$ we obtain

$$m(\mathbf{r}\times\ddot{\mathbf{r}}) = \mathbf{0},$$

that is $$\mathbf{r}\times\ddot{\mathbf{r}} = \mathbf{0}.$$

Integrating, $$\mathbf{r}\times\dot{\mathbf{r}} = \mathbf{h},$$

where **h** is a constant vector. As before this result shows that the particle moves in the plane normal to **h** and passing through O.

Areal velocity

Consider a particle subjected to a central force. Let P be its position at time t and O the centre of force. Using polar co-ordinates and with the unit vectors **i**, **j** as shown in Fig. 13.3, we have

$$\mathbf{r} = r\mathbf{i}.$$

Therefore
$$\dot{\mathbf{r}} = \dot{r}\mathbf{i} + r\frac{d\mathbf{i}}{dt}$$

$$= \dot{r}\mathbf{i} + r\dot{\theta}\mathbf{j}.$$

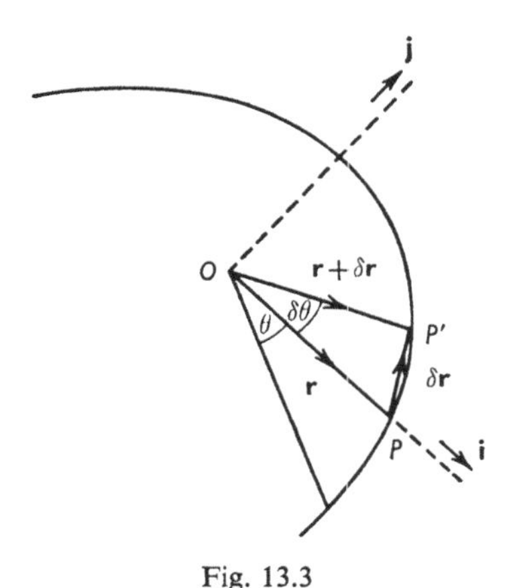

Fig. 13.3

The moment of momentum is

$$\mathbf{H} = m(\mathbf{r}\times\dot{\mathbf{r}})$$

$$= m\{r\mathbf{i}\times(\dot{r}\mathbf{i}+r\dot{\theta}\mathbf{j})\}$$

$$= mr^2\dot{\theta}\mathbf{k}.$$

Since the moment of momentum is constant throughout the motion we can write

$$r^2\dot{\theta} = \text{constant} = h,$$

where $r^2\dot{\theta}$ is the magnitude of the moment of momentum per unit mass.

Now the area δA swept out in time δt by the radius vector **OP** in moving from OP to OP' is given approximately by

$$\delta A = \tfrac{1}{2}|\mathbf{r}|^2\delta\theta$$

where angle $POP' = \delta\theta$. Therefore

$$\frac{\delta A}{\delta t} = \tfrac{1}{2}r^2\frac{\delta\theta}{\delta t}.$$

In the limit as $\delta t \to 0$, we have

$$\frac{dA}{dt} = \tfrac{1}{2}r^2\frac{d\theta}{dt}.$$

Hence $$\frac{dA}{dt} = \tfrac{1}{2}h = \text{constant}.$$

Thus the rate of change of the area swept out by the radius is constant, or briefly, the areal velocity is constant. This may also be stated: equal areas are swept out in equal times by the radius. This fact was observed empirically by Kepler in his study of the motion of the planets round the sun. It forms his second law of planetary motion.

The equation $$r^2\dot{\theta} = h = \text{constant}$$

expressing the constancy of the moment of momentum is of importance.

The direct law of force

Suppose a particle moves under a central attractive force $n^2\mathbf{r}$ per unit mass toward the point O. Its equation of motion is

$$\ddot{\mathbf{r}} = -n^2\mathbf{r}.$$

The general solution of this equation is

$$\mathbf{r} = \mathbf{A}\cos nt + \mathbf{B}\sin nt$$

where $\mathbf{A}, \mathbf{B}$ are arbitrary constant vectors. Taking the usual rectangular Cartesian x-, y-axes in the plane of motion we have

$$x = a_1\cos nt + b_1\sin nt,$$

$$y = a_2\cos nt + b_2\sin nt,$$

which are the parametric equations of the ellipse

$$(a_1 y - a_2 x)^2 + (b_1 y - b_2 x)^2 = (a_1 b_2 - a_2 b_1)^2$$

with O as the centre.

The inverse square law of force

Consider a particle attracted toward a point O by a force varying inversely as the square of the distance of the particle from O. Let m be the mass of the particle, $\mathbf{r}$ be its position vector relative to O and μ/r^2 be the magnitude per unit mass of the central force. The equation of motion is

$$m\ddot{\mathbf{r}} = -\frac{m\mu}{r^2}\hat{\mathbf{r}}$$

where $\hat{\mathbf{r}}$ is the unit vector having the same direction and sense as $\mathbf{r}$. Therefore

$$\ddot{\mathbf{r}} = -\frac{\mu}{r^2}\hat{\mathbf{r}}. \tag{13·3}$$

If $\mathbf{h}$ is the constant moment of momentum per unit mass we have

$$\mathbf{h} = r^2\dot{\theta}\hat{\mathbf{h}}$$

where $\hat{\mathbf{h}}$ is the unit vector having the same direction and sense as $\mathbf{h}$.

Multiplying (13·3) vectorially by $\mathbf{h}$ we have

$$\begin{aligned}\ddot{\mathbf{r}} \times \mathbf{h} &= -\frac{\mu}{r^2}\hat{\mathbf{r}} \times \mathbf{h} \\ &= -\frac{\mu}{r^2}\hat{\mathbf{r}} \times r^2\dot{\theta}\hat{\mathbf{h}} \\ &= \mu\dot{\theta}\hat{\mathbf{h}} \times \hat{\mathbf{r}} \\ &= \mu\frac{d\hat{\mathbf{r}}}{dt},\end{aligned}$$

since $\dot{\theta}$ is the rate of turning of $\hat{\mathbf{r}}$, and $\hat{\mathbf{h}}$ is perpendicular to $\hat{\mathbf{r}}$.

Since $\mathbf{h}$ is constant we have on integrating with respect to time

$$\frac{1}{\mu}\dot{\mathbf{r}} \times \mathbf{h} = \hat{\mathbf{r}} + e\hat{\mathbf{e}} \tag{13·4}$$

where $e\hat{\mathbf{e}}$ is the constant vector of integration. Now $\dot{\mathbf{r}} \times \mathbf{h}$ and $\hat{\mathbf{r}}$ are parallel to the plane of motion. Hence $\hat{\mathbf{e}}$ is parallel to this plane. If

θ is the angle between the variable vector $\mathbf{r}$ and the constant vector $\hat{\mathbf{e}}$ we have

$$\mathbf{r}.\hat{\mathbf{e}} = r\cos\theta.$$

Multiplying (13·4) scalarly by $\mathbf{r}$ we have

$$\frac{1}{\mu}\mathbf{r}.(\dot{\mathbf{r}}\times\mathbf{h}) = \mathbf{r}.(\hat{\mathbf{r}}+e\hat{\mathbf{e}}),$$

giving
$$\frac{1}{\mu}(\mathbf{r}\times\dot{\mathbf{r}}).\mathbf{h} = r+re\cos\theta.$$

Since $\mathbf{h} = \mathbf{r}\times\dot{\mathbf{r}}$, this becomes

$$\frac{1}{\mu}\mathbf{h}.\mathbf{h} = r(1+e\cos\theta).$$

Therefore
$$\frac{h^2/\mu}{r} = 1+e\cos\theta.$$

Thus the orbit is a conic section with one of its foci at the centre of force, with eccentricity e and with semi-latus rectum $l = h^2/\mu$.

From (13·4) we have

$$e\hat{\mathbf{e}} = \frac{1}{\mu}(\mathbf{v}\times\mathbf{h})-\hat{\mathbf{r}}, \tag{13·5}$$

where $\mathbf{v} = \dot{\mathbf{r}}$. Since $\dot{\mathbf{r}}$ is perpendicular to $\mathbf{h}$ we have

$$\dot{\mathbf{r}}.\mathbf{h} = 0,$$

and on squaring (13·5) we have

$$\begin{aligned} e^2 &= \frac{v^2h^2}{\mu^2}-\frac{2}{\mu}\hat{\mathbf{r}}.(\mathbf{v}\times\mathbf{h})+1 \\ &= \frac{v^2h^2}{\mu^2}-\frac{2}{\mu r}(\mathbf{r}\times\mathbf{v}).\mathbf{h}+1 \\ &= \frac{v^2h^2}{\mu^2}-\frac{2h^2}{\mu r}+1. \end{aligned}$$

The orbit is an ellipse, parabola or hyperbola according as $e < 1$, $e = 1$ or $e > 1$, that is, according as

$$v^2 < \frac{2\mu}{r},\ v^2 = \frac{2\mu}{r} \text{ or } v^2 > \frac{2\mu}{r}.$$

Hence the orbit is independent of the direction of motion of the particle, but is dependent only on the speed at a given position. If the particle is projected with a speed V at a distance D from the centre of force, the orbit will be an ellipse, parabola or hyperbola according as

$$V^2 < \frac{2\mu}{D}, \; V^2 = \frac{2\mu}{D} \text{ or } V^2 > \frac{2\mu}{D}.$$

The eccentricity e is given by

$$e^2 = \frac{V^2h^2}{\mu^2} - \frac{2h}{\mu D} + 1.$$

Moment of momentum or angular momentum of a rigid body

A rigid body may be regarded as a system of particles in which the distances between the particles remain constant with time. A consideration of the maintenance of the structure of this system involves an investigation of the internal forces between the particles of the rigid body. However, it is not possible by an analytical discussion to decide on the nature of these internal forces, so it is assumed that the forces which two particles of a rigid body exert on each other are equal and opposite and that the forces act along the line joining the particles. With this assumption we may obtain equations of motion and then use them to predict rigid body motions. The fact that these predictions agree with experience based on practical observations is regarded as justification for the assumption made about the nature of the internal forces.

Regarding a rigid body as a system of particles, the equation of motion of the pth particle is

$$m_p \ddot{\mathbf{r}}_p = \mathbf{F}_p^e + \mathbf{F}_p^i$$

where m_p, $\mathbf{r}_p$ are the mass and position vector relative to a fixed origin O of the particle at time t, and $\mathbf{F}_p^e$, $\mathbf{F}_p^i$ are the external and internal forces acting on the particle.

Multiplying vectorially by $\mathbf{r}_p$ we have

$$\mathbf{r}_p \times m_p \ddot{\mathbf{r}}_p = \mathbf{r}_p \times \mathbf{F}_p^e + \mathbf{r}_p \times \mathbf{F}_p^i,$$

that is,

$$m_p(\mathbf{r}_p \times \ddot{\mathbf{r}}_p) = \mathbf{r}_p \times \mathbf{F}_p^e + \mathbf{r}_p \times \mathbf{F}_p^i.$$

Summing for all particles of the system we write

$$\Sigma m_p(\mathbf{r}_p \times \ddot{\mathbf{r}}_p) = \Sigma \mathbf{r}_p \times \mathbf{F}_p^e + \Sigma \mathbf{r}_p \times \mathbf{F}_p^i.$$

We now consider carefully the contribution of the moment of the internal forces. Let $\mathbf{F}_{pq}^i$ denote the internal force on the pth particle due to the action of the qth particle. Then by Newton's third law, $\mathbf{F}^i_{pq}$ is equal and opposite to $\mathbf{F}^i_{qp}$, that is,

$$\mathbf{F}^i_{pq} = -\mathbf{F}^i_{qp}.$$

The contributions from the interactions between the pth and qth particles are

$$\begin{aligned}\mathbf{r}_p \times \mathbf{F}^i_{pq} + \mathbf{r}_q \times \mathbf{F}^i_{qp} &= (\mathbf{r}_p - \mathbf{r}_q) \times \mathbf{F}^i_{pq} \\ &= \mathbf{0},\end{aligned}$$

since it is assumed that the internal forces between two particles are directed along the line joining them.

Therefore

$$\Sigma \mathbf{r}_p \times \mathbf{F}^i_p = \mathbf{0},$$

and hence

$$\Sigma m_p(\mathbf{r}_p \times \ddot{\mathbf{r}}_p) = \Sigma \mathbf{r}_p \times \mathbf{F}^e_p. \tag{13·6}$$

The moment of momentum of the pth particle about O is

$$\mathbf{r}_p \times m_p \dot{\mathbf{r}}_p = m_p(\mathbf{r}_p \times \dot{\mathbf{r}}_p).$$

Hence the total moment of momentum or angular momentum $\mathbf{H}$ of the system about O is

$$\mathbf{H} = \Sigma m_p(\mathbf{r}_p \times \dot{\mathbf{r}}_p).$$

Differentiating,

$$\begin{aligned}\frac{d\mathbf{H}}{dt} &= \frac{d}{dt}\Sigma m_p(\mathbf{r}_p \times \dot{\mathbf{r}}_p) \\ &= \Sigma m_p(\dot{\mathbf{r}}_p \times \dot{\mathbf{r}}_p + \mathbf{r}_p \times \ddot{\mathbf{r}}_p) \\ &= \Sigma m_p(\mathbf{r}_p \times \ddot{\mathbf{r}}_p).\end{aligned}$$

Hence (13·6) becomes

$$\frac{d\mathbf{H}}{dt} = \Sigma \mathbf{r}_p \times \mathbf{F}^e_p = \mathbf{G} \tag{13·7}$$

where $\mathbf{G}$ is the vector sum of the moments of the external forces. Therefore:

The rate of change of the moment of momentum about a fixed origin of a system of particles is equal to the sum of the moments about the origin of the external forces acting on the system.

This result can also be proved to be true for moments about the centre of mass whether or not this moves. Some writers use the term angular momentum for moment of momentum about the centre of mass.

If $\mathbf{G} = \mathbf{0}$ we have the principle of conservation of moment of momentum for a system of particles, namely:

The total moment of momentum of a system of particles about any fixed origin is constant if the sum of the moments of the external forces about the origin vanishes.

Calculation of angular momentum of a rigid body with a point fixed

Consider a rigid body with one point O fixed. Its angular momentum $\mathbf{H}$ about O is given by

$$\begin{aligned} \mathbf{H} &= \Sigma \mathbf{r}_p \times m_p \mathbf{v}_p \\ &= \Sigma m_p \mathbf{r}_p \times (\boldsymbol{\omega} \times \mathbf{r}_p) \\ &= \Sigma m_p \{\mathbf{r}_p^2 \boldsymbol{\omega} - (\mathbf{r}_p . \boldsymbol{\omega}) \mathbf{r}_p\}. \end{aligned} \tag{13·8}$$

Let $\quad \mathbf{r}_p = x_p \mathbf{i} + y_p \mathbf{j} + z_p \mathbf{k}$

and $\quad \boldsymbol{\omega} = \omega_x \mathbf{i} + \omega_y \mathbf{j} + \omega_z \mathbf{k}.$

Then

$$\begin{aligned} \mathbf{H} &= \Sigma m_p \{(x_p^2 + y_p^2 + z_p^2)\,(\omega_x \mathbf{i} + \omega_y \mathbf{j} + \omega_z \mathbf{k}) \\ &\qquad - (x_p \omega_x + y_p \omega_y + z_p \omega_z)\,(x_p \mathbf{i} + y_p \mathbf{j} + z_p \mathbf{k})\} \\ &= \Sigma m_p \{(y_p^2 + z_p^2)\,\omega_x - x_p y_p \omega_y - x_p z_p \omega_z\} \mathbf{i} \\ &\qquad + \Sigma m_p \{(z_p^2 + x_p^2)\,\omega_y - y_p z_x \omega_z - y_p x_p \omega_x\} \mathbf{j} \\ &\qquad + \Sigma m_p \{(x_p^2 + y_p^2)\,\omega_z - z_p x_p \omega_x - z_p y_p \omega_y\} \mathbf{k}. \end{aligned}$$

The quantities

$$A = \Sigma m_p (y_p^2 + z_p^2), \quad B = \Sigma m_p (z_p^2 + x_p^2), \quad C = \Sigma m_p (x_p^2 + y_p^2),$$

and $\quad D = \Sigma m_p y_p z_p, \qquad E = \Sigma m_p z_p x_p, \qquad F = \Sigma m_p x_p y_p,$

depend only on the distribution of mass within the body. In general their values depend on the choice of the axes of reference. A, B, C are known as the moments of inertia of the body about the axes x, y, z respectively. D, E, F are known as the products of inertia of the body with respect to the yOz, zOx, xOy planes respectively.

The angular momentum $\mathbf{H}$ in terms of these moments and products of inertia is given by

$$\begin{aligned} \mathbf{H} = (A\omega_x - F\omega_y - E\omega_z)\mathbf{i} + (B\omega_y - D\omega_z - F\omega_x)\mathbf{j} \\ + (C\omega_z - E\omega_x - D\omega_y)\mathbf{k}. \end{aligned} \tag{13·9}$$

It can be shown that at any point of a rigid body there is at least one set of mutually perpendicular axes, relative to which all the products of inertia vanish. These axes are known as the principal axes of inertia at the point. Thus if **i**, **j**, **k** have the directions of the principal axes at the fixed point O, (13·9) reduces to

$$\mathbf{H} = A\omega_x\mathbf{i}+B\omega_y\mathbf{j}+C\omega_z\mathbf{k}.$$

In particular, consider a solid of revolution. Take its axis of symmetry as one of the axes of reference. By considering the contributions of appropriate pairs of particles to the sums Σ, it will be seen that the products of inertia vanish and that the moments of inertia about the axes perpendicular to the axis of symmetry are equal. Thus the axis of symmetry and any other two perpendicular axes in the plane perpendicular to the axis of symmetry are principal axes of a solid of revolution.

The gyroscope

The term 'gyroscope' is used for any system in which a body spinning about its axis of symmetry is so mounted that the direction of its angular velocity and its axis may change freely. A well known example of gyroscopic motion is that exhibited by a top. When a top is set spinning on the floor with its axis not verical, its axis will slowly rotate about a vertical axis through the point of contact with the floor. This rotation of the plane containing the axis and the vertical is known as precession.

Consider a simple gyroscope in the form of a solid flywheel of uniform density. Suppose it is so mounted that its centre of mass O is fixed. Then the motion of the flywheel is not affected by the force of gravity. Let the flywheel spin with constant angular speed n along its axis OZ (Fig. 13.4). If no other forces except gravity act on the gyroscope the direction of its axis will remain fixed. Because of symmetry about the z-axis the angular momentum of the flywheel about O is given by

$$\mathbf{H} = In\mathbf{k}$$

where I is its moment of inertia about its axis.

Now suppose the motion is disturbed by applying a force $F\mathbf{i}$ to the axis at the point $a\mathbf{k}$. The moment of this force about O is given by

$$\begin{aligned}\mathbf{G} &= a\mathbf{k}\times F\mathbf{i}\\ &= aF\mathbf{j}.\end{aligned}$$

Using the result of (13·7), namely,

$$\frac{d\mathbf{H}}{dt} = \mathbf{G}$$

we have
$$\frac{d\mathbf{H}}{dt} = aF\mathbf{j}.$$

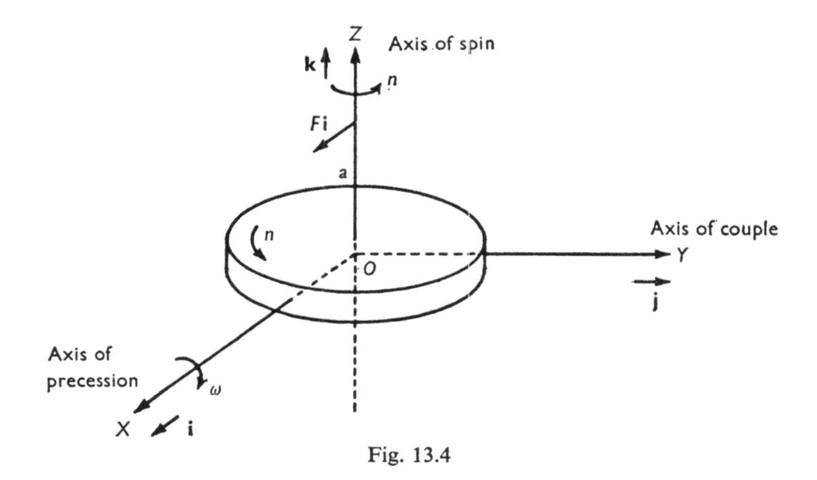

Fig. 13.4

Thus the resulting motion is such that the angular momentum is increased in the OY direction. Therefore the axis of the flywheel will begin to tilt toward OY, that is, the axis of the flywheel will begin to turn about the OX axis.

Suppose that the axis turns through a small angle $\delta\theta$ in the time δt and that the change in the angular momentum is $\delta\mathbf{H}$. Then

$$\delta H = H\delta\theta = In\,\delta\theta,$$

approximately. In the limit as $\delta t \to 0$, we have

$$\frac{d\mathbf{H}}{dt} = In\frac{d\theta}{dt}\mathbf{j}.$$

Hence
$$In\frac{d\theta}{dt} = aF,$$

showing that the axis of the flywheel begins to rotate in the plane perpendicular to the applied force with angular speed

$$\omega = \frac{d\theta}{dt} = \frac{aF}{In}.$$

If a constant couple $G\mathbf{j}$ acts on the flywheel then its axis will rotate with constant angular velocity $-(G/In)\mathbf{i}$.

Kinetic energy of a rigid body with a point fixed

Consider a rigid body moving about a fixed point O with angular velocity $\boldsymbol{\omega}$. Let $\mathbf{r}_p$ be the position vector of the pth particle of mass m_p relative to O. The kinetic energy T of the body is given by

$$\begin{aligned} T &= \tfrac{1}{2}\Sigma m_p \dot{\mathbf{r}}_p^2 \\ &= \tfrac{1}{2}\Sigma m_p(\boldsymbol{\omega}\times\mathbf{r}_p)^2 \\ &= \tfrac{1}{2}\Sigma m_p\{\boldsymbol{\omega}^2\mathbf{r}_p^2-(\boldsymbol{\omega}.\mathbf{r}_p)^2\} \\ &= \tfrac{1}{2}\boldsymbol{\omega}.\Sigma m_p\{\boldsymbol{\omega}\mathbf{r}_p^2-(\boldsymbol{\omega}.\mathbf{r}_p)\mathbf{r}_p\} \\ &= \tfrac{1}{2}\boldsymbol{\omega}.\mathbf{H}, \end{aligned}$$

from (13·8).

An alternative form of the expression for the kinetic energy is available.

Let d_p be the perpendicular distance of the pth particle of mass m_p from the instantaneous axis, (Fig. 13.5). Then

$$\begin{aligned} T &= \tfrac{1}{2}\Sigma m_p \dot{\mathbf{r}}_p^2 \\ &= \tfrac{1}{2}\Sigma m_p(\boldsymbol{\omega}\times\mathbf{r}_p)^2 \\ &= \tfrac{1}{2}\Sigma m_p(\omega r_p \sin\theta)^2 \\ &= \tfrac{1}{2}\Sigma m_p \omega^2 d_p^2 \\ &= \tfrac{1}{2}\omega^2\Sigma m_p d_p^2. \end{aligned}$$

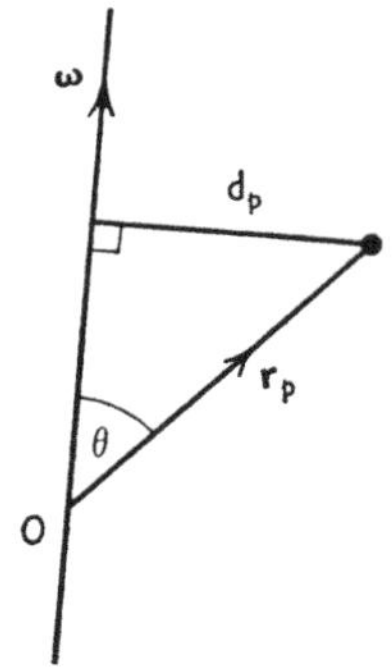

Fig. 13.5

Now $\Sigma m_p d_p^2 = I$ is the moment of inertia of the body about the instantaneous axis and therefore the kinetic energy takes the form

$$T = \tfrac{1}{2}I\omega^2.$$

Electrodynamics

Force on a moving charge in a magnetic field

When a particle of charge e is moving with velocity $\mathbf{v}$ in a constant and uniform magnetic field $\mathbf{H}$, it is found that the force $\mathbf{F}$ on the particle is perpendicular to both $\mathbf{v}$ and $\mathbf{H}$, and that the force is given by

$$\mathbf{F} = e\mathbf{v}\times\mathbf{H},$$

where e is measured in electromagnetic units. If e is measured in electrostatic units the force is given by

$$\mathbf{F} = \frac{e}{c}\mathbf{v}\times\mathbf{H}.$$

Let m be the mass of the particle. Then

$$\ddot{\mathbf{r}} = \frac{e}{mc}\dot{\mathbf{r}}\times\mathbf{H}. \tag{13·10}$$

Multiplying scalarly by $\dot{\mathbf{r}}$ we have

$$\dot{\mathbf{r}}.\ddot{\mathbf{r}} = \frac{e}{mc}\dot{\mathbf{r}}.(\dot{\mathbf{r}}\times\mathbf{H}).$$

Therefore $$\frac{d}{dt}(\dot{\mathbf{r}}^2) = 0,$$

implying $$\dot{\mathbf{r}}^2 = \text{constant}.$$

From this we see that the speed of the particle remains constant and hence the kinetic energy of the particle remains constant. Therefore the magnetic field only changes the direction of motion of the particle.

Multiplying scalarly by $\mathbf{H}$ we have

$$\ddot{\mathbf{r}}.\mathbf{H} = \frac{e}{mc}(\dot{\mathbf{r}}\times\mathbf{H}).\mathbf{H}.$$

Therefore $$\frac{d\dot{\mathbf{r}}}{dt}.\mathbf{H} = 0,$$

implying $$\dot{\mathbf{r}}.\mathbf{H} = \text{constant}.$$

From this we see that the component of the velocity parallel to the field remains constant. Further, since $|\dot{\mathbf{r}}|$ is constant, the direction of motion of the particle makes a constant angle α with the direction of the magnetic field.

Integrating (13·10), $$\dot{\mathbf{r}} = \frac{e}{mc}\mathbf{r}\times\mathbf{H}+\mathbf{A}$$

where $\mathbf{A}$ is a constant vector. Substituting for $\dot{\mathbf{r}}$ in (13·10) we have

$$\ddot{\mathbf{r}} = \frac{e}{mc}\left\{\frac{e}{mc}\mathbf{r}\times\mathbf{H}+\mathbf{A}\right\}\times\mathbf{H}$$

$$= -\frac{e^2}{m^2c^2}\mathbf{H}\times(\mathbf{r}\times\mathbf{H})+\frac{e}{mc}\mathbf{A}\times\mathbf{H}$$

$$= -\frac{e^2}{m^2c^2}\{H^2\mathbf{r}-(\mathbf{r}.\mathbf{H})\mathbf{H}\}+\frac{e}{mc}\mathbf{A}\times\mathbf{H}.$$

Therefore $$\ddot{\mathbf{r}}+\frac{e^2H^2}{m^2c^2}\mathbf{r} = \frac{e^2}{m^2c^2}(\mathbf{r}.\mathbf{H})\mathbf{H}+\frac{e}{mc}\mathbf{A}\times\mathbf{H},$$

showing that the motion of the particle is of an oscillatory nature.

To analyse the motion further, we choose fixed axes Ox, Oy, Oz with Oz parallel to $\mathbf{H}$, that is,

$$\mathbf{H} = (0, 0, H).$$

From (13·10) $$\frac{d\mathbf{v}}{dt} = \frac{e}{mc}(v_yH, \ -v_xH, \ 0),$$

where v_x, v_y, v_z are the components of the velocity $\mathbf{v}$ along the Ox, Oy, Oz axes respectively. Therefore

$$\frac{dv_x}{dt} = \omega v_y, \tag{13·11}$$

$$\frac{dv_y}{dt} = -\omega v_x, \tag{13·12}$$

$$\frac{dv_z}{dt} = 0, \tag{13·13}$$

where $\omega = eH/mc$ is the precessional frequency.

Differentiating (13·11) we obtain

$$\frac{d^2v_x}{dt^2} = \omega\frac{dv_y}{dt}.$$

Substituting for dv_y/dt we obtain

$$\frac{d^2v_x}{dt^2} = -\omega^2v_x.$$

The solution of this equation is

$$v_x = u\sin(\omega t+\epsilon) \tag{13·14}$$

where u, ϵ are constants. From (13·11),

$$v_y = \frac{1}{\omega}\frac{dv_x}{dt}.$$

Therefore
$$v_y = u\cos(\omega t+\epsilon). \tag{13·15}$$

From (13·13)
$$v_z = d \tag{13·16}$$
where d is a constant.

The velocity of the particle may therefore be considered in terms of a constant component $\mathbf{v}_1$ parallel to $\mathbf{H}$, that is,

$$\mathbf{v}_1 = d\mathbf{k},$$

and a component $\mathbf{v}_2$ perpendicular to $\mathbf{H}$, that is,

$$\mathbf{v}_2 = u\sin(\omega t+\epsilon)\mathbf{i}+u\cos(\omega t+\epsilon)\mathbf{j}.$$

Thus $\mathbf{v}_2$ rotates steadily about the direction of the magnetic field, with period $2\pi/\omega = 2\pi mc/eH$.

Integrating (13·14), (13·15), (13·16) we obtain

$$x = a-(u/\omega)\cos(\omega t+\epsilon),$$

$$y = b+(u/\omega)\sin(\omega t+\epsilon),$$

$$z = c+dt,$$

where a, b, c are constants. Thus the particle describes a circular helix whose axis is along the direction of the field.

Exercise 13(*b*)

(1) Prove the following results for a particle moving in a plane, where $\mathbf{r}$ is its position vector at time t, $\hat{\mathbf{r}}$ the unit vector having the same direction and sense as $\mathbf{r}$, $\hat{\mathbf{n}}$ the unit vector perpendicular to the plane and θ the plane polar angle.

$$\dot{\hat{\mathbf{r}}} = \dot{\theta}\hat{\mathbf{n}}\times\hat{\mathbf{r}},$$

$$\dot{\mathbf{r}} = \dot{r}\hat{\mathbf{r}}+r\dot{\theta}\hat{\mathbf{n}}\times\hat{\mathbf{r}},$$

$$\ddot{\mathbf{r}} = (\ddot{r}-r\dot{\theta}^2)\hat{\mathbf{r}}+(2\dot{r}\dot{\theta}+r\ddot{\theta})\hat{\mathbf{n}}\times\hat{\mathbf{r}}.$$

(2) A particle moves so that its position vector $\mathbf{x}$ at time t satisfies the equation

$$\frac{d\mathbf{x}}{dt} = \boldsymbol{\omega} \times \mathbf{x},$$

where $\boldsymbol{\omega}$ is a constant vector. Show that the particle lies on a fixed sphere and also in a fixed plane. Deduce that the particle moves in a circle with constant speed.

(3) The position vector of a particle of mass m at time t is given by $\mathbf{r} = \mathbf{a}\cos nt + \mathbf{b}\sin nt$ where $\mathbf{a}$, $\mathbf{b}$ are constant vectors. Prove that its angular momentum about the origin is given by $mn(\mathbf{a} \times \mathbf{b})$.

(4) A particle moves under a central force of attraction μ/r^2 per unit mass and is projected with a velocity $(\mu/R)^{\frac{1}{2}}$ at an angle 30° to the radius vector when it is a distance R from the centre of force. Show (i) the orbit is an ellipse, (ii) the eccentricity of the orbit is $\sqrt{3}/2$, and (iii) the periodic time of the particle in its orbit is $2\pi(R^3/\mu)^{\frac{1}{2}}$.

(5) A particle moves under a central repulsive force of μ/r^2 per unit mass. The particle approaches from a large distance with speed u and if deflected would pass at a distance d from the centre of force. Show that the orbit is a hyperbola and that the distance of closest approach is $k + (k^2 + d^2)^{\frac{1}{2}}$ where $k = \mu/u^2$.

(6) Prove, by using vector methods, that for two particles, masses m_1, m_2,

(i) the centre of mass moves in the same way as a particle of mass $(m_1 + m_2)$ under the action of the external forces,

(ii) the sum of their angular momenta about the origin O is equal to that about the centre of mass G of the motion relative to G, together with that about O of a particle of mass $(m_1 + m_2)$ moving with G,

(iii) the rate of increase of the angular momentum about G of the motion relative to G is equal to the sum of the moments about G of the external forces.

(7) Prove the following for a system of particles in which m_i is the mass of the ith particle relative to a fixed point O, M is the total mass of the system, $\mathbf{r}_i$ and $\mathbf{r}_i'$ are the position vectors of the ith particle relative to O and to the centre of mass G of the system respectively, $\bar{\mathbf{r}}$ is the position vector of G relative to O, and $\mathbf{F}_i$ is the external force acting on the ith particle.

(i) $$M\ddot{\bar{\mathbf{r}}} = \Sigma \mathbf{F}_i,$$

(ii) $$\Sigma \mathbf{r}_i \times m_i \dot{\mathbf{r}}_i = \Sigma \mathbf{r}'_i \times m_i \dot{\mathbf{r}}'_i + \bar{\mathbf{r}} \times M \dot{\bar{\mathbf{r}}},$$

(iii) $$\frac{d}{dt} \Sigma \mathbf{r}'_i \times m_i \dot{\mathbf{r}}'_i = \Sigma \mathbf{r}'_i \times \mathbf{F}_i.$$

(8) State the principle of conservation of angular momentum.

A uniform rod AB of mass m and length $3a$ can rotate freely about a fixed pivot at A, and hangs in its position of stable equilibrium. One end of a light inextensible string of length $2a$ is attached to the pivot, and the other end is attached to a particle of mass $3m$. The particle is held with the string taut and horizontal, and is then released. If, after colliding, the particle and the rod stick together, show that the rod comes instantaneously to rest when it has turned through an angle $\cos^{-1}(9/25)$. (C.)

(9) From the vector equation for the motion of a particle of mass m in a plane, namely

$$m\ddot{\mathbf{r}} = \mathbf{P}$$

show how to derive the equations

$$\text{(i)}\ \tfrac{1}{2}m(\dot{\mathbf{r}}\,.\,\dot{\mathbf{r}}) = \int \mathbf{P}\,.\,d\mathbf{r} + \text{constant},$$

$$\text{(ii)}\ m(\mathbf{r} \times \dot{\mathbf{r}}) = \int (\mathbf{r} \times \mathbf{P})\,dt + \text{constant},$$

and explain the significance of these equations.

Show that, if $\mathbf{P} = -(m\mu\mathbf{r}/r^3)$, equation (i) gives

$$\dot{r}^2 + r^2\dot{\theta}^2 = 2\mu/r + c,$$

where (r, θ) are polar co-ordinates and c is a constant. (M.A.)

(10) A particle of mass m and charge e (e.m.u.) is projected in the plane $z = 0$ with velocity $\mathbf{v}$, and is subject to a constant magnetic field $\mathbf{H}$ whose direction is that of the z-axis. Show that the particle remains in the plane $z = 0$, and that it describes a circle of radius $m|\mathbf{v}|/(e|\mathbf{H}|)$ at a constant speed of $|\mathbf{v}|$. Also show that the period of the circular orbit is $2\pi m/(e|\mathbf{H}|)$.

(11) A particle of mass m moving along a helical path has its position vector $\mathbf{r}$ at time t given by

$$\mathbf{r} = a\cos\omega t\,\mathbf{i} + a\sin\omega t\,\mathbf{j} + bt\,\mathbf{k},$$

where a, b, ω are constants. Show that the total force acting on it is constant in magnitude and acts in a radial direction towards the axis of the helix. Also show that the angular momentum of the particle about the origin is

$$mab\{(\sin\omega t - \omega t\cos\omega t)\mathbf{i} - (\cos\omega t + \omega t\sin\omega t)\mathbf{j}\} + ma^2\omega\mathbf{k}.$$

(12) A particle of mass m and charge e moves under the influence of gravity $\mathbf{g}$ and of a constant magnetic field $\mathbf{H}$ which produces a force $e(\mathbf{v}\times\mathbf{H})$, where $\mathbf{v}$ is the velocity of the particle at time t. If $\mathbf{H}$ is in the horizontal direction, show that the particle will move parallel to $\mathbf{H}$ with constant speed, and in the plane perpendicular to $\mathbf{H}$ the particle has a drift velocity of magnitude g/ω in the horizontal direction together with circular motion.

(13) If $\mathbf{r}$ is the position vector of a particle of mass m relative to a fixed origin O and $\mathbf{F}$ is the resultant force on the particle show that

$$\frac{d}{dt}(\mathbf{r}\times m\dot{\mathbf{r}}) = \mathbf{r}\times\mathbf{F},$$

and interpret this equation.

A gyroscope of mass M is mounted on an axis OP, its centre of mass is on OP at a distance h from O and its moments of inertia are A about OP and B about an axis through O perpendicular to OP. An electric motor keeps the gyroscope spinning about OP with angular velocity n and OP rotates about the vertical through O with angular velocity ω, making a fixed angle θ with the upward vertical. Show that the horizontal component of the angular momentum of the gyroscope about O is

$$\{(A-B)\omega\cos\theta+An\}\sin\theta,$$

and hence prove that

$$(A-B)\omega^2\cos\theta+An\omega = Mgh. \qquad \text{(M.A.)}$$

14

CURVES IN SPACE

Introduction

The application of the methods of the calculus to the geometry of curves and surfaces in space is known as *differential geometry*. We shall in this chapter give an introduction to the study of curves in space.

By a curve in space is meant a twisted curve, that is, a curve which does not lie in one plane; for example, the curve drawn on a cylinder or a sphere. The equation of a curve can be expressed as a single vector equation such as

$$\mathbf{r} = \mathbf{f}(u)$$

where $\mathbf{r}$ is the position vector of any point on the curve and $\mathbf{f}(u)$ is a given vector function involving the parameter u.

There are two important properties of a space curve which we shall consider. The first is known as the curvature, and speaking loosely this may be regarded as the rate of change of direction with the distance traversed along the curve, that is, a measure of the 'tightness or sharpness of the bend'. The second property is known as the torsion and this may be regarded as the measure of the departure of the curve from a plane curve, that is, the 'twistiness'. In plane curves we have to consider only the curvature. But in space or skew curves we have to consider both curvature and torsion.

We remind the reader that the derivative of a vector of constant length is perpendicular to the vector. This may be quickly proved as follows. Let $\mathbf{a}$ be a vector of constant length. Therefore

$$\mathbf{a}.\mathbf{a} = a^2 = \text{constant}.$$

Differentiating with respect to a parameter u we get

$$\mathbf{a}.\frac{d\mathbf{a}}{du} = 0.$$

Hence $d\mathbf{a}/du$ is perpendicular to $\mathbf{a}$.

Tangent and normal to a curve

Let A be a fixed point on a curve and the position of every point on the curve be given in terms of its distance s measured along the curve from A. Then the position vector of any point is a function of the scalar parameter s. Let P, Q be points with parameters s, $s+\delta s$ and their position vectors be $\mathbf{r}$, $\mathbf{r}+\delta\mathbf{r}$ respectively (Fig. 14.1). Then

$$\frac{d\mathbf{r}}{ds} = \underset{\delta s\to 0}{\mathrm{Lt}}\ \frac{\delta\mathbf{r}}{\delta s} = \underset{\delta s\to 0}{\mathrm{Lt}}\ \frac{\mathbf{PQ}}{\delta s} = \hat{\mathbf{t}}, \tag{14·1}$$

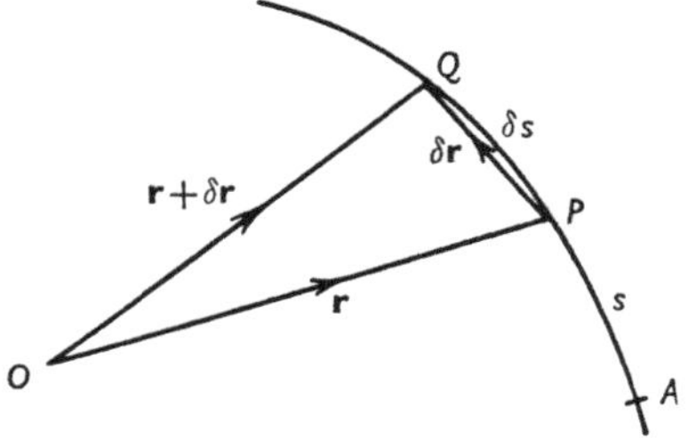

Fig. 14.1

where $\hat{\mathbf{t}}$ is the unit vector in the direction of the tangent to the curve at P in the sense of increasing s, that is, in the forward positive sense. We shall refer to $\hat{\mathbf{t}}$ as the *unit tangent* to the curve at P.

Suppose the space curve is given as a function of the parameter u, that is, $\mathbf{r} = \mathbf{f}(u)$. The unit tangent vector $\hat{\mathbf{t}}$ may be found in the following way.

$$\hat{\mathbf{t}} = \frac{d\mathbf{r}}{ds} = \frac{d\mathbf{r}}{du}\frac{du}{ds}.$$

Squaring,
$$\hat{\mathbf{t}}^2 = \left(\frac{d\mathbf{r}}{du}\right)^2\left(\frac{du}{ds}\right)^2.$$

But
$$\hat{\mathbf{t}}^2 = 1.$$

Therefore
$$\frac{du}{ds} = 1\Big/\left|\frac{d\mathbf{r}}{du}\right|.$$

Hence
$$\hat{\mathbf{t}} = \frac{d\mathbf{r}}{du}\Big/\left|\frac{d\mathbf{r}}{du}\right|.$$

Alternatively, by definition, $d\mathbf{r}/du$ is the vector whose direction is that of the tangent at the point $P(\mathbf{r})$. Hence

$$\hat{\mathbf{t}} = \frac{d\mathbf{r}}{du}\Big/\left|\frac{d\mathbf{r}}{du}\right|.$$

Example

Find the unit tangent to the space curve $x = 2\cos\theta$, $y = 2\sin\theta$, $z = \theta$.

We have $$\mathbf{r} = 2\cos\theta\mathbf{i}+2\sin\theta\mathbf{j}+\theta\mathbf{k}.$$

Differentiating $$\frac{d\mathbf{r}}{d\theta} = -2\sin\theta\mathbf{i}+2\cos\theta\mathbf{j}+\mathbf{k}.$$

Therefore $$\left|\frac{d\mathbf{r}}{d\theta}\right| = \sqrt{(4\sin^2\theta+4\cos^2\theta+1)} = \sqrt{5}.$$

Now $$\hat{\mathbf{t}} = \frac{d\mathbf{r}}{ds} = \frac{d\mathbf{r}}{d\theta}\frac{d\theta}{ds}$$
$$= (-2\sin\theta\mathbf{i}+2\cos\theta\mathbf{j}+\mathbf{k})\frac{d\theta}{ds}.$$

Squaring, $$1 = 5\left(\frac{d\theta}{ds}\right)^2.$$

Therefore $$\frac{d\theta}{ds} = \frac{1}{\sqrt{5}}.$$

Hence $$\hat{\mathbf{t}} = (-2\sin\theta\mathbf{i}+2\cos\theta\mathbf{j}+\mathbf{k})/\sqrt{5}.$$

Every line through P perpendicular to the tangent at P is called a normal to the curve at P. There is an infinite number of these normals and they all lie in the plane through P perpendicular to $\hat{\mathbf{t}}$. This plane is called the normal plane to the curve at P, and any straight line in this plane is normal to the curve at P and perpendicular to the tangent at P.

If $\mathbf{r}_1$ is a given point on the curve and $\hat{\mathbf{t}}_1$ is the unit tangent at this point then the equation of the tangent is

$$\mathbf{r} = \mathbf{r}_1+\lambda\hat{\mathbf{t}}_1,$$

and the equation of the normal plane is

$$(\mathbf{r}-\mathbf{r}_1).\hat{\mathbf{t}}_1 = 0.$$

Principal normal, curvature and osculating plane

The direction of the tangent at a point on a curve varies with the distance of the point from the point A, (Fig. 14.1). Let P, Q be two close points on a curve such that the arc $PQ = \delta s$ (Fig. 14.2(i)).

Let the angle between the tangents at P and Q be $\delta\psi$. Then $\delta\psi/\delta s$ is the average arc-rate of rotation of the tangent of the arc PQ. The *curvature* κ of the curve at the point P is defined as the limiting value of $\delta\psi/\delta s$ as $\delta s \to 0$, that is,

$$\kappa = \operatorname*{Lt}_{\delta s \to 0} \frac{\delta\psi}{\delta s} = \frac{d\psi}{ds} = \psi'.$$

Let $\hat{\mathbf{t}}$ and $\hat{\mathbf{t}}+\delta\hat{\mathbf{t}}$ be the unit tangents at P and Q. Let $\mathbf{CX} = \hat{\mathbf{t}}$, $\mathbf{CY} = \hat{\mathbf{t}}+\delta\hat{\mathbf{t}}$ and angle $XCY = \delta\psi$ (Fig. 14.2(ii)). Since $\mathbf{CX}$, $\mathbf{CY}$ are unit vectors we have

$$|\mathbf{CX}| = |\mathbf{CY}| = 1,$$

and hence

$$\text{arc } XY = \delta\psi.$$

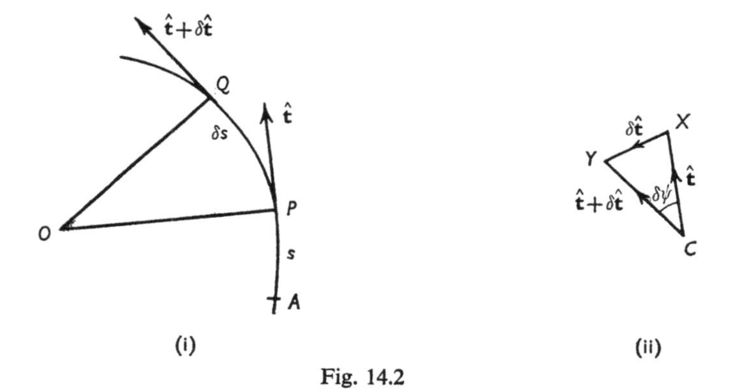

Fig. 14.2

Consider the vector $d\hat{\mathbf{t}}/ds$. Its direction is perpendicular to $\hat{\mathbf{t}}$ since it is the derivative of a vector of constant length. Its magnitude is $d\psi/ds$ since

$$\left|\frac{d\hat{\mathbf{t}}}{ds}\right| = \operatorname*{Lt}_{\delta s \to 0} \left|\frac{\delta\hat{\mathbf{t}}}{\delta s}\right| = \operatorname*{Lt}_{\delta s \to 0} \left(\frac{XY}{\delta\psi}\frac{\delta\psi}{\delta s}\right) = \frac{d\psi}{ds}.$$

If $\hat{\mathbf{n}}$ is the unit vector along $d\hat{\mathbf{t}}/ds$ we can write

$$\frac{d\hat{\mathbf{t}}}{ds} = \frac{d\psi}{ds}\hat{\mathbf{n}} = \kappa\hat{\mathbf{n}}. \tag{14·2}$$

Since $\hat{\mathbf{n}}$ is perpendicular to $\hat{\mathbf{t}}$ it is known as the *unit principal normal.*

The straight line through P in the direction of $\hat{\mathbf{n}}$ is called the *principal normal* to the curve at P. The plane containing P, $\hat{\mathbf{t}}$, $\hat{\mathbf{n}}$ is called the *osculating plane* of the curve at P. This plane is the limiting position of the plane containing P and two adjacent points Q, R on the curve, as $Q \to P$, $R \to P$. For a plane curve the osculating plane is the plane of the curve. If $\hat{\mathbf{t}}_1$ and $\hat{\mathbf{n}}_1$ are the unit tangent and unit normal at the point $\mathbf{r}_1$ on the curve, the equation of the osculating plane at $\mathbf{r}_1$ is

$$[\mathbf{r}-\mathbf{r}_1, \hat{\mathbf{t}}_1, \hat{\mathbf{n}}_1] = 0,$$

and the equation of the principal normal at $\mathbf{r}_1$ is

$$\mathbf{r} = \mathbf{r}_1 + a\hat{\mathbf{n}}_1,$$

where a is a parameter.

The reciprocal of the curvature κ is known as the radius of curvature ρ, that is

$$\rho = 1/\kappa.$$

The centre of curvature is the point C on the principal normal at P such that $\mathbf{PC} = \rho\hat{\mathbf{n}}$.

The circle of curvature for the point P is the circle with centre C and radius ρ, lying in the osculating plane.

Example

Find the principal normal and the curvature of the space curve $x = 2\cos\theta$, $y = 2\sin\theta$, $z = \theta$.

We have shown (Example on p. 277) that the unit tangent $\hat{\mathbf{t}}$ is given by

$$\hat{\mathbf{t}} = (-2\sin\theta\mathbf{i} + 2\cos\theta\mathbf{j} + \mathbf{k})/\sqrt{5}.$$

Differentiating with respect to s,

$$\frac{d\hat{\mathbf{t}}}{ds} = \frac{(-2\cos\theta\mathbf{i} - 2\sin\theta\mathbf{j})}{\sqrt{5}}\frac{d\theta}{ds}.$$

We have also shown

$$\frac{d\theta}{ds} = \frac{1}{\sqrt{5}}.$$

Therefore $$\frac{d\hat{\mathbf{t}}}{ds} = (-2\cos\theta\mathbf{i} - 2\sin\theta\mathbf{j})/5.$$

But $$\frac{d\hat{\mathbf{t}}}{ds} = \kappa\hat{\mathbf{n}}.$$

Hence $$\kappa = \left|\frac{d\hat{\mathbf{t}}}{ds}\right| = \frac{2}{5},$$

and $$\hat{\mathbf{n}} = -\cos\theta\mathbf{i} - \sin\theta\mathbf{j}.$$

Binormal and torsion

We define the *binormal* at P to be the straight line through P, perpendicular to the osculating plane.

Consider the vector $\hat{\mathbf{b}} = \hat{\mathbf{t}} \times \hat{\mathbf{n}}$. Since $\hat{\mathbf{t}}$, $\hat{\mathbf{n}}$ are perpendicular to one another and are unit vectors, we see that the vector $\hat{\mathbf{b}}$ is also a unit vector perpendicular to both $\hat{\mathbf{t}}$ and $\hat{\mathbf{n}}$. Hence $\hat{\mathbf{b}}$ is normal to the curve and to the osculating plane at P. We refer to $\hat{\mathbf{b}}$ as the *unit binormal* at P. The orthogonal triad $\hat{\mathbf{t}}$, $\hat{\mathbf{n}}$, $\hat{\mathbf{b}}$ forms a right-handed system. For a plane curve, $\hat{\mathbf{b}}$ is perpendicular to the plane of the curve. For a curve in space as P moves along the curve the triad $\hat{\mathbf{t}}$, $\hat{\mathbf{n}}$, $\hat{\mathbf{b}}$ twists in direction

In (14·2) the scalar κ measures the arc-rate of turning of the unit tangent $\hat{\mathbf{t}}$ at the point P. We shall now obtain a scalar τ which measures the arc-rate of turning of the unit binormal $\hat{\mathbf{b}}$ at the point P, and we shall define τ to be the *torsion* of the curve at P. Since the binormal is perpendicular to the osculating plane, the arc-rate of rotation of the osculating plane is the same as the arc-rate of turning of the binormal. We shall also regard the torsion to be positive when the rotation of the binormal as s increases is right-handed relative to $\hat{\mathbf{t}}$.

Since

$$\begin{aligned}
\hat{\mathbf{b}} &= \hat{\mathbf{t}} \times \hat{\mathbf{n}}, \\
\frac{d\hat{\mathbf{b}}}{ds} &= \frac{d}{ds}(\hat{\mathbf{t}} \times \hat{\mathbf{n}}) \\
&= \frac{d\hat{\mathbf{t}}}{ds} \times \hat{\mathbf{n}} + \hat{\mathbf{t}} \times \frac{d\hat{\mathbf{n}}}{ds} \\
&= \kappa\hat{\mathbf{n}} \times \hat{\mathbf{n}} + \hat{\mathbf{t}} \times \frac{d\hat{\mathbf{n}}}{ds} \\
&= \hat{\mathbf{t}} \times \frac{d\hat{\mathbf{n}}}{ds}.
\end{aligned}$$

But $\hat{\mathbf{n}}$ is perpendicular to $\hat{\mathbf{t}}$ and also perpendicular to $d\hat{\mathbf{n}}/ds$. Therefore $\hat{\mathbf{n}}$ is parallel to $d\hat{\mathbf{b}}/ds$ provided that $(d\hat{\mathbf{b}}/ds) \neq \mathbf{0}$. We may now write

$$\frac{d\hat{\mathbf{b}}}{ds} = -\tau\hat{\mathbf{n}}, \tag{14·3}$$

where the negative sign in (14·3) indicates that the torsion τ is positive when the rotation of $\hat{\mathbf{b}}$ as s increases is right-handed relative to $\hat{\mathbf{t}}$, in which case, by referring to Fig. 14.3, it is seen that $d\hat{\mathbf{b}}/ds$ is opposite in sense to $\hat{\mathbf{n}}$.

Since $$\hat{\mathbf{b}} = \hat{\mathbf{t}}\times\hat{\mathbf{n}}$$
it follows that $$\hat{\mathbf{n}} = \hat{\mathbf{b}}\times\hat{\mathbf{t}}.$$
Differentiating with respect to s

$$\begin{aligned}\frac{d\hat{\mathbf{n}}}{ds} &= \frac{d\hat{\mathbf{b}}}{ds}\times\hat{\mathbf{t}}+\hat{\mathbf{b}}\times\frac{d\hat{\mathbf{t}}}{ds}\\ &= -\tau\hat{\mathbf{n}}\times\hat{\mathbf{t}}+\hat{\mathbf{b}}\times\kappa\hat{\mathbf{n}}\\ &= -\tau(\hat{\mathbf{n}}\times\hat{\mathbf{t}})+\kappa(\hat{\mathbf{b}}\times\hat{\mathbf{n}})\\ &= \tau(\hat{\mathbf{t}}\times\hat{\mathbf{n}})-\kappa(\hat{\mathbf{n}}\times\hat{\mathbf{b}})\\ &= \tau\hat{\mathbf{b}}-\kappa\hat{\mathbf{t}}. \end{aligned} \tag{14·4}$$

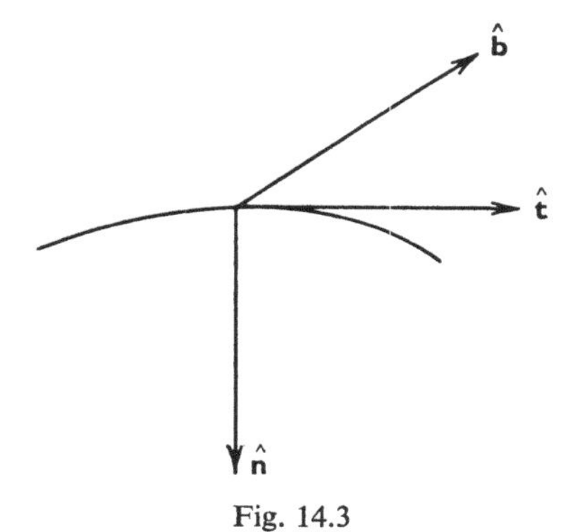

Fig. 14.3

The formulae

$$\frac{d\hat{\mathbf{t}}}{ds} = \kappa\hat{\mathbf{n}},\quad \frac{d\hat{\mathbf{b}}}{ds} = -\tau\hat{\mathbf{n}},\quad \frac{d\hat{\mathbf{n}}}{ds} = -\kappa\hat{\mathbf{t}}+\tau\hat{\mathbf{b}},$$

are called the Frenet–Serret formulae, and they are the vector equivalent of nine scalar equations.

Formulae for curvature and torsion

Let $\mathbf{r}$ be a function of a parameter u. We shall denote differentiation with respect to u by dots above the symbol. We have

$$\dot{\mathbf{r}} = \frac{d\mathbf{r}}{ds}\dot{s} = \dot{s}\hat{\mathbf{t}}, \tag{14·5}$$

$$\ddot{\mathbf{r}} = \ddot{s}\hat{\mathbf{t}}+\dot{s}\frac{d\hat{\mathbf{t}}}{ds}\dot{s} = \ddot{s}\hat{\mathbf{t}}+\kappa\dot{s}^2\hat{\mathbf{n}}. \tag{14·6}$$

Therefore $$\dot{\mathbf{r}}\times\ddot{\mathbf{r}} = \kappa\dot{s}^3\hat{\mathbf{t}}\times\hat{\mathbf{n}} = \kappa\dot{s}^3\hat{\mathbf{b}}. \tag{14·7}$$

If s is measured in the sense of increasing u, $\dot{s}$ is positive and we have from (14·5) and (14·7)

$$\dot{s} = |\dot{\mathbf{r}}|, \tag{14·8}$$

and

$$\kappa = \frac{|\dot{\mathbf{r}}\times\ddot{\mathbf{r}}|}{\dot{s}^3}. \tag{14·9}$$

Also $\hat{\mathbf{t}}$ has the direction and sense of $\dot{\mathbf{r}}$ and $\hat{\mathbf{b}}$ has the direction and sense of $\dot{\mathbf{r}}\times\ddot{\mathbf{r}}$.

Differentiating (14·7) we have

$$\ddot{\mathbf{r}}\times\ddot{\mathbf{r}}+\dot{\mathbf{r}}\times\dddot{\mathbf{r}} = \kappa\dot{s}^3\frac{d\hat{\mathbf{b}}}{ds}\,\dot{s}+\hat{\mathbf{b}}\,\frac{d}{du}\,(\kappa\dot{s}^3),$$

and hence using (14·3) we obtain

$$\dot{\mathbf{r}}\times\dddot{\mathbf{r}} = \kappa\dot{s}^4(-\tau\hat{\mathbf{n}})+\hat{\mathbf{b}}\,\frac{d}{du}\,(\kappa\dot{s}^3).$$

Multiplying scalarly by $\ddot{\mathbf{r}}$ and using (14·6) we obtain

$$\begin{aligned}\ddot{\mathbf{r}}\,.\,(\dot{\mathbf{r}}\times\dddot{\mathbf{r}}) &= -\kappa\dot{s}^4\ddot{s}\tau\hat{\mathbf{n}}\,.\,\hat{\mathbf{t}}+\ddot{s}\hat{\mathbf{b}}\,.\,\hat{\mathbf{t}}\,\frac{d}{du}\,(\kappa\dot{s}^3)-\kappa^2\dot{s}^6\tau\hat{\mathbf{n}}\,.\,\hat{\mathbf{n}}+\kappa\dot{s}^2\hat{\mathbf{b}}\,.\,\hat{\mathbf{n}}\,\frac{d}{du}\,(\kappa\dot{s}^3)\\ &= -\kappa^2\dot{s}^6\tau\\ &= -\frac{|\dot{\mathbf{r}}\times\ddot{\mathbf{r}}|}{\dot{s}^6}\,\dot{s}^6\tau.\end{aligned}$$

Hence

$$\tau = \frac{[\dot{\mathbf{r}},\,\ddot{\mathbf{r}},\,\dddot{\mathbf{r}}]}{|\dot{\mathbf{r}}\times\ddot{\mathbf{r}}|^2}. \tag{14·10}$$

The helix

A helix is a curve drawn on a cylinder (not necessary circular) cutting the generators at a constant angle α, that is the tangents are inclined at a constant angle α to the generators. Let $\hat{\mathbf{g}}$ be a unit vector having the same direction as the generators. Then $\hat{\mathbf{t}}\,.\,\hat{\mathbf{g}} = \cos\alpha$. The following properties hold.

(*a*) $\hat{\mathbf{n}}$ and $\hat{\mathbf{g}}$ are perpendicular.

Since

$$\hat{\mathbf{t}}\,.\,\hat{\mathbf{g}} = \cos\alpha,$$

we have on differentiating with respect to s,

$$\frac{d\hat{\mathbf{t}}}{ds}\,.\,\hat{\mathbf{g}} = 0,$$

and by Frenet's formula (14·2)

$$\kappa\hat{\mathbf{n}}\,.\,\hat{\mathbf{g}} = 0.$$

Hence $\hat{\mathbf{n}}$ is perpendicular to $\hat{\mathbf{g}}$.

(*b*) $\hat{\mathbf{g}}$ is parallel to the plane of $\hat{\mathbf{t}}$ and $\hat{\mathbf{b}}$, and is inclined at a constant angle with $\hat{\mathbf{b}}$.

Since
$$\hat{\mathbf{b}} = \hat{\mathbf{t}} \times \hat{\mathbf{n}},$$

$\hat{\mathbf{n}}$ is perpendicular to the plane of $\hat{\mathbf{b}}$ and $\hat{\mathbf{t}}$. Thus from (*a*) $\hat{\mathbf{g}}$ is parallel to the plane of $\hat{\mathbf{b}}$ and $\hat{\mathbf{t}}$, and the angle made by $\hat{\mathbf{g}}$ with $\hat{\mathbf{b}}$ is $\frac{1}{2}\pi \pm \alpha$.

(*c*) τ/κ is constant.

Since
$$\hat{\mathbf{n}}.\hat{\mathbf{g}} = 0$$
we have on differentiation,
$$\frac{d\hat{\mathbf{n}}}{ds}.\hat{\mathbf{g}} = 0,$$

and by Frenet's formula (14·4)
$$(\tau\hat{\mathbf{b}} - \kappa\hat{\mathbf{t}}).\hat{\mathbf{g}} = 0,$$

that is,
$$\tau\hat{\mathbf{b}}.\hat{\mathbf{g}} - \kappa\hat{\mathbf{t}}.\hat{\mathbf{g}} = 0.$$

Hence
$$\frac{\tau}{\kappa} = \frac{\hat{\mathbf{t}}.\hat{\mathbf{g}}}{\hat{\mathbf{b}}.\hat{\mathbf{g}}}.$$

Hence using (*a*) and (*b*),
$$\frac{\tau}{\kappa} = \pm\cot\alpha. \qquad (14\cdot 11)$$

(*d*) If τ/κ is constant on a curve then the curve is a helix. This is the converse of (*c*).

Suppose $\tau/\kappa = c$, where c is a constant. Then by Frenet's formulae (14·3) and (14·2) we have
$$\frac{d\hat{\mathbf{b}}}{ds} = -\tau\hat{\mathbf{n}} = -c\kappa\hat{\mathbf{n}} = -\frac{d}{ds}(c\hat{\mathbf{t}}).$$

Integrating with respect to s we have
$$\hat{\mathbf{b}} + c\hat{\mathbf{t}} = \mathbf{a},$$

where $\mathbf{a}$ is a constant. Multiplying scalarly by $\hat{\mathbf{t}}$ we obtain
$$\mathbf{a}.\hat{\mathbf{t}} = c,$$

showing that $\hat{\mathbf{t}}$ is inclined at a constant angle to the fixed direction of $\mathbf{a}$. Hence the curve is a helix.

(*e*) κ and τ are constant multiples of the curvature κ_0 of the plane section of the cylinder perpendicular to the generators.

Take as origin a point O on the surface of the cylinder, (Fig. 14.4). Let P be a point on the helix and P' be the orthogonal projection of P on to the normal section of the cylinder whose plane passes through O. Let $\mathbf{OP} = \mathbf{r}$, $\mathbf{OP}' = \mathbf{r}'$. Then

$$\mathbf{r}' = \mathbf{r} - (\hat{\mathbf{g}}.\mathbf{r})\hat{\mathbf{g}}. \qquad (14\cdot12)$$

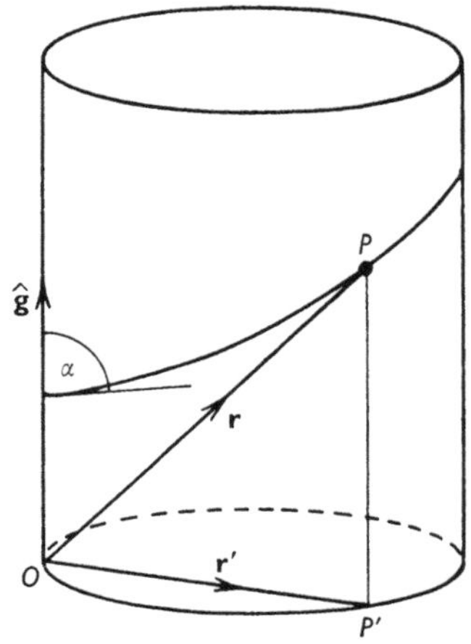

As P moves, P' also moves. Let s, s' be arc parameters for the loci of P, P'. From (14·8) we have

Fig. 14.4

$$\begin{aligned}\frac{ds'}{ds} &= \left|\frac{d\mathbf{r}'}{ds}\right| \\ &= \left|\frac{d\mathbf{r}}{ds} - \left(\hat{\mathbf{g}}.\frac{d\mathbf{r}}{ds}\right)\hat{\mathbf{g}}\right| \\ &= |\hat{\mathbf{t}} - (\hat{\mathbf{g}}.\hat{\mathbf{t}})\hat{\mathbf{g}}| \\ &= |\{\hat{\mathbf{t}}^2 - 2(\hat{\mathbf{g}}.\hat{\mathbf{t}})^2 + (\hat{\mathbf{g}}.\hat{\mathbf{t}})^2\hat{\mathbf{g}}^2\}^{\frac{1}{2}}| \\ &= |(1 - 2\cos^2\alpha + \cos^2\alpha)^{\frac{1}{2}}| \text{ (since } \hat{\mathbf{g}}.\hat{\mathbf{t}} = \cos\alpha,\ \hat{\mathbf{t}}^2 = \hat{\mathbf{g}}^2 = 1) \\ &= |(1 - \cos^2\alpha)^{\frac{1}{2}}| \\ &= |\sin\alpha|.\end{aligned}$$

The curvature κ_0 is given by

$$\begin{aligned}\kappa_0 &= \left|\frac{d\hat{\mathbf{t}}'}{ds'}\right| \\ &= \left|\frac{d}{ds'}\left(\frac{d\mathbf{r}'}{ds'}\right)\right| \\ &= \left|\frac{d^2\mathbf{r}'}{ds'^2}\right| \\ &= \left|\frac{d^2\mathbf{r}'}{ds^2}\right|\operatorname{cosec}^2\alpha.\end{aligned}$$

But since $\hat{\mathbf{g}}.\hat{\mathbf{t}}$ is constant we have from differentiating (14·12) and using (14·1)

$$\left|\frac{d^2\mathbf{r}'}{ds^2}\right| = \frac{d\hat{\mathbf{t}}}{ds} = \kappa.$$

Therefore $$\kappa_0 = \kappa\,\text{cosec}^2\alpha.$$

Hence $\kappa = \kappa_0 \sin^2\alpha$ and from (14·11) we obtain $\tau = \pm\kappa_0 \sin\alpha\cos\alpha$.

(*f*) A curve having constant curvature κ and constant torsion τ is a helix on a circular cylinder.

Since τ/κ is constant, it follows from (*d*) that the curve is a helix. Since κ is constant it follows from (*e*) that κ_0 is constant, so the cylinder on which the helix is drawn is a circular cylinder.

Exercise 14

(1) $P(x, y, z)$ and $P'(x', y', z')$ are two points on the space curve

$$\mathbf{r} = p^3\mathbf{i} + 2p\mathbf{j} + 3p^2\mathbf{k}.$$

The tangent vector at the point P is given by

$$\mathbf{t} = \underset{p'\to p}{\text{Lt}} \left(\frac{x'-x}{p'-p}\mathbf{i} + \frac{y'-y}{p'-p}\mathbf{j} + \frac{z'-z}{p'-p}\mathbf{k}\right)$$

where p, p' are the parameters at the points P, P'.

By substituting for x, y, z and x', y', z' in terms of p and p' respectively, show that

$$\mathbf{t} = 3p^2\mathbf{i} + 2\mathbf{j} + 6p\mathbf{k}.$$

Hence obtain the tangent vector and the equation of the tangent line for the point $y = 6$.

(2) Find the radius of curvature of the following plane curves:

(i) $\mathbf{r} = a\cos\theta\mathbf{i} + a\sin\theta\mathbf{j}$ (circle),

(ii) $\mathbf{r} = at^2\mathbf{i} + 2at\mathbf{j}$ (parabola),

(iii) $\mathbf{r} = cp\mathbf{i} + \dfrac{c}{p}\mathbf{j}$ (rectangular hyperbola).

(3) Show that the curvature κ of the plane curve $\mathbf{r} = x\mathbf{i} + y\mathbf{j}$ where $x = f(t)$, $y = \phi(t)$, t being a scalar parameter, is given by

$$\kappa = |\dot{x}\ddot{y} - \dot{y}\ddot{x}|/(\dot{x}^2 + \dot{y}^2).$$

(4) The equation of a space curve is given by

$$\mathbf{r} = t\mathbf{i} + t^2\mathbf{j} + t^3\mathbf{k},$$

where $\mathbf{r}$ is the position vector of a point on the curve and t is a scalar parameter. Find

(i) the point in which the tangent at the point $x = 3$ meets the xOy plane,

(ii) the angle between the tangents at the two points on the curve for which $y = 1$.

(5) Find the radius of curvature at the point t for the twisted curve whose equation is

$$\mathbf{r} = 3t\mathbf{i}+3t^2\mathbf{j}+2t^3\mathbf{k}$$

where t is a scalar parameter.

(6) Find the unit tangent, unit normal and unit binormal vectors at the point θ for the space curve

$$\mathbf{r} = a\cos\theta\mathbf{i}+a\sin\theta\mathbf{j}+c\theta\mathbf{k}$$

where θ is a scalar parameter.

(7) Find the curvature and torsion of the circular helix given by

$$x = a\cos\theta, \quad y = a\sin\theta, \quad z = a\theta\cot\alpha.$$

(8) Prove that the radii of curvature and torsion at any point of the curve $6x = z^3$, $2y = z^2$ are equal.

(9) In the equation $\hat{\mathbf{t}} = d\mathbf{r}/ds$ prove that, for any curve, $\hat{\mathbf{r}}.\hat{\mathbf{t}} = dr/ds$, where $\mathbf{r} = r\hat{\mathbf{r}}$, $\hat{\mathbf{r}}$ being the unit vector in the direction and sense of $\mathbf{r}$.

The length p of the perpendicular from the origin to the tangent is given by $p = -\hat{\mathbf{n}}.\mathbf{r}$ where $\hat{\mathbf{n}}$ is the unit vector normal to the curve. Prove that for a plane curve

$$\frac{dp}{ds} = \kappa r \frac{dr}{ds}$$

and therefore

$$\kappa = \frac{1}{r}\frac{dp}{dr},$$

where κ is the curvature of the curve.

ANSWERS TO EXERCISES

Exercise 5 (p. 67)

(2) $\frac{7}{8}$, $-\frac{5}{8}$. (5) 9; $\frac{1}{9}$, $-\frac{4}{9}$, $\frac{8}{9}$.

(6) 7, $\frac{1}{7}(2\mathbf{i}+3\mathbf{j}+6\mathbf{k})$. (12) 120°.

Exercise 6 (p. 86)

(1) 5, 36° 52′ with x axis; $\sqrt{229}$, 352° 24′ with x axis.

(2) 9 units, bearing $\tan^{-1}\frac{4}{7}$ N. of E., elevation $\sin^{-1}\frac{4}{9}$;
9 units, bearing $\tan^{-1}\frac{4}{7}$ N. of E. depression $\sin^{-1}\frac{4}{9}$.

(3) $4\sqrt{2}$ km/h from N.W.

(4) $\sqrt{58}$ m/s, 336° 48′ with horizontal.

(5) 5 m/s at $\tan^{-1}(-\frac{4}{3})$ with AB; 16 m; 2·4 s.

(8) $4\mathbf{i}-4\mathbf{j}+2\mathbf{k}$. (9) $\sqrt{\frac{43}{3}}$ newtons.

(10) 5 newtons; direction cosines $\dfrac{3}{5\sqrt{2}}, \dfrac{4}{5\sqrt{2}}, \dfrac{1}{\sqrt{2}}$.

(18) 1·7, 0·2, 1·2.

Exercise 7 (p. 106)

(4) 6, $2\sqrt{2}$. (10) $-\dfrac{v^2d}{a^2}, \dfrac{2uv}{a}$.

Exercise 8 (p. 131)

(1) $\dfrac{-5}{3}, \dfrac{10}{3}, \dfrac{5}{3}, -5$. (3) 5, -10, $25\mathbf{j}-40\mathbf{k}$.

(6) AC, DE are perpendicular. (7) $\sqrt{65}$.

(11) $\pm\frac{1}{7}(3\mathbf{i}+6\mathbf{j}+2\mathbf{k})$.

(17) $\cos^{-1}\dfrac{1}{3\sqrt{2}}, \cos^{-1}\dfrac{7}{3\sqrt{13}}, \cos^{-1}\dfrac{3}{\sqrt{26}}$; $AB = 3$, $BC = \sqrt{13}$,
$CA = 2\sqrt{2}$; $\sqrt{17}$.

(24) 9 joules. (25) $\dfrac{-4}{3}$. (26) 0.

Miscellaneous exercises (p. 134)

(1) (*a*) $\dfrac{3\sqrt{3}+2}{2}, \dfrac{2\sqrt{3}-3}{3}$.   (2) (*b*) $-2\mathbf{i}+\frac{4}{3}\mathbf{j}$.

(3) (*a*) $a = 3, b = 2$. (*b*) $\mathbf{OM} = \dfrac{2-\lambda}{2}\mathbf{i}+\dfrac{3+5\lambda}{2}\mathbf{j}$, $10x+2y = 13$.

(5) $\left(\dfrac{p}{1+l}-\dfrac{mq}{1+m}\right)\mathbf{AB}$, etc., $\dfrac{2p(1-l)\triangle}{1+l}$.

7 (ii) $c = a\cos B+b\cos A$, $c^2 = a^2+b^2-2ab\cos C$.

Exercise 9(*a*) (p. 145)

(1) (*a*) $\mathbf{r} = 2(1+2t)\mathbf{i}-(1-t)\mathbf{j}+3(1-t)\mathbf{k}$,

(*b*) $\dfrac{x-2}{4} = \dfrac{y+1}{1} = \dfrac{z-3}{-3}$.

(2) $(-3, 4, -7)$.

(3) $\dfrac{x-2}{1} = \dfrac{y+3}{-9} = \dfrac{z-4}{5}$. $(6/5, 21/5, 0)$.

(4) Yes.

(5) $\dfrac{x-6}{2} = \dfrac{y-2}{-1} = \dfrac{z+4}{3}$. (i) $\sqrt{126}$, (ii) $\sin^{-1}(\sqrt{14}/7)$.

(6) $(-3, 1, 2)$, $(1, 7, 0)$.

(7) $A(2, 1, 3)$, $B(-1, 4, 5)$, $C(3, 2, -2)$;

$A = \pi-\cos^{-1}(10/\sqrt{594})$, $B = \cos^{-1}(32/\sqrt{1518})$,
$C = \cos^{-1}(37/\sqrt{1863})$.

(8) $p = -2$, $\mathbf{r} = 2(1+t)\mathbf{i}+(2+3t)\mathbf{j}+3\mathbf{k}$.

(9) (i) $(-2, 3, -5)$, (ii) $2\sqrt{17}$, (iii) $\dfrac{x-4}{3} = \dfrac{y-7}{2} = \dfrac{z+9}{-2}$.

Exercise 9(*b*) (p. 149)

(1) $\sqrt{59}$, $\dfrac{x-8}{5} = \dfrac{y-1}{-3} = \dfrac{z-3}{5}$.

(2) $5x = 24$, $2y = z$. $4\sqrt{5}$.

Exercise 9(*c*) (p. 157)

(1) $10x+9y-2z = 22$.

(2) $3x-2y+z+1 = 0$.

(3) $x-2y-z = 6$.

(4) $x-2y-4z = 12$.

(6) $2x+3y-2z = 2$. $(0, -4, -7)$.

(7) 3.

(9) $\mathbf{b}.\mathbf{n} = 0$. $\sqrt{21}/7$.

(11) $\dfrac{x+4}{5} = \dfrac{y}{3} = \dfrac{z-4}{-4}$. $(1/6, 7/6, 4/3)$.

Exercise 9(*d*) (p. 161)

(1) $\cos^{-1}(3/\sqrt{84})$, (ii) $\cos^{-1}(1/\sqrt{30})$.

(2) $\sin^{-1}(18/91)$.

Exercise 10 (p. 179)

(3) $\mathbf{r} = a\sin 2\theta\mathbf{i}+2a\sin^2\theta\mathbf{j}$.

(4) $\mathbf{r} = (1+12\cos\theta+24\sin\theta,\ 2+18\cos\theta-12\sin\theta,\ 3-6\cos\theta+12\sin\theta)$.

(5) $\mathbf{r} = (2+2t^2+12t,\ -4+8t^2,\ 1-6t^2+4t)$.

(7) $(1, 0, 2)$, 8, $6x+5y-3z = 0$.

(8) $$\mathbf{r} = \left(1+\frac{\sqrt{110}}{10}p^2+\frac{2\sqrt{2}}{3},\ 2-\frac{3\sqrt{110}}{10}p^2+\frac{5\sqrt{2}}{3},\ 3+\frac{\sqrt{110}}{10}p^2+\frac{13\sqrt{2}}{3}\right).$$

(12) $\mathbf{r} = 2ap\mathbf{i}+2a\mathbf{j}$.

(13) (i) $\mathbf{r} = \mathbf{i}-2\mathbf{j}+\lambda(-\mathbf{i}+\mathbf{j})$, i.e. $x+y+1 = 0$,
$\mathbf{r} = 9\mathbf{i}+6\mathbf{j}+\lambda(3\mathbf{i}+\mathbf{j})$, i.e. $x-3y+9 = 0$,
(ii) $\cos^{-1}(1/\sqrt{5})$.

(14) $\mathbf{r} = at^2\mathbf{i}+2at\mathbf{j}+\lambda(t\mathbf{i}+\mathbf{j})$,

$\mathbf{r} = at^2\mathbf{i}+2at\mathbf{j}+\lambda(\mathbf{i}-t\mathbf{j})$.

(17) $4a(\sin^2\phi\cos^2\phi+\phi^2)^{\frac{1}{2}}$.

Exercise 11(*b*) (p. 206)

(1) $5\mathbf{i}+7\mathbf{j}+11\mathbf{k}$, $\sqrt{195}/14$.

(2) $\pm(4\mathbf{i}+5\mathbf{j}+7\mathbf{k})/(3\sqrt{10})$.

(3) (i) -2, (ii) $(-7, -3, 8)$, (iii) $(-9, -1, 8)$, (iv) 8, (v) $(-4, 4, -2)$.

(7) $11\sqrt{6}/2$. $(\mathbf{i}-\mathbf{j}+2\mathbf{k})/\sqrt{6}$.

Exercise 11(*c*) (p. 210)

(1) $10x+9y-2z = 22$. $3x-2y+z+1 = 0$. $x-2y-z = 6$.

(2) $2x-y+z = 5$.

(3) (i) Line in plane, (ii) line parallel to plane, (iii) line intersects plane at point $(2, -6, -16)$, (iv) line normal to plane at point $(4, 0, -10)$.

(5) $\pm(1, -1, -1)/\sqrt{3}$. $5/\sqrt{3}$.

(6) (i) $\cos^{-1}4/5$, (ii) $\frac{1}{3}\pi$, (iii) $3/\sqrt{2}$, (iv) $\sqrt{3}$.

Exercise 11(*d*) (p. 213)

(1) $3(-9t^2, 2t, 2)$.

(2) $\mathbf{r}\times\dfrac{d^2\mathbf{r}}{dt^2}$.

(5) $-\left(\dfrac{t^2}{2}\mathbf{i}+t^3\mathbf{j}+\dfrac{5t^4}{4}\mathbf{k}\right)$.

(10) $2\pi a^2\mathbf{k}$.

Exercise 11(*e*) Miscellaneous (p. 221)

(1) $\frac{8}{7}(2\mathbf{i}+\mathbf{j})$Nm.

(3) $2\pi(30\mathbf{i}+7\mathbf{j}-2\mathbf{k})$.

(4) $\cos^{-1}\sqrt{35}/7$, $\cos^{-1} 2/\sqrt{13}$, $\cos^{-1} 9/(2\sqrt{35})$.

(5) (i) 0, (ii) 0.

(6) (i) OA perpendicular to BC, (ii) OA parallel to BC.

(9) $\cos^{-1}(-5/7)$.

(12) $\mathbf{i}+2\mathbf{j}$. 5/3.

(14) $\left(\dfrac{23-20\sqrt{3}}{25}, \dfrac{10+\sqrt{3}}{10}, \dfrac{53+30\sqrt{3}}{50}\right)$.

(15) $\dfrac{1}{\omega^2}\,\boldsymbol{\omega}\times\mathbf{u}$.

Exercise 12(*a*) (p. 233)

(1) 6. 6, -6.

(2) Positive.

(3) $\mathbf{a}.(\mathbf{a}+\mathbf{b})$.

(4) 12.

(6) 25.

(9) $45x-32y-7z = 166$.

(10) $11x+7y-5z = 0$.

(11) $\frac{35}{3}$, 14, $-\frac{112}{3}$, $\frac{16}{3}$.

(12) 0, $\dfrac{Fa}{\sqrt{6}}$.

Exercise 12(*b*) (p. 240)

(7) (ii) 1, $\dfrac{1}{\{\mathbf{a}, \mathbf{b}, \mathbf{c}\}}$; $\mathbf{a} = \dfrac{\mathbf{b}'\times\mathbf{c}'}{\{\mathbf{a}', \mathbf{b}', \mathbf{c}'\}}$, $\mathbf{b} = \dfrac{\mathbf{c}'\times\mathbf{a}'}{\{\mathbf{b}', \mathbf{c}', \mathbf{a}'\}}$,

$\mathbf{c} = \dfrac{\mathbf{a}'\times\mathbf{b}'}{\{\mathbf{c}', \mathbf{a}', \mathbf{b}'\}}$.

(8) $\mathbf{r} = \dfrac{\mathbf{F}\times\mathbf{G}}{\mathbf{F}^2}$.

Exercise 12(c) (p. 248)

(1) $p = 2, \quad q = 5$.

(2) $(5, -1, 1)$.

(5) $\dfrac{\mathbf{b}\times\mathbf{a}+k\mathbf{a}}{\mathbf{a}^2}$.

(6) $p\mathbf{a}$.

(7) $\mathbf{x} = \frac{1}{6}(1, 7, 3), \quad \lambda = \frac{1}{6}$.

(8) $\mathbf{x} = \mathbf{b}+\lambda\mathbf{a}$, where λ is an arbitrary scalar.

(11) (i) $\dfrac{\mathbf{a}\times\mathbf{b}}{\mathbf{a}^2}+\lambda\mathbf{a}$, where λ is a variable scalar.

(ii) $\{\lambda^2\mathbf{a}+(\mathbf{a}.\mathbf{c})\mathbf{c}+\lambda(\mathbf{c}\times\mathbf{a})\}/\{\lambda(\lambda^2+\mathbf{c}^2)\}$.

(iii) $\{a\mathbf{w}+(\mathbf{v}\times\mathbf{w})\times\mathbf{u}\}/\{a(a+\mathbf{u}.\mathbf{v})\}$, when $a+\mathbf{u}.\mathbf{v} \neq 0$;
$\lambda\mathbf{v}+\mathbf{w}/a$, where λ is a variable scalar, when $a+\mathbf{u}.\mathbf{v} = 0$.

(12) $\mathbf{x} = \{\mathbf{b}-\mathbf{a}\times\mathbf{c}+(\mathbf{a}.\mathbf{b})\mathbf{a}\}/(1+\mathbf{a}^2)$,
$\mathbf{y} = \{\mathbf{c}-\mathbf{a}\times\mathbf{b}+(\mathbf{a}.\mathbf{c})\mathbf{a}\}/(1+\mathbf{a}^2)$.

Exercise 13(a) (p. 254)

(1) (i) In equilibrium. (ii) Not in equilibrium.

(2) (i) Not collinear; resultant $= \frac{1}{3}\mathbf{Y}$, acting through point $(10, 5, 3)$.

(ii) Collinear; resultant $= 3\mathbf{X}$, acting through point $(3, 0, 6)$.

(4) Resultant $= 3\mathbf{i}+3\mathbf{j}+7\mathbf{k}$, acting along line
$7x-35 = 7y = 3z-22$.

(5) $-5(\mathbf{i}+2\mathbf{j}+\mathbf{k})$.

(7) (i) Force is $5\mathbf{i}+3\mathbf{j}-\mathbf{k}$; couple is $-11\mathbf{i}+7\mathbf{j}-4\mathbf{k}$.

(ii) Force is $5\mathbf{i}+3\mathbf{j}-\mathbf{k}$; couple is $-9\mathbf{j}+3\mathbf{k}$.

Exercise 14 (p. 285)

(1) $27\mathbf{i}+2\mathbf{j}+18\mathbf{k}, \dfrac{x-27}{27} = \dfrac{y-6}{2} = \dfrac{z-27}{18}$.

(2) (i) a, (ii) $2a(1+t^2)^{\frac{3}{2}}$, (iii) $\dfrac{c}{2}\left(p^2+\dfrac{1}{p^2}\right)^{\frac{3}{2}}$.

(4) (i) (2, 2, 0), (ii) $\cos^{-1}3/7$.

(5) $\dfrac{3(1+2t^2)^2}{2}$.

(6) $\dfrac{1}{\sqrt{(a^2+c^2)}}(-a\sin\theta\mathbf{i}+a\cos\theta\mathbf{j}+c\mathbf{k})$, $-\cos\theta\mathbf{i}-\sin\theta\mathbf{j}$,

$\dfrac{1}{\sqrt{(a^2+c^2)}}(c\sin\theta\mathbf{i}-c\cos\theta\mathbf{j}+a\mathbf{k})$.

(7) $\dfrac{1}{a}\sin^2\alpha$, $\dfrac{1}{a}\sin\alpha\cos\alpha$.

BIBLIOGRAPHY

R. H. Atkin. *Classical Dynamics*. Heinemann.

R. A. Becker. *Introduction to Theoretical Mechanics*. McGraw-Hill.

P. M. Cohn. *Solid Geometry*. Routledge and Kegan Paul.

C. J. Eliezer. *Concise Vector Analysis*. Pergamon Press.

M. B. Glauert. *Principles of Dynamics*. Routledge and Kegan Paul.

A. M. Macbeath. *Elementary Vector Algebra*. Oxford University Press.

A. S. W. Massey and H. Kestelman. *Ancillary Mathematics*. Pitman.

Mathematical Association. *A Second Report on the Teaching of Mechanics in Schools*. Bell.

E. A. Maxwell. *Coordinate Geometry with Vectors and Tensors*. Oxford University Press.

E. A. Milne. *Vectorial Mechanics*. Methuen.

D. E. Rutherford. *Classical Mechanics*. Oliver and Boyd.

D. E. Rutherford. *Vector Methods*. Oliver and Boyd.

S. Schuster. *Elementary Vector Geometry*. Wiley.

J. L. Synge and B. A. Griffiths. *Principles of Mechanics*. McGraw-Hill.

C. E. Weatherburn. *Differential Geometry*. Cambridge University Press.

C. E. Weatherburn. *Elementary Vector Analysis*. Bell.

C. Wexler. *Analytic Geometry: A Vector Approach*. Addison-Wesley.

INDEX

acceleration, 74, 88, 93, 96, 97, 99, 102
  as a vector, 74, 94
  Coriolis, 103
  normal and tangential, 101, 102
  radial and transverse, 98, 102, 257
  relative, 75
  uniform, 74
addition
  of displacements, 5, 8, 72
  of vectors, 5, 13, 18, 20, 23, 25, 27, 29, 58, 61
angle
  between line and plane, 160
  between planes, 159
  between straight lines, 145
  between vectors, 54, 63, 64, 115, 118, 202
angular momentum
  of particle, 237, 257
  of rigid body, 237, 263–6
angular velocity, 77, 214
Apollonius's theorem
  vector proof of, 120
areal velocity, 259
Associative Law, 23, 35, 191, 197

binormal, 280

Cartesian co-ordinates, 57, 58, 60, 96
centre of gravity, 85
centre of mass, 84
central orbit, 258–63
  areal velocity, 259
  direct law of force, 260
  inverse square law of force, 261
centroid
  of a tetrahedron, 51
  of a triangle, 50
  of points, 47, 65
charged particle in magnetic field, 174, 269
circle
  definition of, 162
  vector equation of, 162–5
circular helix, 174
circular motion, 99, 101, 256
collinear points, 45
Commutative Law, 23, 35, 111, 191, 197
components of a vector, 16, 30, 57, 61
concurrent forces, 81, 207, 255
co-ordinate system
  change of origin, 177
  rotation of axes, 178
coplanar vectors, 228
cosine rule
  vector proof of, 121
couple, 253
cross product, 187
curve
  binormal of, 280
  curvature of, 275, 278
  normal plane to, 277
  normal to, 276
  principal normal to, 278
  tangent to, 276
  torsion of, 275, 280
  unit principal normal to, 278
  unit tangent to, 276

decomposition of vectors, 58, 60
differentiation
  of a vector, 88–91
  of scalar product, 125
  of scalar triple product, 238
  of unit vector, 95
  of vector of constant length, 125, 275
  of vector product, 211
  of vector triple product, 238
directed line segment, 2, 12, 14
direct law of force, 260
direction
  cosines, 59, 62, 64, 139, 152
  of displacements, 2
  of vectors, 12
  of vector product, 188, 197
  ratios, 139, 152
direction-vector, 138, 140
displacement, 1, 69, 73
  free, 2
  localized, located, 2
  on earth's surface, 69

distance
  between points, 59, 62, 64
Distributive Law, 36, 111, 192, 198
division
  of straight line, 44
  of vectors, 33
dot product, 110

electrodynamics, 269
ellipse
  vector equation of, 172
equal vectors, 13
equations
  vector, 218–20, 241–8
equilibrium of forces, 251
equivalent systems of forces, 250–1
equivalent vectors, 14, 250

finite rotations, 70
force
  as a vector, 80
  moment about a point, 205, 216
  moment about a straight line, 231
  principle of transmissibility of, 250
  work done by, 127
forces
  concurrent, 81, 207, 255
  equilibrium of, 251
  equivalent system of, 250
  resultant of, 81, 252, 253
frame of reference, 220
  invariance of, 220
free displacement, 2
free vector, 14
Frenet–Serret formulae, 281

gyroscope, 266–8

helix
  circular, 174, 271
  definition of, 174
  properties of, 282–5

integration
  of vectors, 104
  involving scalar product, 126
  involving vector product, 212
intrinsic co-ordinates, 99
invariance of frames of reference, 220
inverse square law of force, 261
inverse vector, 11, 23

kinetic energy of rigid body, 268

left- and right-handed systems, 183–5
length
  of a displacement, 2
  of a vector, 12, 21, 59, 62, 64
like vectors, 13, 21
line
  direction cosines of, 139
  direction ratios of, 139
  direction-vector of, 138
  gradient of, 138
  segment, 2, 12, 14
  vector, 15
localized, located
  displacement, 2
  vector, 14
locus
  circle, 162–6
  circular helix, 174–6
  ellipse, 172
  of charged particle in magnetic field, 174, 269
  of projectile, 170
  parabola, 166
  plane, 139–56
  rectangular hyperbola, 173
  sphere, 173–4
  straight line, 138–45
  vector equation of, 137

magnitude
  of a displacement, 2
  of a vector 12, 13, 21, 59, 62, 64
  of a vector product, 193, 196
mean centre, 47
  weighted, 47
median of a triangle, 50
modulus
  of a vector, 13, 21, 59, 62, 64
  of a vector product, 193, 196
moment of a force
  about a line, 231
  about a point, 205, 216
moment of inertia, 265
moment of momentum of particle, 237, 257
moment of momentum of rigid body, 237, 263–6
motion of charged particle in magnetic field, 269–71
motion of particle
  in a circle, 99, 256–7

motion of a particle (*cont.*)
  in a plane, 96–102
  on a rotating plane, 102–3
  under gravity, 170–2
multiplication of vectors
  by a scalar, 32, 61
  by a vector, 108, 183

negative vector, 23
Newton's Laws, 80
normal,
  plane, 277
  to curve, 276
normal-vector of plane, 152
null vector, 11
  see zero vector
number pair, 1, 57
number triple, 15, 60

osculating plane, 278

parabola
  definition of, 166
  vector equation of, 166–8
parallel
  displacements, 3
  vectors, 34, 111, 116, 118, 190, 197, 201
parallelogram law, 19
particle
  moment of momentum, 237, 257
  motion in two dimensions, 96–102
  motion on a rotating plane, 102–3
perpendicular vectors, 108, 111, 116, 118, 183, 185, 201
plane
  angle made with another plane, 159
  angle made with a straight line, 160
  direction cosines of, 152, 153
  direction ratios of, 152, 153
  equation of, 149–54, 228
  normal-vector of, 152
  osculating, 278
polar co-ordinates, 97
polygon
  of forces, 82
  of vectors, 29
position vectors, 43
principal normal to curve, 278
principle of transmissibility of forces, 250
product
  of inertia, 265
  of number and vector, 32
  of three vectors, 224–6, 234–7
  of two vectors, 108–18, 183–201
projectile, 170
  vector equation of path of, 171
projection of vectors, 54, 56, 62, 116, 195
Pythagoras's theorem
  vector proof of, 122

radius of curvature, 279
rectangular hyperbola,
  vector equation of, 173
relative
  acceleration, 75
  velocity, 75
representation
  of displacements, 1
  of vectors, 12
  of vector quantities, 71
resolution of vectors, 57
resultant
  acceleration, 74
  force, 82
  of two forces on a rigid body, 252
  of two parallel forces on a rigid body, 253
  of vectors, 29
  velocity, 73
rigid body
  kinetic energy of, 268
  moment of momentum of, 237, 263–6
rotation
  finite, 70
  of axes, 178
  of vectors, 108

scalar product, 110–18
  comparison with vector product, 201
  definition, 110, 117
  differentiation of, 125–6
  geometrical meaning of, 116
  integration involving, 126
  properties of, 111–12, 116–18
scalar triple product, 224–6
  definition of, 224
  differentiation of, 238
  geometrical interpretation of, 226
  properties of, 224–5

INDEX

sense
  of angular velocity, 77, 214
  of displacements, 2
  of vectors, 12
  of vector product, 187, 197
sine rule
  vector proof of, 203
skew lines
  definition, 147
  shortest distance between, 147, 209
sliding vector, 15, 251
speed, 69
sphere
  definition of, 173
  vector equation of, 173
straight line
  angle between straight line and plane, 160
  angle between two straight lines, 145
  direction cosines of, 138–40
  direction ratios of, 138–40
  direction-vector of, 138
  equations of, 140–5
  gradient of, 138
subtraction of vectors, 26
successive displacements, 4
symbols, 6
systems
  left- and right-handed, 183

tangent to a curve, 88, 92, 276
tetrahedron, 51, 123, 229
tied vector, 15
triangle,
  law, 5, 13, 18, 69
  of accelerations, 74
  of velocities, 74

uniform
  acceleration, 74
  velocity, 73
unit principal normal, 278
unit tangent, 276
unit vector, 13, 58, 59, 60, 62, 63, 95, 111, 118, 193, 200
unlike vectors, 13, 21

Varignon's theorem, 207, 217
vector equations, 218–20, 241–8
vector product, 183–202
  comparison with scalar product, 201
  definition of, 188, 196
  differentiation of, 211
  geometrical meaning of, 201
  integration involving, 212
  modulus of, 193, 196
  properties of, 190–3, 197–200
vector triple product, 224, 234–7
  definition of, 224
  differentiation of, 238
  properties of, 236–7
velocity, 69, 73, 88, 91, 96, 97, 99
  angular, 77, 214
  as a vector, 73, 93
  radial and transverse, 98, 102, 256
  relative, 75
  tangential 100, 102
  uniform, 73

weighted mean centre, 47
work done by force, 127

zero vector, 11, 12, 20, 22, 117, 190, 197

www.ingramcontent.com/pod-product-compliance
Ingram Content Group UK Ltd.
Pitfield, Milton Keynes, MK11 3LW, UK
UKHW032325190125
453752UK00011B/151